Developments in Statistical Evaluation of Clinical Trials

Kees van Montfort • Johan Oud •
Wendimagegn Ghidey
Editors

Developments in Statistical Evaluation of Clinical Trials

Springer

Editors
Kees van Montfort
Department of Biostatistics
Erasmus Medical Center
Rotterdam
The Netherlands

Johan Oud
Behavioural Science Institute
University of Nijmegen
Nijmegen
The Netherlands

Wendimagegn Ghidey
Department of Hematology
Erasmus Medical Center
Rotterdam
The Netherlands

ISBN 978-3-662-52211-0 ISBN 978-3-642-55345-5 (eBook)
DOI 10.1007/978-3-642-55345-5
Springer Heidelberg New York Dordrecht London

Printed on acid-free paper

Springer is part of Springer Science+Business Media (www.springer.com)

Preface

Over the last few decades the role of statistics in the evaluation and interpretation of clinical data has become more and more important. As a result the standards of clinical study design, conduct and interpretation have been advanced. In this book statistical considerations in data analysis as a basis for deriving an accurate clinical interpretation are elaborated.

Most often it is the physician who decides whether to prescribe a specific drug for a specific patient in a specific situation. The decisions of the physicians are largely based on the interpretation of data they have read and heard. This book describes various ways of approaching and interpreting the data that result from a clinical trial study.

The book reemphasizes the essential role that biostatistics plays in clinical trials. The book contains 18 carefully reviewed chapters on recent developments in trials and statistics. The chapters in this book are generally autonomous and may be read in any order. Each chapter is written by one or more experts in the specific approach. Starting from (a) some background information about the specific approach (short history and main publications), the chapter (b) describes the type of research questions the approach is able to answer and the kind of data to be collected, (c) gives the statistical and mathematical explanation of the model(s) used in the analysis of the data, (d) discusses the input and output of the software used in the analysis, and (e) provides one or more examples with typical data sets enabling the readers to apply the programs themselves. The chapters are worked out in a homogeneous style to enhance comparability between the approaches. The data sets and the computer code for the analysis with various softwares are a very important component of the book. They are available upon request (by emailing the authors of the chapters).

Each chapter is self-contained in this edited volume. The chapters are written and reviewed by experts in the specific approach. Although an authored volume could have advantages, because of the rapid changes in the field, an edited book written by people who are in the middle of the latest developments in the specific approach is preferable. In addition, the authors of the chapters use a shared notation to enable the reader to compare methods more easily.

The book addresses the great majority of researchers in the field of clinical trials. Included are biostatisticians, medical researchers and physicians. It is meant as a reference work for all those actually doing and using research in the field of clinical trials. To reach this vast audience, knowledge of statistics as taught at master degree level in medical and biomedical sciences is required. However, the restricted number of chapters gives each of the chapters the opportunity to go into sufficient details to enable the readers to understand and apply the methods. In addition, the book addresses biostatisticians and physicians, who are professionally dealing with research in the field of clinical trials, to provide standards for state-of-the-art practices. Furthermore, the book offers researchers new ideas about the use of biostatistical analysis in solving their research problems. Finally the book is suitable as obligated literature for courses in clinical trial evaluation given at university medical and epidemiological research schools.

We thank the authors of the chapters for their willingness to contribute to the book, the anonymous reviewers for their expertise and time invested and Springer Publishers for their decision to publish the book in their Statistics book program.

Rotterdam, The Netherlands — Kees van Montfort
Nijmegen, The Netherlands — Johan Oud
Rotterdam, The Netherlands — Wendimagegn Ghidey

Contents

List of Contributors

Ariel Alonso Department of Methodology and Statistics, Maastricht University, Maastricht, The Netherlands

Gerard van Breukelen Department of Methodology and Statistics, Maastricht University, Maastricht, The Netherlands

Yuki Choai Department of Statistical Science, School of Multidisciplinary Sciences, The Graduate University for Advanced Studies, Tachikawa, Tokyo, Japan

Richard J. Cook Department of Statistics and Actuarial Science, University of Waterloo, Waterloo, ON, Canada

P. J. Devereaux Faculty of Health Sciences, McMaster University, Hamilton, ON, Canada

Adélaide Doussau USMR, Bordeaux University-Hospital, ISPED Centre INSERM U897-Epidemiologie-Biostatistique, Bordeaux, France

Peter Fennema Advanced Medical Research, Männedorf, Switzerland

Frank Fleischer Department of Biostatistics, Boehringer Ingelheim Pharma GmbH and Co. KG, Biberach, Germany

Birgit Gaschler-Markefski Department of Biostatistics, Boehringer Ingelheim Pharma GmbH and Co. KG, Biberach, Germany

Wendimagegn Ghidey Department of Hematology, Erasmus Medical Center, The Netherlands

Akihiro Hirakawa Center for Advanced Medicine and Clinical Research, Nagoya University Graduate School of Medicine, Showa-ku, Nagoya, Japan

Julia Hocke Department Biostatistics, Boehringer Ingelheim Pharma GmbH and Co. KG, Biberach, Germany

Afisi S. Ismaila Department of Clinical Epidemiology and Biostatistics, McMaster University, Hamilton, ON, Canada

Thomas Jaki Medical and Pharmaceutical Statistics Research Unit, Department of Mathematics and Statistics, Lancaster University, Lancaster, United Kingdom

Meinhard Kieser Institute of Medical Biometry and Informatics, University of Heidelberg, Heidelberg, Germany

Kazuhiko Kuribayashi Pfizer Japan Inc., Tokyo, Japan

Ker-Ai Lee Department of Statistics and Actuarial Science, University of Waterloo, Waterloo, ON, Canada

Shigeyuki Matsui Department of Biostatistics, Graduate School of Medicine, Nagoya University, Showa-ku, Nagoya, Japan

Michael A. McIsaac Department of Public Health Sciences, Queen's University, Kingston, ON, Canada

Geert Molenberghs I-BioStat, Universiteit Hasselt, Diepenbeek, Belgium

KU Leuven, Leuven, Belgium

Kees van Montfort Department of Biostatistics, Erasmus Medical Center, Rotterdam, The Netherlands

Takahiro Nonaka Pharmaceuticals and Medical Devices Agency, Chiyoda-ku, Tokyo, Japan

Johan H.L. Oud Behavioural Science Institute, Radboud University Nijmegen, Nijmegen, The Netherlands

Xavier Paoletti Department of Biostatistics/INSERM U900, Institut Curie, Paris, France

Janice Pogue Department of Clinical Epidemiology and Biostatistics and Faculty of Health Sciences, McMaster University, Hamilton, ON, Canada

Hiroyuki Saeki FUJIFILM RI Pharma Co. LTD., Chuo-ku, Tokyo, Japan

Karin Schiefele Department of Epidemiology and Medical Biometry, University Ulm, Ulm, Germany

Toshiro Tango Center for Medical Statistics, Minato-ku, Tokyo, Japan

Jos W.R. Twisk Department of Epidemiology and Biostatistics, VU Medical Centre, Amsterdam, The Netherlands

Stephen D. Walter Department of Medicine, McMaster University, Hamilton, ON, Canada

Jixian Wang Novartis Pharma AG, Basel, Switzerland

James M.S. Wason MRC Biostatistics Unit Hub for Trials Methodology Research, Institute of Public Health, Cambridge, United Kingdom

Andrew R. Willan SickKids Research Institute, Dalla Lana School of Public Health, University of Toronto, Toronto, ON, Canada

Salim Yusuf Faculty of Health Sciences, McMaster University, Hamilton, ON, Canada

Chapter 1
Statistical Models and Methods for Incomplete Data in Randomized Clinical Trials

Michael A. McIsaac and Richard J. Cook

Abstract In this chapter we discuss several models by which missing data can arise in clinical trials. The likelihood function is used as a basis for discussing different missing data mechanisms for incomplete responses in short-term and longitudinal studies, as well as for missing covariates. We critically discuss common ad hoc strategies for dealing with incomplete data, such as complete-case analyses and naive methods of imputation, and we review more broadly appropriate approaches for dealing with incomplete data in terms of asymptotic and empirical frequency properties. These methods include the EM algorithm, multiple imputation, and inverse probability weighted estimating equations. Simulation studies are reported which demonstrate how to implement these procedures and examine performance empirically.

1.1 Introduction

In well-conducted randomized clinical trials, randomization eliminates the possible effect of confounding variables in the assessment of treatment effects. That is, when the assignment of the treatment to patients is carried out by random allocation, different treatment groups will have similar distributions of demographic and clinical features, so any differences seen in the distribution of responses between the treatment groups are attributable to the different treatments they receive. There are a number of other rationale put forward for use of randomization in health research [40], but it is the elimination of the effect of confounding variables and facilitation of causal inference that has had the most profound impact in advancing scientific understanding.

Following recruitment and randomization, however, participants in clinical trials often withdraw before completion of follow-up, leading to incomplete outcome

M.A. McIsaac (✉)
Department of Public Health Sciences, Queen's University, Kingston, ON, Canada
e-mail: mcisaacm@queensu.ca

R.J. Cook
Department of Statistics and Actuarial Science, University of Waterloo, Waterloo, ON, Canada
e-mail: rjcook@uwaterloo.ca

K. van Montfort et al. (eds.), *Developments in Statistical Evaluation of Clinical Trials*, DOI 10.1007/978-3-642-55345-5_1

data. Incomplete data can of course arise for a variety of reasons; many illustrative examples can be seen in the second chapter of Molenberghs and Kenward [26]. Depending on the reasons for withdrawal, the individuals who remain in the study may no longer form groups with similar distributions of the demographic and clinical features, which compromises the validity of causal inferences. The purpose of this article is to discuss models and mechanisms by which incomplete data can arise in clinical trials, the consequences missing data can have on the interpretation of study results, and methods which can be employed to minimize the effect of these consequences. A clear understanding of the practical and statistical issues involved with incomplete response data will improve ability to critically appraise the clinical literature.

The remainder of this chapter is organized as follows. In Sect. 1.2 we discuss the problem of incomplete binary responses. We restrict attention to the case of a binary treatment indicator and a single binary confounding variable to simplify the discussion, calculations, and empirical studies, but we remark on practical issues with more complex settings at the end of this section. We discuss the case of incomplete longitudinal data in Sect. 1.3, and the problem of incomplete covariates in Sect. 1.4. Concluding remarks are made in Sect. 1.5.

1.2 Incomplete Binary Response Data

1.2.1 Models and Measures of Treatment Effect

Consider a balanced two-arm clinical trial in which patients are randomized to receive either an experimental treatment or standard care. Let $X = 1$ indicate that a patient was allocated to receive experimental therapy and $X = 0$ otherwise, where $P(X = 1) = 0.5$. Suppose the outcome of interest is whether the patient had a successful response; we let $Y = 1$ if this is the case and $Y = 0$ otherwise. We illustrate the problem of dependently missing data by considering a situation with a single additional binary variable V, where $V = 1$ indicates the presence of a particular feature and $V = 0$ otherwise; $P(V = 1) = p$. Suppose that the variable V is an effect modifier [33] so that the treatment has a different effect for individuals with and without the feature. This may be represented by the logistic model

$$P(Y = 1|X, V; \gamma) = \text{expit}(\gamma_0 + \gamma_1 X + \gamma_2 V + \gamma_3 XV) \,, \tag{1.1}$$

where $\gamma = (\gamma_0, \gamma_1, \gamma_2, \gamma_3)'$. In most situations there will be sub-populations between which there is variation in the event rate and the effect of treatment; (1.1) is the simplest model which accommodates this phenomenon.

While (1.1) may reflect reality, in clinical trials we typically aim to assess treatment effects based on marginal models (i.e. models that do not condition on prognostic variables such as V); indeed provided X is independent of V, the causal effect of treatment is typically defined in terms of such a model. Thus the logistic model used for treatment comparisons is formulated as

$$P(Y = 1|X; \beta) = \text{expit}(\beta_0 + \beta_1 X) , \tag{1.2}$$

where $\beta = (\beta_0, \beta_1)'$. Of course,

$$P(Y = 1|X; \beta) = E_V [P(Y = 1|X, V; \gamma); p] , \tag{1.3}$$

since V is independent of X due to randomization, and so it is possible to obtain the functional form of β in terms of $(\gamma', p)'$.

The resulting response rates in the control and treatment arms are $p_C = P(Y = 1|X = 0) = \text{expit}(\beta_0)$ and $p_T = P(Y = 1|X = 1) = \text{expit}(\beta_0 + \beta_1)$, respectively. Some common measures of treatment effect include the absolute difference AD $= p_T - p_C$, the number needed to treat NNT $= (p_T - p_C)^{-1}$, the relative risk RR $= p_T / p_C$, and the odds ratio OR $= [p_T/(1 - p_T)]/[p_C/(1 - p_C)]$ [16, 22]. When the experimental treatment has a higher response rate, the AD and NNT measures are positive and the RR and OR are larger than one.

Let $I(A)$ be an indicator function such that $I(A) = 1$ if A is true and $I(A) = 0$ otherwise. If response data are incomplete, in order to thoroughly discuss modeling issues it is necessary to introduce a new random variable $R = I(Y \text{ observed})$, so $R = 1$ if Y is observed and $R = 0$ otherwise. The biases that result from incomplete data arise if there is an association between the response (Y) and whether we observe it or not (R). There are a variety of ways of introducing an association between Y and R including through bivariate binary models [6] and shared random effect models [1]. Here we consider the setting in which both Y and R are associated with the covariates X and V. When V is unknown, an association between Y and R exists because of the omission of V from the analysis. We adopt this framework because when V is known, there are a variety of approaches to incorporating information about V into the analyses to mitigate problems, as we discuss in the following sections.

Suppose that the missing data model is

$$P(R = 1|X, V; \alpha) = \text{expit}(\alpha_0 + \alpha_1 X + \alpha_2 V + \alpha_3 XV) , \tag{1.4}$$

where $\alpha = (\alpha_0, \alpha_1, \alpha_2, \alpha_3)'$. This model accommodates a different dependence on V in the two treatment arms. We assume in this idealized setting that $R \perp Y|X, V$. Since $X \perp V$ by randomization, the marginal proportion of missing data is

$$\begin{aligned} p_R = P(R = 1; \alpha, p) &= E_X \{E_V [P(R = 1|X, V)]\} \\ &= \sum_{x=0}^{1} \sum_{v=0}^{1} P(R = 1|X = x, V = v; \alpha)\ P(V = v; p)\ P(X = x) , \end{aligned}$$

where $P(V = v; p) = p^v (1 - p)^{1-v}$, and $P(X = x) = 1/2$ if randomization is balanced. The joint probability mass function for $Y, R|X$ is

$$\begin{aligned} P(Y, R|X;\theta) &= E_V\left[P(Y|X, V;\gamma)\ P(R|X, V;\alpha)\right] \\ &= \sum_{v=0}^{1} P(Y|X, V = v;\gamma)\ P(R|X, V = v;\alpha)P(V = v; p)\ , \end{aligned} \tag{1.5}$$

where $\theta = (\alpha', \gamma', p)'$. From (1.5) we can derive the conditional odds ratio for the association between Y and R given X as

$$OR_{Y,R|X} = \frac{P(Y = 1, R = 1|X;\theta)}{P(Y = 1, R = 0|X;\theta)} \Big/ \frac{P(Y = 0, R = 1|X;\theta)}{P(Y = 0, R = 0|X;\theta)}\ ,$$

and we can calculate the conditional probability

$$P(Y|X, R;\theta) = \frac{P(Y, R|X;\theta)}{P(R|X;\theta)} = \frac{P(Y, R|X;\theta)}{\sum_{y=0}^{1} P(Y = y, R|X;\theta)}\ . \tag{1.6}$$

So, thus far we have defined a simple model for $Y|X, V$ and $R|X, V$ under the assumption that Y and R are conditionally independent given (X, V). When we condition on X but not V, the response Y and the missing data indicator R are associated (i.e. dependent). We have mentioned that this setting was problematic, but here we will explore why this is the case.

1.2.2 *Parameter Estimation with Incomplete Response Data*

1.2.2.1 Complete-Case Analyses

Complete-Case Analyses when Covariate V Is Unknown

The likelihood function is perhaps the most fruitful starting point when considering inference based on parametric models [39]. When response data may be incomplete, the availability of the response of interest is stochastic, and hence the observed data likelihood is

$$L \propto P(Y, R = 1|X)^R\ P(R = 0|X)^{1-R}\ .$$

Noting that $P(Y, R = 1|X) = P(Y|R = 1, X)P(R = 1|X)$, this may be re-expressed as $L_{Y|R=1,X} \cdot L_{R|X}$ where

$$L_{Y|R=1,X} = \left[P(Y = 1|R = 1, X)^Y\ P(Y = 0|R = 1, X)^{1-Y}\right]^R \tag{1.7}$$

is obtained from $P(Y|R = 1, X)^R$ by considering the two possible realizations of Y, and

$$L_{R|X} = P(R = 1|X)^R \; P(R = 0|X)^{1-R} \, . \tag{1.8}$$

When responses are not available from all individuals in a sample, it is tempting to restrict attention to individuals with complete data and base analyses on this subset. This restriction, however, implicitly conditions on $R = 1$ so that a complete-case maximum likelihood analysis actually maximizes the partial likelihood (1.7). It appears that (1.8) does not contain information about the parameters we are interested in because it relates to the missing data process alone. Note however that while (1.7) is indexed by θ, the quantities estimated by standard analyses based on available data (i.e. the sub-sample of individuals with $R = 1$) are

$$\beta_0^\dagger = \text{logit}\; P(Y = 1|X = 0, R = 1; \theta)$$

and

$$\beta_1^\dagger = \text{logit}\; P(Y = 1|X = 1, R = 1; \theta) - \beta_0^\dagger \, .$$

These parameters differ from β_0 and β_1 whenever $P(Y|X, R = 1) \neq P(Y|X)$, which will occur here if $P(Y|X, V) \neq P(Y|X)$ and $P(R|X, V) \neq P(R|X)$. Using (1.6), we can compute the naive measures of treatment effect which are actually being estimated from complete-case analyses: $\text{AD}^\dagger = P(Y=1|X=1, R=1) - P(Y = 1|X = 0, R = 1)$, $\text{NNT}^\dagger = 1/\text{AD}^\dagger$, $\text{RR}^\dagger = P(Y = 1|X = 1, R = 1)/P(Y = 1|X = 0, R = 1)$, and $\text{OR}^\dagger = [P(Y = 1|X = 1, R = 1)/P(Y = 0|X = 1, R = 1)]/[P(Y = 1|X = 0, R = 1)/P(Y = 0|X = 0, R = 1)]$.

To explore this more fully, we consider here some specific parameter configurations. Let $P(X = 1) = 0.5$ and $P(V = 1) = 0.5$. In the response model (1.1), we let $\gamma_2 = 0$ and $\gamma_3 = \log\, 2$ so the odds ratio characterizing the treatment effect is twice as big for those with $V = 1$ compared to those with $V = 0$. We set $\beta_1 = \log\, 1.5$ in (1.2), so the marginal odds ratio of the treatment effect is 1.5, and we solve for γ_0 and γ_1 so that $P(Y = 1|X = 0) = \text{expit}(\beta_0) = 0.5$ (i.e. the probability of response is 0.5 in the control arm). The marginal relative risk is therefore 1.2. In the missing data model (1.4) we set $\alpha_1 = \alpha_2 = 0$ and for each α_3 we solve for α_0 so that $P(R = 1) = 0.5$.

Figure 1.1 displays a plot of $RR^\dagger$ and $OR^\dagger$, the limiting values of complete-case estimators of *RR* and *OR*, as a function of α_3. When $\alpha_3 = 0$, the probability of the response being missing is the same for all individuals regardless of their covariates (data are *missing completely at random*, in the terminology of Little and Rubin [20]), so $P(R|X, V) = P(R|X) = P(R)$. In this case, $RR^\dagger = RR = 1.2$ and $OR^\dagger = OR = 1.5$. When $\alpha_3 < 0$, complete-case estimators of these effect measures will be too small and hence correspond to a understatement of the effect

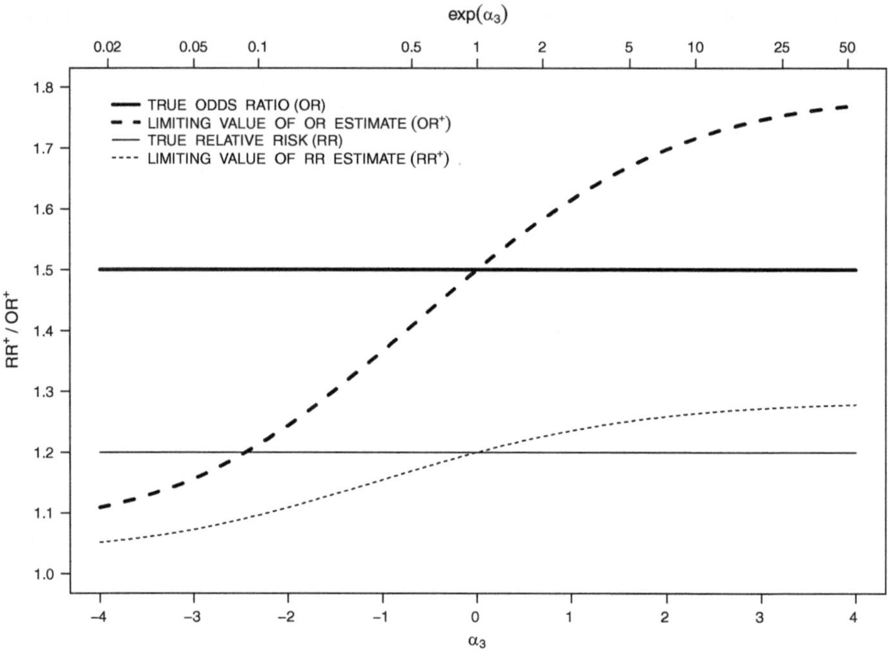

Fig. 1.1 Limiting values of naive complete-case estimators of the relative risk ($RR^{\dagger}$) and odds ratio ($OR^{\dagger}$) as a function of α_3

of treatment. Conversely, when $\alpha_3 > 0$, the inferences regarding the benefit of treatment are anti-conservative.

Complete-Case Analyses when Covariate V Is Known

If we are able to identify the variable V which renders Y and R conditionally independent (i.e., $Y \perp R|X, V$), another option is to write the observed data likelihood based on the conditional model as

$$L \propto P(Y, R = 1|X, V)^R \, [P(R = 0|X, V)]^{1-R} \, .$$

Since $P(Y, R = 1|X, V) = P(Y|X, V)P(R = 1|Y, X, V)$ and $P(R = 1|Y, X, V) = P(R = 1|X, V)$ this can in turn be written as $L_{Y|X,V} \cdot L_{R|X,V}$ where $L_{Y|X,V} \propto P(Y|X, V)$ and $L_{R|X,V} \propto P(R|X, V)$. In practice one would naturally restrict attention to the partial likelihood $L_{Y|X,V}$, since we are not typically interested in modeling the missing data process unless it is necessary. As seen above, a complete-case analysis with restriction to individuals with $R = 1$ yields inconsistent estimators of β when we just condition on X, however when we condition on V as well, a complete-case analysis gives consistent estimators for γ. Identification of variables like V which are prognostic for Y and associated with the missing data process is therefore key to ensure consistent estimation of parameters. It is not sufficient for these variables to be associated with the response alone or the missing data status alone since in either case such variables cannot render Y and R conditionally independent.

While conditioning on a suitable V seems to have solved our problem, the catch is that we did not want to condition on V in our assessment of the treatment effect – we are estimating γ instead of β, so we are estimating the wrong thing! We do have the option of modeling $V|X$, which amounts to modeling the marginal distribution of V since X was determined by randomization, and given an estimate of p as $\hat{p}$, we can compute a crude estimate by solving for β in

$$\tilde{P}(Y = 1|X; \tilde{\beta}) = \sum_{v=0}^{1} P(Y = 1|X, V = v; \hat{\gamma})\, \hat{p}^{v}(1 - \hat{p})^{1-v} .$$

Due to the so-called curse of dimensionality, this process is considerably more challenging and undesirable when V is high dimensional (i.e. a vector) [30]. A very convenient and more direct approach to estimating β is obtained using inverse probability weights as we describe in the next sub-section.

1.2.2.2 Use of Inverse Probability Weights

Suppose we have a sample of n independent subjects giving data $\{(Y_i, X_i, V_i), i = 1, 2, \ldots, n\}$. The score function for the logistic regression model in (1.2) resulting from (1.7) can be written as

$$S(\beta) = \sum_{i=1}^{n} R_i \;(Y_i - E(Y_i|X_i; \beta)) \begin{pmatrix} 1 \\ X_i \end{pmatrix} .$$

With complete data (i.e. if $P(R_i = 1) = 1, i = 1, 2, \ldots, n$) this has expectation zero and hence yields a consistent estimator for β [23]. With incomplete data however,

$$\begin{aligned} E[S(\beta)] &= E_X \left\{ E_{Y|X} \left\{ E_{R|Y,X} \left[S(\beta) \right] \right\} \right\} \\ &= \sum_{i=1}^{n} E_X \left\{ E_{Y|X} \left[P(R_i = 1|Y_i, X_i)\, (Y_i - E(Y_i|X_i; \beta)) \begin{pmatrix} 1 \\ X_i \end{pmatrix} \right] \right\} , \end{aligned}$$

which does not in general equal zero. If the probability of a response being missing depends on Y given X, then inconsistent estimators are obtained for β; the corresponding limiting values are the $\beta^{\dagger}$ given in the previous section.

Now again suppose we are able to identify V as a covariate which renders $Y \perp R|X, V$. In this case we can employ the model for $P(R = 1|Y, X, V) = P(R = 1|X, V; \alpha)$ in an *inverse probability weighted* estimating function defined as

$$U(\beta) = \sum_{i=1}^{n} \frac{R_i}{P(R_i = 1|X_i, V_i; \alpha)} (Y_i - E(Y_i|X_i; \beta)) \begin{pmatrix} 1 \\ X_i \end{pmatrix} \tag{1.9}$$

[32]. Taking the expectation of (1.9) as before yields

$$\begin{aligned} E[U(\beta)] &= \sum_{i=1}^{n} E_{X,V} \left\{ E_{Y|X,V} \left[E_{R|Y,X,V} \left(\frac{R_i}{P(R_i = 1|X_i, V_i)} (Y_i - E(Y_i|X_i; \beta)) \begin{pmatrix} 1 \\ X_i \end{pmatrix} \right) \right] \right\} \\ &= \sum_{i=1}^{n} E_{X,V} \left\{ E_{Y|X,V} \left[(Y_i - E(Y_i|X_i; \beta)) \begin{pmatrix} 1 \\ X_i \end{pmatrix} \right] \right\} \\ &= \sum_{i=1}^{n} E_X \left\{ E_{V|X} \left\{ (E(Y_i|X_i, V_i) - E(Y_i|X_i; \beta)) \begin{pmatrix} 1 \\ X_i \end{pmatrix} \right\} \right\} \\ &= \sum_{i=1}^{n} E_X \left[(E(Y_i|X_i; \beta) - E(Y_i|X_i; \beta)) \begin{pmatrix} 1 \\ X_i \end{pmatrix} \right] = 0 \end{aligned} \tag{1.10}$$

and so a consistent estimator of β is obtained from (1.9) [11].

Note that in practice the parameters in the model $P(R|X, V; \alpha)$ must be estimated and this can easily be carried out via logistic regression since R is a binary variable. Naive standard errors which do not recognize that the weights have been estimated can lead to invalid tests (with incorrect type I error rates) and invalid confidence intervals (with coverage rates not compatible with the nominal level). Large sample theory for correct variance estimation is beyond the scope of this note, but see Robins et al. [32] for general results or Chen and Cook [3] for simpler results corresponding to the present formulation.

1.2.2.3 Multiple Imputation

Multiple imputation is, in its simplest implementation, a simulation-based approach to creating complete data from an incomplete dataset. Again suppose that we have identified a covariate V which renders $Y \perp R|X, V$, and the model for $Y|X, V$ is given by (1.1). A multiple imputation approach involves fitting a model to $Y|X, V$ based on individuals with complete data, even though $Y|X$ is the model of interest. The fitted model would give a consistent maximum likelihood estimator $\hat{\gamma}$, along with the asymptotic covariance matrix for $\hat{\gamma}$, $\mathscr{I}^{-1}(\hat{\gamma})$, where $\mathscr{I}(\gamma)$ is the expected information matrix from an analysis based on (1.1). Since γ is not of interest, this fitted model is simply used to generate complete data which are then analyzed with the model of interest. The particular steps in such analyses are described in the following paragraphs.

The approach has a Bayesian flavour in that after fitting $Y|X, V$ we sample from $\text{MVN}(\hat{\gamma}, \mathscr{I}^{-1}(\hat{\gamma}))$ to obtain another realization of the 4×1 parameter vector $\hat{\gamma}$ which we denote by $g^{(1)}$. If the response for any individual is missing, then we simulate the binary response as a Bernoulli variate with probability $\text{expit}(g_0^{(1)} + g_1^{(1)}X + g_2^{(1)}V + g_3^{(1)}XV)$ using the respective covariate values. This yields *the first imputed value* for each individual with missing data, and we label the realized response $y^{(1)}$. After each individual with incomplete data in the dataset has a response simulated based on $g^{(1)}$, a second sample is drawn from $\text{MVN}(\hat{\gamma}, \mathscr{I}^{-1}(\hat{\gamma}))$ and labelled $g^{(2)}$. Using this value, one samples a second value $Y^{(2)} \sim \text{Bern}(\text{expit}(g_0^{(2)} + g_1^{(2)}X + g_2^{(2)}V + g_3^{(2)}XV))$ for each person with a missing response data. This procedure is repeated m times until we have m "complete" datasets. For each of the m "complete" datasets we then fit the model of interest given by (1.2).

Let $\hat{\beta}_1^{(r)}$ denote the estimate of β_1 from the rth imputed data set and $\omega^{(r)} = \widehat{\text{var}}(\hat{\beta}_1^{(r)})$ be the naive variance estimate ignoring the fact that some data had been imputed by simulation. The combined estimate of β_1 obtained by multiple imputation is simply the average, so $\bar{\hat{\beta}}_1 = \sum_{r=1}^{m} \hat{\beta}_1^{(r)}/m$ is the reported point estimate from multiple imputation. Let $\bar{\omega} = \sum_{r=1}^{m} \omega^{(r)}/m$ denote the average of the naive (within imputation) variance estimates, and let $\omega^* = (m-1)^{-1}\sum_{r=1}^{m}(\hat{\beta}_1^{(r)} - \bar{\hat{\beta}}_1)^2$ denote the variation between imputation samples. Rubin [36] argues that the asymptotic variance of $\bar{\hat{\beta}}_1$ is $\text{var}(\bar{\hat{\beta}}_1) = \bar{\omega} + (1+m^{-1})\omega^*$ and

$$\frac{\bar{\hat{\beta}}_1 - \beta_1}{\sqrt{\text{var}\left(\bar{\hat{\beta}}_1\right)}} \sim t_{u_m}$$

approximately, where the degrees of freedom are given by

$$u_m = (m-1)\left[1 + \frac{m\bar{\omega}}{(1+m)\omega^*}\right]^2 .$$

Wang and Robins [42] prove consistency and derive the large sample properties of estimators arising from multiple imputation under correct model specification. More refinements to the estimated degrees of freedom have since been made [2] and are implemented in SAS. We will not get into these issues here, but remark simply that one appeal of multiple imputation is the ability to make use of auxiliary variables such as V when constructing the imputation model. In the context of longitudinal data with missing at random processes (see Sect. 1.3), this can be achieved by adopting a joint model for the responses over time (e.g., a mixed model) and, while the primary analysis is to be based only on a final response, intermediate values can ensure a more suitable imputation process which may translate to more precise estimates of treatment effects and more powerful tests.

1.2.3 An Illustrative Simulation Study

Here we report on a simple simulation study to illustrate these methods. We let $p_C = 0.5$, $P(V = 1) = p = 0.5$, $\beta_1 = \log 1.5$, $\gamma_2 = \log 0.5$ and $\gamma_3 = \log 2$. These specifications can be used to obtain values for γ_0 and γ_1. Note that the true odds ratio $\exp(\beta_1)$, which would be consistently estimated in the absence of missing data, is 1.5 in this formulation ($\beta_1 \approx 0.4055$). We then specify the missing data model as $\alpha_1 = 0$, $\alpha_2 = \log 2$, $\alpha_3 = \log 4$, and ensure that $P(R = 1) = p_R = 0.5$, so 50 % of subjects will have incomplete response data and there is a differential degree of association between Y and R in the control and treatment arms. The limiting value of a naive estimate of β_1 is 0.4831 based on the earlier calculations, giving an asymptotic bias of approximately 0.0777.

Two thousand datasets of $n = 500$ individuals were simulated and the following analyses were carried out: (i) a complete-case likelihood analysis using (1.7), (ii) an inverse weighted analysis using (1.9) with weights known, (iii) an inverse weighted analysis with weights estimated via logistic regression, and (iv) multiple imputation with $m = 20$ and the imputation model based on $Y|X, V$. In all cases the response model was simply based on $Y|X$. The empirical biases, empirical standard errors (ESE), average asymptotic standard errors (ASE), and empirical coverage of nominal 95 % confidence intervals (ECP) are reported in Table 1.1.

The empirical biases of the complete-case analyses (expected since $\gamma_3 \neq 0$ and $\alpha_3 \neq 0$) are apparent, and this leads to empirical coverage probabilities less than the nominal 95 % level. The bias from the inverse weighted analyses

Table 1.1 Simulation results of naive and adjusted analyses using inverse weighting (known and estimated weights) and multiple imputation; $P(X = 1) = 0.5$; $P(V = 1) = 0.5$; $p_C = 0.5$; $\beta_0 = 0$, $\beta_1 = \log 1.5$, $\gamma_0 = 0.347$, $\gamma_1 = 0.059$, $\gamma_2 = \log 0.5$, $\gamma_3 = \log 2$, $p_R = 0.5$; $\alpha_0 = -0.654$, $\alpha_1 = 0$, $\alpha_2 = \log 2$, $\alpha_3 = \log 4$, Number of subjects = 500; Number of simulations = 2,000

Method of analysis	Parameter	Bias	ESE	ASE	ECP
Complete-case analysis	β_0	−0.072	0.201	0.196	93.3
	β_1	0.076	0.268	0.260	93.1
Weighted analysis	β_0	−0.005	0.204	0.199	95.1
(Known weights)	β_1	0.009	0.278	0.274	94.1
Weighted analysis	β_0	−0.004	0.203	0.200	95.2
(Estimated weights)	β_1	0.008	0.279	0.275	94.3
Multiple imputation[a]	β_0	−0.004	0.203	0.195	94.2
($m = 20$)	β_1	−0.004	0.281	0.277	94.2

[a] m indicates the number of complete pseudo-datasets created for multiple imputation

with known and estimated weights are negligible and the empirical coverage probabilities are compatible with the 95 % level. The biases are similarly small for the estimators based on multiple imputation and the empirical coverage probabilities are compatible with the 95 % level for these as well. Also noteworthy is the similarity in the standard errors of the estimates based on inverse weighting and multiple imputation.

1.2.4 Further Remarks

In many clinical settings there are a number of ad hoc alternative approaches for dealing with missing response data. In dermatology trials, for example, it is common to use so-called *non-responder* imputation [12, 28]. If, as we have described here, the response $Y = 1$ indicates a successful response to treatment (e.g. alleviation of symptoms), then in non-responder imputation (NRI), individuals who do not provide a response are assigned a value $Y = 0$ (i.e. they did not remain in the trial and report an alleviation of symptoms). The rationale for this crude form of imputation may arise from the notion that anything other than completing the course of treatment and exhibiting a good clinical response is undesirable and hence should be treated as a failure. An intuitively appealing aspect of this form of imputation is that all patients randomized are utilized in the analysis. However with NRI, a naive estimator of the probability of a successful response given X is, in fact, consistent for the joint probability $P(Y = 1, R = 1|X)$; this reflects that individuals must both provide a response and the response must be successful. The validity of estimates achieved through this method depends, therefore, on the process giving rise to the missing data. If $R \perp (Y, X)$, estimates of response rates within treatment arms (and therefore also estimates of AD) are conservative in that they are down-weighted by the probability of a response being observed (in fact, we are consistently estimating $P(Y = 1|X) \cdot P(R = 1)$). When data are not missing completely at random, NRI analyses will not yield consistent estimates of RR, OR, or AD. Depending on the mechanism giving rise to the missing data (which is generally unknown), NRI analyses can lead to conservative (too small) or anti-conservative (too large) estimates of treatment effect [25]. Despite this, NRI is commonly assumed to be a conservative method of analysis [37].

When responses are continuous, the calculations discussed in previous sections can be carried out following similar principles; to make this clear we wrote the expressions in a general form using expectations and explicit probability statements in key places. With continuous responses, however, another common crude method of imputation is often used called *mean value* imputation. In this case the average value of the response (perhaps for that particular treatment arm, or overall) is assigned to individuals with missing responses. This strategy can also lead to conservative or anti-conservative estimates of treatment effect depending on the particular setting, and naive standard errors will not typically reflect the effect of imputation.

The discussion of multiple imputation given earlier is often referred to as *parametric* multiple imputation since it relies on the explicit specification of a parametric model to simulate the imputed data for each data set. Other versions of multiple imputation are often adopted which employ implicit models to exploit the data observed in the sample [15, 21]. *Nonparametric* multiple imputation involves finding a set of completely observed individuals who are "similar" to an individual with a missing response (with respect to key attributes or a summary measure) and randomly selecting the responses from this set of similar individuals [29, 38]. This sampling is done with replacement to make up multiple complete datasets. Here judgement is not required to specify a probability model for imputation of the response, but rather to identify the set of "similar" individuals for each individual with a missing response [36]. Matching, stratification or use of propensity scores are useful for this goal, and several procedures are available in common statistical packages to facilitate this.

1.3 Incomplete Longitudinal Data

1.3.1 Notation and Terminology

Consider a longitudinal study in which the plan is to assess each of n individuals over K distinct assessment times. Let $Y_i = (Y_{i1}, \ldots, Y_{iK})'$ denote the random variable corresponding to the response vector for individual i over the K assessments. Suppose that every individual under study has measurements taken on p baseline covariates so that subject i has baseline covariate vector $X_i = (X_{i1}, \ldots, X_{ip})'$. We assume X_i is completely observed, and let $P(Y_i|X_i)$ denote the probability model of interest.

We restrict attention to incomplete longitudinal data due to drop-out, and suppose that the last time an observation for individual i occurred was at time K_i; this is a random variable and we let k_i denote its realization, as illustrated in Fig. 1.2. We can then partition the response vector as $Y_i = (\bar{Y}_i, Y_i^-)$, where $\bar{Y}_i = (Y_{i1}, \ldots, Y_{iK_i})'$ is observed and $Y_i^- = (Y_{i,K_i+1}, \ldots, Y_{iK})$ is missing. Let $R_i = (R_{i1}, \ldots, R_{iK})'$ be the corresponding vector of missing data indicators, where $R_{ik} = I(k \leq K_i)$, $k = 1, \ldots, K$. We can therefore equivalently think of R_i as a random vector or K_i as a random variable. Little and Rubin [20] and Rubin [35] define three classes of missing data mechanisms for this context.

Data are said to be *missing completely at random (MCAR)* if missingness (failing to observe a value) does not depend on any observed or unobserved measurements, i.e. $P(R_i|Y_i, X_i) = P(R_i)$. Data are said to be *missing at random (MAR)* if, conditional on the observed data, missingness does not depend on the data that are unobserved; that is, $P(R_i|Y_i, X_i) = P(R_i|\bar{Y}_i, X_i)$. Data are said to be *not missing at random* or, equivalently, *missing not at random (MNAR)* if missingness

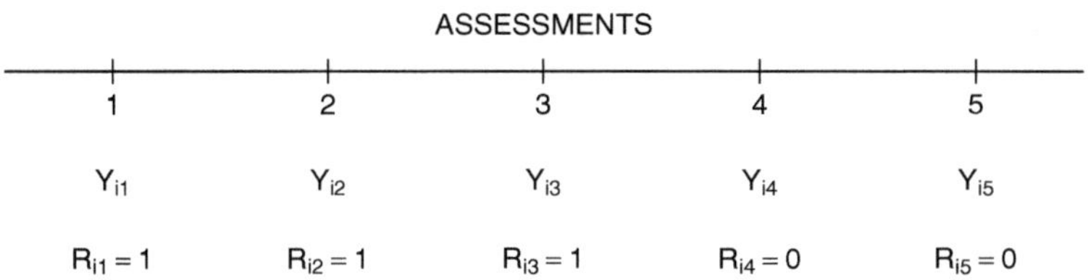

Fig. 1.2 Schematic of schedule of assessments in longitudinal study with $K = 5$ for an individual with $k_i = 3$

depends on the value of the realized (but unobserved) response, i.e. $P(R_i|Y_i, X_i)$ cannot be simplified. It is perhaps worth emphasizing that these terms must be used and interpreted in the context of the available information (or at least the information being used); MNAR mechanism can become a MAR mechanism in light of additional information used judiciously.

1.3.2 Likelihood-Based Methods of Estimation and Inference

As in the univariate case, the likelihood for incomplete longitudinal data is developed by specifying the joint distribution of response variable Y_i and the missing data indicators R_i (or equivalently K_i), given the covariates X_i. Two classes of models have been proposed based on alternative factorizations of the joint distribution of $(Y_i, R_i)|X_i$ [19]: one is based on *selection models* [20], the other is based on *pattern mixture models* [10, 18].

With selection models, the joint distribution of Y_i and R_i is factored as

$$P(R_i, Y_i|X_i; \beta, \alpha) = P(R_i|Y_i, X_i; \alpha)\ P(Y_i|X_i; \beta)\ , \tag{1.11}$$

where the distribution of R_i, $P(R_i|Y_i, X_i; \alpha)$, is indexed by a vector of parameters α and the distribution of Y_i, $P(Y_i|X_i; \beta)$, is indexed by a vector of β.

With pattern-mixture models, the factorization of the joint distribution is

$$P\ (R_i, Y_i|X_i; \beta, \alpha) = P\ (Y_i|X_i, R_i; \zeta)\ P\ (R_i|X_i; \theta)\ , \tag{1.12}$$

where in $P(Y_i|X_i, R_i; \zeta)$, the distribution of Y_i, is defined separately for each missing data configuration and indexed by parameters ζ, and the distribution of R_i, $P(R_i|X_i; \theta)$, is known up to parameters θ.

When we are concerned with the parameters of the marginal distribution of Y, averaged over the missing data patterns, it is in many senses more natural to use selection models, because people do not want to make inference conditional on the missing data indicators. In the following, we focus on selection models.

To describe the likelihood based approach we derive the joint density of the observed data $(\bar{Y}_i, R_i)$ by integrating out the missing data Y_i^- in the selection model of the joint distribution as

$$P(R_i, \bar{Y}_i | X_i; \alpha, \beta) = \int P(R_i | \bar{Y}_i, Y_i^-, X_i; \alpha)\ P(\bar{Y}_i, Y_i^- | X_i; \beta)\, dY_i^- .$$

Let $\bar{Y} = \{\bar{Y}_i, i = 1, 2, \ldots, n\}$ and $R = \{R_i, i = 1, 2, \ldots, n\}$ for a sample of n independent subjects. Then the observed-data joint likelihood for $(\alpha', \beta')'$ is

$$L(\alpha, \beta; \bar{Y}, R) = \prod_{i=1}^{n} \int P(R_i | \bar{Y}_i, Y_i^-, X_i; \alpha)\ P(\bar{Y}_i, Y_i^- | X_i; \beta)\, dY_i^- . \quad (1.13)$$

When the missing data mechanism is MAR, $P(R_i | \bar{Y}_i, Y_i^-, X_i) = P(R_i | \bar{Y}_i, X_i)$ and (1.13) becomes

$$L(\alpha, \beta; \bar{Y}, R) = \prod_{i=1}^{n} \left\{ P(R_i | \bar{Y}_i, X_i; \alpha) \int P(\bar{Y}_i, Y_i^- | X_i; \beta)\, dY_i^- \right\} \quad (1.14)$$

$$= \prod_{i=1}^{n} \{ P(R_i | \bar{Y}_i, X_i; \alpha)\ P(\bar{Y}_i | X_i; \beta) \} .$$

If the parameters α and β are functionally independent, then likelihood inference for β from (1.14) is the same as a likelihood inference for β from the observed "partial" likelihood simply using the available data

$$L(\beta; \bar{Y}) = \prod_{i=1}^{n} P(\bar{Y}_i | X_i; \beta) . \quad (1.15)$$

Thus likelihood functions are unaffected by MAR mechanisms and this has contributed in part to the popularity of mixed effects models for the analysis of longitudinal data. If data are MNAR, then the simplification in (1.14) is not possible and we must use (1.13). This likelihood may lead to identifiability problems and so sensitivity analyses are often advocated for this case [31].

We remark that, as in the univariate case, one can sometimes identify an auxiliary covariate V_i which renders $R_i \perp Y_i^- | \bar{Y}_i, X_i, V_i$, so that inclusion of V_i in the analysis causes the missing data mechanism to be MAR. In this case, consider

$$P(R_i, \bar{Y} | X_i, V_i) = \int P(R_i | \bar{Y}_i, Y_i^-, X_i, V_i)\ P(\bar{Y}_i, Y_i^- | X_i, V_i)\, dY_i^-$$

$$= P(R_i | \bar{Y}_i, X_i, V_i)\ P(\bar{Y}_i | X_i, V_i) .$$

This is only useful if we aim to estimate the effect of both X_i and V_i on the distribution of Y_i. Again, however, V_i may be useful for multiple imputation (as in Sect. 1.2.2.3) or for inverse weighting as we discuss in the next section.

1.3.3 Generalized Estimating Equations

Using standard notation for generalized linear models of binary data, we let $E(Y_{ik}|x_i) = P(Y_{ik} = 1|x_i) = \mu_{ik}$ and $\text{var}(Y_{ik}|x_i) = \mu_{ik}(1 - \mu_{ik})$, $k = 1, \ldots, K$. Furthermore, we let $\Sigma_i(\beta, \rho) = \text{cov}(Y_i|x_i) = \mathbb{A}_i^{\frac{1}{2}} Q(\rho) \mathbb{A}_i^{\frac{1}{2}}$ where $\mathbb{A}_i = \text{diag}\{\mu_{ik}(1 - \mu_{ik}), k = 1, \ldots, K\}$ and $Q(\rho)$ is a $K \times K$ working correlation matrix with (k, k') entry, $Q_{kk'}(\rho)$, parameterized in terms of a vector of association parameters ρ. A marginal generalized linear model is formed by letting $g(\mu_{ik}) = x'_{ik}\beta$ where $g(\cdot)$ is a known link function and $\beta = (\beta_0, \ldots, \beta_p)'$ is a $(p + 1) \times 1$ vector of regression coefficients.

Generalized estimating equations for β take the form

$$U(\beta, \rho) = \sum_{i=1}^{n} U_i(\beta, \rho) = 0 \tag{1.16}$$

where $U_i(\beta, \rho) = G_i'(\beta)\Sigma_i^{-1}(\beta, \rho)(Y_i - \mu_i)$, with $\mu_i = (\mu_{i1}, \ldots, \mu_{iK})'$ and $G_i(\beta) = \partial\mu_i(\beta)/\partial\beta'$ a $K \times (p + 1)$ matrix of derivatives [17]. If $\hat{\beta}$ is the solution for fixed $\rho = \rho_o$, then asymptotically $\sqrt{n}(\hat{\beta} - \beta) \sim N(0, \text{var}(\sqrt{n}(\hat{\beta} - \beta)))$ with

$$\text{var}(\sqrt{n}(\hat{\beta} - \beta)) = [A^{-1}(\beta, \rho_o)][B(\beta, \rho_o)][A^{-1}(\beta, \rho_o)]', \tag{1.17}$$

where $A(\beta, \rho) = E(\partial U_i(\beta, \rho)/\partial\beta')$ and $B(\beta, \rho) = E(U_i(\beta, \rho)U_i'(\beta, \rho))$. When ρ is not specified, estimation of β is facilitated by iteratively replacing ρ with a $\sqrt{n}$-consistent moment-type estimate based on estimates of β at successive iterations of a scoring algorithm [17].

The functional form of $Q_{kk'}(\rho)$, $k \neq k'$, $k, k' = 1, \ldots, K$, is typically unknown, but even if the correlation structure is misspecified, consistent estimators of β arise from solving (1.16), and (1.17) will still hold. However, misspecification of the correlation structure in (1.16) can lead to inefficient estimators of β and, in more extreme cases, problematic asymptotic properties arise for the solution [7]. In many cases, the working independence assumption can yield quite efficient estimators [41], so we set $Q_{kk'}(\rho) = \rho_o = 0$ for $k \neq k'$ in what follows. An estimate of (1.17) is obtained in this case by computing

$$\widehat{\text{var}}(\sqrt{n}(\hat{\beta} - \beta)) = [\hat{A}^{-1}(\hat{\beta}, \rho_o)][\hat{B}(\hat{\beta}, \rho_o)][\hat{A}^{-1}(\hat{\beta}, \rho_o)]', \tag{1.18}$$

where

$$\hat{A}(\hat{\beta}, \rho_o) = -n^{-1} \sum_{i=1}^{n} G_i'(\hat{\beta}) \mathbb{A}_i^{-1}(\hat{\beta}, \rho_o) G_i(\hat{\beta}),$$

and

$$\hat{B}(\hat{\beta}, \rho_o) = n^{-1} \sum_{i=1}^{n} G_i'(\hat{\beta}) \mathbb{A}_i^{-1}(\hat{\beta}, \rho_o)(y_i - \hat{\mu}_i)(y_i - \hat{\mu}_i)' \mathbb{A}_i^{-1}(\hat{\beta}, \rho_o) G_i(\hat{\beta}) .$$

As in the univariate case, however, this estimating equation approach is not appropriate when data are incomplete and not missing completely at random.

Selection models provide a natural framework for characterizing factors which affect the risk of attrition in longitudinal studies. Let $R_{ik} = I(k \leq K_i)$ and $\bar{R}_{ik} = \{R_{i1}, \ldots, R_{ik}\}$, $k = 1, \ldots, K_i$. Selection models involve modeling the conditional probability of drop-out at each visit, which we denote here as $\eta_{ik} = P(R_{ik} = 0 | R_{i1} = \cdots = R_{i,k-1} = 1, y_i, x_i)$. As mentioned in Sect. 1.3.1, the nature of the relation between this conditional probability of drop-out, covariates, and (possibly missing) responses determines the impact that drop-outs have on inferences regarding the regression coefficients in the response model. We restrict attention here to settings in which data are MAR, with any covariate dependence based only on previously observed covariates or responses. In this case, η_{ik} may be a function of $\bar{Y}_i$ and X_i, but not of Y_i^-. Let $H_{ik}^y = \{y_{i1}, \ldots, y_{i,k-1}\}$ be the history of response Y up to time k. In practice, we typically let η_{ik} depend on H_{ik}^y and X_i.

Since R_{ik} is a binary variable it is convenient to formulate logistic regression models for the conditional probability of drop-out given by

$$\log(\eta_{ik}/(1 - \eta_{ik})) = w_{ik}' \alpha^{(k)} , \tag{1.19}$$

where $\alpha^{(k)} = (\alpha_0^{(k)}, \ldots, \alpha_{q_k}^{(k)})'$ is a $(q_k + 1) \times 1$ vector of regression coefficients characterizing the nature of the relationship between w_{ik} and η_{ik}, and w_{ik} is a covariate vector containing relevant observed information in H_{ik}^y and X_i.

The inverse-weighted estimating equations under the working independence assumption take the form

$$U(\beta, \alpha) = \sum_{i=1}^{n} U_i(\beta, \alpha) = 0 \tag{1.20}$$

where under cluster-specific weights as discussed by Fitzmaurice [9],

$$U_i(\beta, \alpha) = G_i'(\beta) \Sigma_i^{-1}(\beta) \Delta_i(\alpha)(Y_i - \mu_i) ,$$

$\Sigma_i(\beta) = \text{diag}\{\eta_{ik}(1 - \eta_{ik}), k = 1, \ldots, K_i\}$, $\Delta_i(\alpha) = I(K_i = k_i)/\pi_i(\alpha)$, and $\pi_i(\alpha) = P(K_i = k_i | \bar{Y}_i, x_i; \alpha)$. We often assume all subjects are available for the first assessment, so $\pi_i(\alpha) = \eta_{i2}(\alpha)$ if $k_i = 1$, $\pi_i(\alpha) = (1 - \eta_{i2}(\alpha))\eta_{i3}(\alpha)$ if $k_i = 2$, $\pi_i(\alpha) = (1 - \eta_{i2}(\alpha))(1 - \eta_{i3}(\alpha))$ if $k_i = 3$, etc. In practice, an estimate of α can be obtained by fitting ordinary logistic regression models to the missing data indicators as appropriate. Inserting $\hat{\alpha}$ into (1.20) gives estimating equations which can be solved for β in the usual fashion [32].

1.3.4 Naive Methods of Imputation

The "last observation carried forward" (LOCF) imputation approach for dealing with missing values due to drop-outs operates as follows: if $k_i < K$, missing observations at visits $k = k_i + 1, \ldots, K$ are replaced with the value of the most recently observed response (i.e. y_{ik_i}). To distinguish the actual (possibly latent) responses from the pseudo-responses used under this imputation scheme, we use Y_i^* to denote the response vector under LOCF imputation. Therefore $Y_{ik}^* = Y_{ik}$ for $k \leq k_i$ and $Y_{ik}^* = Y_{ik_i}$ for $k > k_i$, $k = 1, 2, \ldots, K$. Assumptions made for the response Y_i are adopted for the pseudo-response Y_i^* since analyses are typically carried out under the assumption that they are in some sense equivalent. In fact, in most situations for which the assumptions regarding Y_i are true, they will not be true for Y_i^*, implying that the estimating equation (1.16) is misspecified for the pseudo response. The frequency properties of estimators of β based on Y_i^* have been investigated under a wide range of settings by several authors [5, 27] based on the theory of misspecified models [34, 43]. As with the other naive imputation approaches discussed earlier, LOCF leads to inconsistent estimators in a wide variety of settings and can result in either conservative or anti-conservative estimates of treatment effect.

1.4 Missing Covariates

1.4.1 Likelihood Analyses

Now consider a setting of a clinical trial in which the secondary analyses are directed at fitting a regression model which controls for a variable Z in addition to the treatment indicator; for the sake of simplicity we again suppose Z is a binary variable. One might simply specify a model with the main effects, but we consider a model of the form

$$P(Y = 1|X, Z; \lambda) = \text{expit}\,(\lambda_0 + \lambda_1 X + \lambda_2 Z + \lambda_3 XZ)\ . \tag{1.21}$$

This would be of interest if there are questions about whether the effect of treatment was significantly different in different subgroups defined by a binary covariate Z, for example, in which case λ_3 is parameter of primary interest. Such questions arise frequently when the goal is to examine the robustness and generalizability of findings; in cancer trials, for example, the aim may be to investigate whether the effect of chemotherapy varies according to tumour type. Some centers may not collect complete histological data and in such circumstances covariate data on tumour type will be incomplete.

Let $C = I(Z \text{ observed})$ indicate whether the covariate value was recorded. The observed data likelihood can then be written as

$$L \propto P(Y, Z, C = 1|X)^C \; P(Y, C = 0|X)^{1-C} , \tag{1.22}$$

where we can marginalize over Z with $\sum_z P(Y, Z = z, C = 0|X)$ to obtain $P(Y, C = 0|X)$, the contribution from individuals for whom Z is unobserved.

As in the case of incomplete responses, the tendency is to focus on simple analyses such as those restricted to individuals with complete covariate data. In this case the adopted likelihood would be based on the response model with the implicit condition $C = 1$ and so is proportional to

$$\begin{aligned} P(Y|Z, X, C = 1) &= \frac{P(C = 1|Y, Z, X) \; P(Y|Z, X)}{\sum_y P(C = 1|Y = y, Z, X) \; P(Y = y|Z, X)} \\ &= \frac{P(C = 1|Y, Z, X)}{P(C = 1|Z, X)} \; P(Y|Z, X) . \end{aligned} \tag{1.23}$$

If $C \perp Y|Z, X$, then (1.23) reduces to $P(Y|Z, X)$ and a complete-case analysis will yield consistent estimators of λ, but otherwise inconsistent estimators are obtained; we show this by example in the simulation studies that follow. Note that with incomplete covariate data, missingness can depend on the potentially missing variable (Z) and a complete-case analysis remains valid because it involves conditioning on this covariate; this is in contrast to the setting of missing responses where the missing data must be modelled. However even when valid, this complete-case analysis ignores the information contained in the responses from individuals with incomplete data, and therefore may result in less than optimal efficiency.

1.4.2 An EM Algorithm

If one makes assumptions regarding the distribution of the incomplete covariate in likelihood analyses based on (1.22), one can exploit information from individuals with $C = 0$ and improve efficiency. To see this note that the second term in (1.22),

$$P(Y, C = 0|X) = \sum_{z=0}^{1} P(Y|Z = z, X) \; P(Z = z|X) \; P(C = 0|Y, Z = z, X) ,$$

is indexed by λ (as well as the parameters in $P(Z|X)$ and those of the missing data process). If $P(C|Y,Z,X) = P(C|Y,X)$ or $P(C|X)$, then the missing data process can be modelled using observed data (Y and X). If $P(C|Y,Z,X) = P(C|Z,X)$, then while this is a desirable missing data process for complete-case analysis (see (1.23)), in this setting there is a need to make uncheckable assumptions about the missing data process, since the dependence between C and Z given X cannot be modelled in general. Progress can be made here if an auxiliary variable V can be found which satisfies $C \perp Z|X,V,Y$ (see Sects. 1.4.3 and 1.4.4).

The assumptions that are needed to exploit information from individuals with $C = 0$ could include the fully specified conditional covariate distribution, or simply its parametric form. In the latter case, the EM algorithm offers a convenient method for estimation [8]. The complete data likelihood L_C corresponding to (1.22) is proportional to

$$[P(C|Y,Z,X)\ P(Y|Z,X)\ P(Z|X)]^C\ \ [P(C|Y,Z,X)\ P(Y|Z,X)\ P(Z|X)]^{1-C}\,.$$

We typically work with the "partial" complete data likelihood

$$L_C \propto [P(Y|Z,X)\ P(Z|X)]^C\ \ [P(Y|Z,X)\ P(Z|X)]^{1-C} \tag{1.24}$$

under the assumption that the information regarding λ in the missing-data model is negligible. Working with (1.24) then requires an expression for

$$P(Z|C=0,Y,X) = \frac{P(C=0|Y,Z,X)\ P(Y|Z,X)\ P(Z|X)}{\sum_z P(C=0|Y,Z=z,X)\ P(Y|Z=z,X)\ P(Z=z|X)} \tag{1.25}$$

for the expectation step of the EM algorithm, which if $C \perp Z|Y,X$ gives simply

$$\frac{P(Y|Z,X)\ P(Z|X)}{\sum_z P(Y|Z=z,X)\ P(Z=z|X)}\,. \tag{1.26}$$

It is clear from (1.26) that, provided $P(C|Y,Z,X) = P(C|Y,X)$, the partial complete data likelihood (1.24) can be used if assumptions are made regarding the distribution of $Z|X$. In fact, when treatment is randomly assigned, only the marginal distribution of Z is required since $Z \perp X$. However, if C depends on Z given Y and X, then there is an identifiability problem and (1.25) cannot be evaluated without strong assumptions regarding the missing data process.

1.4.3 Multiple Imputation with Missing Covariates

Suppose now that there exists a completely observed covariate V which renders $C \perp Z|Y,X,V$. Again for simplicity we assume V is binary with $P(V=1) = p$

and $P(V = 0) = 1 - p$. Multiple imputation can be carried out using a model for $P(Z|Y, X, V, C) = P(Z|Y, X, V)$ and because $Z \perp C|Y, X, V$, the model for $Z|Y, X, V$ can be fitted based on individuals with complete data. For illustration here, we adopt a simpler model whereby $P(Z|V, X, Y) = P(Z|V, X)$ which can be easily fitted using a saturated logistic regression model,

$$P(Z = 1|X, V) = \text{expit}(\delta_0 + \delta_1 X + \delta_2 V + \delta_3 XV) . \tag{1.27}$$

Suppose the missing data model is

$$P(C = 1|X, V) = \text{expit}(\alpha_0 + \alpha_1 X + \alpha_2 V + \alpha_3 XV) , \tag{1.28}$$

and the response is generated according to

$$P(Y = 1|X, Z, V) = \text{expit}(\lambda_0^* + \lambda_1^* X + \lambda_2^* Z + \lambda_3^* XZ + \lambda_4^* V) . \tag{1.29}$$

The response model of interest (1.21) can be recovered by noting that $P(Y = 1|X, Z) = E_{V|X,Z}[P(Y = 1|X, Z, V)]$.

The association between Y and C given X and Z is determined by the joint model

$$\begin{aligned} P(Y, C|X, Z) &= \sum_v P(Y|X, Z, V = v, C)P(C|X, Z, V = v)P(V = v|X, Z) \\ &= \sum_v P(Y|X, Z, V = v)P(C|X, V = v)P(V = v|Z) . \end{aligned}$$

If we simply fit the response model in (1.21), a complete-case analysis is generally invalid in this setting because $C \not\perp Y|X, Z$ due to the omission of the variable V in (1.21).

Following the same arguments as given earlier, for any given data set we may carry out multiple imputation of Z based on the model $P(Z|Y, X, V)$. If this model is fit and an estimate of δ is obtained, by standard large sample theory $\hat{\delta} \sim \text{MVN}(\delta, \mathscr{I}^{-1}(\hat{\delta}))$.

We proceed by letting $d^{(r)}$ denote the rth realization from $\text{MVN}(\hat{\delta}^*, \mathscr{I}^{-1}(\hat{\delta}^*))$, and using $d^{(r)}$ to generate values for all missing Z according to $P(Z|Y, X, V; d^{(r)})$. Then based on this "complete" data set, we fit $P(Y|Z, X; \lambda)$ to obtain $\hat{\lambda}^{(r)}$. This is repeated m times, and we let $\bar{\hat{\lambda}} = \sum_{r=1}^{m} \hat{\lambda}^{(r)}/m$ and compute the standard errors as described in Sect. 1.2.2.3.

1.4.4 Inverse Probability Weighted Estimating Functions

Inverse probability weighting can be used to obtain unbiased estimating functions for a complete-case analysis. If $P(C_i|Y_i, X_i, V_i, Z_i) = P(C_i|Y_i, X_i, V_i)$, then we can write the inverse weighted estimating function as

$$U(\beta) = \sum_{i=1}^{n} \frac{C_i}{P(C_i = 1|Y_i, X_i, V_i)} (Y_i - E(Y_i|X_i, Z_i; \lambda))\ W_i\ , \tag{1.30}$$

where $W_i = (1, X_i, Z_i, X_i Z_i)'$, and this can be shown to have expectation zero. Since the model in the weight indicates a dependence on (Y_i, X_i, V_i) which are always observed, then it can be fit and a $\sqrt{n}$-consistent estimator of α in (1.28) inserted; a consistent estimator of λ will then be obtained by setting (1.30) equal to zero and solving for λ.

1.4.5 A Simulation Study

Here we report on a simulation study designed to demonstrate the performance of several methods of dealing with missing covariates. We consider the response model (1.21) with $\lambda_4^* = 0$ and log 4 in (1.29) and find the parameters of the covariate distribution to ensure these parameter values were obtained. We set $\lambda_1 = 0$, $\lambda_2 = \log 1.5$, $\lambda_3 = \log 0.5$, $P(X = 1) = 0.5$, $P(V = 1) = 0.5$, and $P(Z = 1) = 0.25$ so $P(Y = 1) = 0.5$. We set $\delta_1 = 0$, $\delta_2 = 0$, $\delta_3 = \log 4$ in (1.27) to ensure that, as desired, $P(Z = 1) = 0.25$ based on (1.27). Finally, setting $\alpha_0 = -0.151$, $\alpha_1 = \log 0.8$, $\alpha_2 = \log 1.2$ and $\alpha_3 = \log 2$ in (1.28) yields $P(C = 1) = 0.5$; so for 50 % of subjects we would expect the covariate to be missing. We generated data for sample sizes of 500 and 2,000 individuals in 2,000 simulated datasets. The analyses conducted included a complete-case analysis, inverse probability weighted analyses with known and estimated weights, an EM algorithm for which the correct covariate distribution was assumed, and multiple imputation. The imputation model adopted was a saturated logistic regression model for Z given (Y, X, V), involving eight parameters: the intercept, three main effects, three two way interactions and a three way interaction. The empirical biases, empirical standard errors, average asymptotic standard errors, and empirical coverage probabilities are reported in Table 1.2 for sample sizes of 500 (left column) and 2,000 (right column). The top half of the table corresponds to the case where $C \perp Y|X, Z$; in the bottom half, $C \not\perp Y|X, Z$ but $C \perp Y|X, Z, V$ where V is the auxiliary covariate used for inverse weighting with $P(C|X, V)$, and multiple imputation via $P(Z|X, V)$.

The results where $Y \perp C|X, Z$ (top half) indicate all methods yield approximately unbiased estimates, close agreement between the empirical and average asymptotic standard errors, and empirical coverage that is compatible with the

Table 1.2 Simulation results of naive and adjusted analyses using inverse weighting (known and estimated weights), EM and multiple imputation; $P(X = 1) = 0.5$; $P(V = 1) = 0.5$; $P(Z = 1) = 0.25$, $\delta_1 = 0$, $\delta_2 = 0$, $\delta_3 = \log(4) = 1.3862$; $P(Y = 1) = 0.5$, $\lambda_1 = 0$, $\lambda_2 = \log(1.5) = 0.405$, $\lambda_3 = \log(0.5) = -0.693$; $P(C = 1) = 0.5$, $\alpha_0 = -0.151$, $\alpha_1 = \log(0.8) = -0.223$, $\alpha_2 = \log(1.2) = 0.182$, $\alpha_3 = \log(2) = 0.693$; Number of simulations $= 2,000$

Method	Parameter	Sample size: 500				Sample size: 2,000			
		Bias	ESE	ASE	ECP	Bias	ESE	ASE	ECP
		$Y \perp C \mid X, Z$ ($\lambda_4^* = 0$)							
Complete-case analysis	λ_0	0.001	0.203	0.202	95.2	0.001	0.101	0.100	95.1
	λ_1	−0.001	0.301	0.300	95.2	−0.005	0.152	0.149	94.3
	λ_2	0.017	0.531	0.505	95.2	0.015	0.248	0.244	94.9
	λ_3	−0.010	0.655	0.632	94.8	−0.015	0.315	0.307	94.7
Weighted analysis (Known weights)	λ_0	0.001	0.203	0.203	95.0	0.001	0.101	0.100	95.1
	λ_1	0.000	0.304	0.304	95.5	−0.006	0.152	0.151	94.7
	λ_2	0.017	0.532	0.506	95.5	0.015	0.248	0.244	95.0
	λ_3	−0.013	0.659	0.637	94.8	−0.015	0.317	0.310	94.6
Weighted analysis (Estimated weights)	λ_0	0.002	0.204	0.203	95.0	0.001	0.101	0.101	95.2
	λ_1	0.000	0.305	0.304	95.1	−0.006	0.152	0.151	94.8
	λ_2	0.018	0.534	0.507	95.5	0.015	0.248	0.244	95.2
	λ_3	−0.013	0.662	0.638	94.8	−0.016	0.317	0.310	94.5
EM	λ_0	−0.016	0.167	0.165	95.0	−0.016	0.081	0.082	94.9
	λ_1	0.010	0.240	0.238	94.6	0.011	0.118	0.118	95.0
	λ_2	−0.001	0.515	0.495	95.7	0.002	0.244	0.240	94.7
	λ_3	0.010	0.641	0.622	94.8	−0.001	0.310	0.304	94.5
Multiple imputation[a] ($m = 20$)	λ_0	0.007	0.157	0.158	95.4	0.002	0.075	0.077	95.6
	λ_1	−0.009	0.239	0.239	94.4	−0.003	0.116	0.116	95.0
	λ_2	−0.035	0.516	0.523	96.0	0.003	0.248	0.248	95.2
	λ_3	0.043	0.646	0.647	95.0	−0.010	0.315	0.311	94.5

		$Y \not\perp C \mid X, Z$ ($\lambda_4^* = \log(4)$)							
Complete-case analysis	λ_0	0.031	0.203	0.202	94.9	0.029	0.101	0.100	93.7
	λ_1	0.102	0.295	0.300	94.0	0.112	0.149	0.149	87.8
	λ_2	0.027	0.524	0.506	95.8	0.001	0.245	0.244	95.1
	λ_3	−0.084	0.647	0.633	95.8	−0.051	0.309	0.307	94.5
Weighted analysis	λ_0	0.000	0.203	0.203	95.1	−0.003	0.101	0.100	94.7
(Known weights)	λ_1	−0.004	0.300	0.304	95.6	0.005	0.150	0.151	95.5
	λ_2	0.027	0.525	0.507	95.7	0.001	0.245	0.245	95.1
	λ_3	−0.039	0.651	0.637	95.3	−0.008	0.311	0.309	95.0
Weighted analysis	λ_0	0.000	0.199	0.203	95.7	−0.003	0.099	0.101	95.0
(Estimated weights)	λ_1	−0.006	0.297	0.304	95.7	0.006	0.147	0.151	96.3
	λ_2	0.028	0.526	0.508	95.5	0.001	0.245	0.245	95.2
	λ_3	−0.037	0.651	0.638	95.2	−0.008	0.311	0.309	95.1
EM	λ_0	−0.009	0.166	0.165	94.8	−0.014	0.082	0.082	95.0
	λ_1	0.007	0.236	0.239	95.2	0.012	0.119	0.119	95.2
	λ_2	−0.007	0.512	0.496	95.4	−0.029	0.241	0.240	94.8
	λ_3	−0.005	0.638	0.622	95.3	0.026	0.305	0.303	95.0
Multiple imputation[a]	λ_0	0.010	0.155	0.163	95.9	0.000	0.076	0.077	95.8
($m = 20$)	λ_1	−0.009	0.233	0.250	96.1	0.002	0.116	0.116	94.8
	λ_2	−0.024	0.500	0.544	96.8	−0.010	0.245	0.248	95.3
	λ_3	0.015	0.622	0.680	96.2	0.001	0.309	0.310	95.3

[a] m indicates the number of pseudo-complete datasets created for multiple imputation

nominal 95 % level. The efficiency gains realized by modeling the covariate distribution are apparent by comparing the standard errors from the complete-case analysis with those of the EM algorithm. The standard errors of the estimates from the EM and MI algorithms are in close agreement. For the bottom half of the table, the empirical biases from the complete-case analyses expected due to (1.23) are apparent. The weighted analyses yielded estimators with much smaller empirical biases and better performance with the larger sample size. Smaller biases and smaller standard errors are seen with the EM algorithm. The multiple imputation analyses yielded small empirical biases as well and their standard errors are in close agreement with those of the EM algorithm. The empirical coverage probabilities for all valid methods are compatible with the nominal 95 % level. Simulations and analyses were carried out in R version 2.14.0 and SAS 9.2 on the Sun Solaris 10 platform.

1.5 Discussion

Incomplete data can arise in a number of settings for a variety of different reasons. Key factors influencing the extent of the impact on standard analyses are the proportion of missing data, and as demonstrated in this chapter, the nature of the stochastic mechanism which causes the data to be incomplete. Even when analyses are valid, loss of efficiency and decreased power are always issues. When possible, the extent of missing data should always be minimized.

Likelihood methods which have been developed and applied to minimize the effect of incomplete data are often directed at retrieving information about parameters of interest and improving power, but these come at the cost of making modeling assumptions beyond those typically made in analyses with complete data. These additional model assumptions are explicit, for example, when a parametric multiple imputation approach is adopted for incomplete response data. When covariates are missing and the EM algorithm is applied, one must make assumptions regarding the covariate distribution, which is not customary in routine analyses. When inverse probability weights are used, a model for the missing data process must be specified, which again is not something that is routinely done in standard analyses. The specified models should be checked carefully since consistent estimators only result if these are correct.

Throughout this chapter we have emphasized simple models with binary data, primarily for transparency and so that explicit results would be easy to obtain. When responses are continuous, inverse probability weighting changes very little; this approach requires modeling the missing data indicator which remains binary. Multiple imputation can be carried out in this case based on a linear regression model. The methods for longitudinal data can be similarly adapted. When incompletely observed covariates are continuous or categorical, the necessary model assumptions for the EM algorithm or multiple imputation may become more involved and robustness of inferences becomes more of a concern. When multiple covariates are missing, high-dimensional joint models for the covariates are required and these can be challenging to specify and check. These challenges, in part, are reasons for the appeal of inverse probability weighted analyses of individuals with complete data [24].

We have considered the cases of a missing response or a single missing covariate separately. Frequently both responses and covariates can be missing in a given dataset and hybrid methods can be employed [4].

We have emphasized the setting in which interest lies in a regression model for a marginal mean parameter. In some settings, association parameters (e.g. correlations or odds ratios) are viewed as of comparable importance. This occurs when scientific interest lies in the nature of the association structure, or if concerns lie in optimizing efficiency. In this case, regression models can be formulated for the association parameters and appropriate likelihood functions can be formed [13, 14]. Zhao and Prentice [45] describe how to do this using second order estimating equations. In the likelihood setting, the EM algorithm can be adopted and the idea of using inverse weighting for estimating association parameters can be adapted [44].

Acknowledgements This work was supported by a Post-Graduate Scholarship to Michael McIsaac from the Natural Sciences and Engineering Research Council (NSERC) of Canada and grants to Richard Cook from NSERC (Grant No. 101093) and the Canadian Institutes of Health Research (Grant No. 105099). Richard Cook is a Tier I Canada Research Chair in Statistical Methods for Health Research.

References

1. Albert, P.S., Follmann, D.: Shared-parameter models. In: Fitzmaurice, G., Davidian, M., Verbeke, G., Molenberghs, G. (eds.) *Longitudinal Data Analysis*, Chapter 18, 433–452. CRC Press, Boca Raton, FL. (2009)
2. Barnard, J., Rubin D.B.: Miscellanea. small-sample degrees of freedom with multiple imputation. *Biometrika* **86(4)**, 948–955 (1999)
3. Chen, B., Cook, R.J.: Strategies for bias reduction in estimation of marginal means with data missing at random. *Optimization and Data Analysis on Biomedical Informatics*. Ed: Panos Pardalos. American Mathematics Society (2011)
4. Chen, B., Yi, G.Y., Cook, R.J.: Weighted generalized estimating functions for longitudinal response and covariate data that are missing at random. *Journal of the American Statistical Associaton* **105**, 336–353 (2010)
5. Cook, R.J., Zeng, L., Yi, G.Y.: Marginal analysis of incomplete longitudinal binary data: a cautionary note on LOCF imputation. *Biometrics* **60(3)**, 820–828 (2004)
6. Cox, D.R.: The analysis of multivariate binary data. *Applied Statistics* **21**, 113–120 (1972)
7. Crowder, M.: On the use of a working correlation matrix in using generalized linear models for repeated measures. *Biometrika* **82**, 407–410 (1995)
8. Dempster, A.P., Laird, N.M., Rubin, D.B.: Maximum likelihood from incomplete data via the EM algorithm (with discussion). *Journal of the Royal Statistical Society: Series B* **39**, 1–38 (1977)
9. Fitzmaurice, G.M., Molenberghs, G., Lipsitz, S.R.: Regression models for longitudinal binary responses with informative drop-outs. *Journal of the Royal Statistical Society. Series B* (Methodological) **57(4)**, 691–704 (1995)

10. Glynn, R.J., Laird, N.M., Rubin, D.B.: Multiple imputation in mixture models for nonignorable nonresponse with followups. *Journal of the American Statistical Associaton* **88**, 984–993 (1993)
11. Godambe, V.P.: *Estimating Functions*. Oxford University Press, USA (1991)
12. Gordon, K.B., Langley, R.G., Leonardi, C., Toth, D., Menter, M.A., Kang, S., Hefferman, M., Miller, B., Hamlin, R., Lim, L., Zhong, J., Hoffman, R., Okun, M.M.: Clinical response to adalimumab treatment in patients with moderate to severe psoriasis: Double-blind, randomized controlled trial and open-label extension study. *Journal of the American Academy of Dermatology* **55**, 598–606 (2006)
13. Heagerty, P.J.: Marginalized transition models and likeliood inference for longitudinal categorical data. *Biometrics* **58**, 342–351 (2002)
14. Heagerty, P.J., Zeger, S.L.: Marginalized multilevel models and likelihood inference. *Statistical Science* **15**, 1–19 (2000)
15. Herzog, T., and Rubin, D.B.: Using multiple imputations to handle nonresponse in sample surveys. In: Madow, W.G., Olkin, I., Rubin, D.B. (eds.) *Incomplete Data in Sample Surveys, Volume 2: Theory and Bibliography*, 209–245. New York: Academic Press (1983)
16. Laupacis, A., Sackett, D.L., Roberts, R.S. : An assessment of clinically useful measures of the consequences of treatment. *New England Journal of Medicine* **318**, 1728–1733 (1988)
17. Liang, K.Y., Zeger, S. : Longitudinal data analysis using generalized linear models. *Biometrika* **73**, 13–22 (1986)
18. Little, R.J.A. : Pattern-mixture models for multivariate incomplete data. *Journal of the American Statistical Associaton* **88**, 125–134 (1993)
19. Little, R.J.A. : Modeling the drop-out mechanism in repeated-measures studies. *Journal of the American Statistical Associaton* **90**, 1112–1121 (1995)
20. Little, R.J.A., Rubin, D.B.: *Statistical Analysis with Missing Data*. Wiley, New York (1987)
21. Little, R.J.A., Rubin, D.B.: *Statistical Analysis with Missing Data*, Second Edition. Wiley, New York (2002)
22. Matthews, D.E., Farewell, V.T.: *Using and Understanding Medical Statistics*, 3rd Revised Edition. Karger, Basel, Switzerland (1996)
23. McCullagh, P., Nelder, J.A. : *Generalized Linear Models*, Second Edition. Chapman & Hall/CRC., London, UK (1989)
24. McIsaac, M.A., Cook, R.J.: Response-Dependent Sampling with Clustered and Longitudinal Data. In: *ISS-2012 Proceedings Volume On Longitudinal Data Analysis Subject to Measurement Errors, Missing Values, and/or Outliers*, 157–181. New York: Springer (2013)
25. McIsaac, M.A., Cook, R.J., Poulin-Costello, M.: Incomplete data in randomized dermatology trials: Consequences and statistical methodology. *Dermatology* **226**(1), 19–27 (2013). DOI 10.1159/000346247
26. Molenberghs, G., Kenward M. : *Missing Data in Clinical Studies*. John Wiley & Sons Ltd, West Sussex, England, UK (2007)
27. Prakash, A., Risser, R. C., Mallinckrodt, C. H.: The impact of analytic method on interpretaion of outcomes in longitudinal clinical trials. *International Journal of Clinical Practice* **62**, 1147–1158 (2008)
28. Reich, K., Nestle, F.O., Papp, K., Ortonne, J.P., Evans, R., Guzzo, C., Dooley, L.T., Griffiths, C.E.M. for the EXPRESS Study Investigators : Infliximab induction and maintenance therapy for moderate-to-severe psoriasis: a phase III, multicentre, double-blind trial. *Lancet* **366**, 1367–1374 (2005)
29. Reilly, M., Pepe, M.: The relationship between hot-deck multiple imputation and weighted likelihood. *Statistics in Medicine* **16**, 5–19 (1997)
30. Robins, J.M., Ritov, Y.: Toward a curse of dimensionality approximate (CODA) asymptotic theory for semiparameric models. *Statistics in Medicine* **16**, 285–319 (1997)
31. Robins, J.M., Hernan, M.A., Brumback, B. Marginal structural models and causal inference in epidemiology. *Epidemiology* **11(5)**, 550–560 (2000)
32. Robins, J.M., Rotnitzky, A., Zhao, L.P.: Analysis of semiparametric regression models for repeated outcomes in the presence of missing data. *Journal of the American Statistical Association* **90(429)**, 106–121 (1995)

33. Rothman, K.J., Greenland, S., eds: *Modern Epidemiology*, Second Edition. Lippincott Williams & Wilkins, Philadelphia (1998)
34. Rotnitzky, A., Wypij, D.: A note on the bias of estimators with missing data. *Biometrics* **50**, 1163–1170 (1994)
35. Rubin, D.B. : Inference and missing data. *Biometrika* **63**, 581–592 (1976)
36. Rubin, D.B. : *Multiple Imputation for Nonresponse in Surveys*. Wiley, New York (1987)
37. Saurat, J.H., Stingl, G., Dubertret, L., Papp, K., Langley, R.G., Ortonne, J.P., Unnebrink, K., Kaul, M., Camez, A., for the CHAMPION Study Investigators: Efficacy and safety results from the randomized controlled comparative study of adalimumab vs. methotrexate vs. placebo in patients with psoriasis (CHAMPION). *British Journal of Dermatology* **158**, 558–566 (2007)
38. Schenker, N., Welsh, A.H.: Asymptotic results for multiple imputation. *The Annals of Statistics* **16(4)**, 1550–1566 (1988)
39. Sprott, D.A. : *Statistical Inference in Science*. Springer, New York (2000)
40. Sprott, D.A., Farewell, V.T.: Randomization in experimental science. *Statistical Papers* **34**, 89–94 (1993)
41. Sutradhar, B.C., Das, K.: On the efficiency of regression estimators in generalised linear models for longitudinal data. *Biometrika* **86**, 459–465 (1999)
42. Wang, N., Robins, J. M. : Large-sample theory for parametric multiple imputation procedures. *Biometrika* **85**, 935–948 (1998)
43. White, H.A. : Maximum likelihood estimation of misspecified models. *Econometrica* **50**, 1–25 (1982)
44. Yi, G.Y., Cook, R.J. : Marginal methods for incomplete longitudinal data arising in clusters. *Journal of the American Statistical Association* **97(460)**, 1071–1080 (2002)
45. Zhao, L.P., Prentice, R.L. : Correlated binary regression using a quadratic exponential model. *Biometrika* **77**, 642–648 (1990)

Chapter 2
Bayesian Decision Theory and the Design and Analysis of Randomized Clinical Trials

Andrew R. Willan

Abstract Traditional approaches to the analyses and sample size determinations for randomized clinical trials are based on tests of hypotheses and rely on arbitrarily set error probabilities and smallest clinically important differences. Recently Bayesian methods have been proposed as an alternative. In particular, many authors have argued that Bayesian decision theory and associated value of information methods can be used to the determine if current evidence in support of a new health care intervention is sufficient for adoption and, if not, the optimal sample size for a future trial. Value of information methods incorporate current knowledge, the value of health outcome, the incidence and accrual rates, time horizon and trial costs, while maximizing the expected net benefit of future patients and providing an operational definition of equipoise. In this chapter value of information methods are developed in detail and illustrated using a recent example from the literature.

2.1 Introduction

The standard approach to the analysis and sample size determination for a randomized clinical trial (RCT) is based on the use of tests of hypotheses and the frequentists definition of probability. Consider a randomized clinical trial in which a new health care intervention, referred to as *Treatment* and labeled T, is compared to an existing intervention, referred to as *Standard* and labeled as S. The trial is conducted for the purpose of considering the adoption of *Treatment* if it is superior to *Standard*. This type of trial is often referred to as a superiority trial. Let Y be the random variable representing the primary outcome where larger values of Y are preferred, such as survival (where $Y = 1$ if the patient survives, 0 otherwise), survival time, quality-adjusted survival time or net benefit. Let $E(Y|i), i = T, S$ be the expected value of the outcome for a patient randomized to i, and let $\theta = E(Y|T) - E(Y|S)$. Thus, larger values of θ favour *Treatment*. Typically, in a superiority trial the data is used to test the null hypothesis $H : \theta \leq 0$

A.R. Willan (✉)
SickKids Research Institute, Dalla Lana School of Public Health, University of Toronto, Toronto, ON, Canada
e-mail: andy@andywillan.com

K. van Montfort et al. (eds.), *Developments in Statistical Evaluation of Clinical Trials*, DOI 10.1007/978-3-642-55345-5_2

versus the alternative hypothesis $A : \theta > 0$. *Treatment* is considered for adoption if, and only if, H is rejected in favour of A. The probability of falsely rejecting H, referred to as the Type I error probability, is set to some relatively small value. Sample size is determined by specifying the smallest clinically important (positive) difference for θ, labelled as θ_{SCID}, and requiring that the probability of failing to reject H when $\theta \geq \theta_{\text{SCID}}$ is less that some relatively small value, referred to as the Type II error probability.

There are many problems with this approach. Firstly, the value selected for the Type I error probability is somewhat arbitrary and is almost always set to 0.05. Using the same value for the probability of a Type I error for every trial ignores the seriousness of the error, which clearly varies from trial to trial. Thus, a trial that randomizes patients with age-related macular degeneration between two different wavelengths of laser coagulation [42] uses the same probability of falsely declaring *Treatment* superior, as does a trial of Caesarean section versus vaginal delivery for women presenting in the breech position [23]. Declaring one wavelength superior to another when they are the same is not a serious error since selecting the wavelength is a matter of simply dialing the appropriate frequency and the only difference to patients is the colour of the light observed during the procedure. However, in the latter, declaring Caesarean section superior when it is the same as vaginal delivery is a serious error. Assigning the same probability to the two errors makes no sense, quite apart from the fact that the value of 0.05 is somewhat arbitrary in the first place. Also somewhat arbitrary is the typical choice of 0.2 for the probability of a Type II error. It means that there is a 20 % chance that the effort and money invested in the trial will be wasted, even if a clinically important difference between the treatments exists. Again, it fails to reflect the seriousness of making the error. The choice of θ_{SCID} can be less arbitrary and can be estimated by polling clinicians and decision makers. However, in practice it is often back-solved from the sample size equation after substituting in a sample size that reflects constraints relating to patient recruitment and budget. Even if θ_{SCID} is a reasonable, clinically determined estimate of the smallest clinically important difference, there is a range of values for the true treatment difference that is less than the smallest clinically important difference, for which the probability of rejecting the null hypothesis and adopting *Treatment* is greater than 50 %. This sometimes referred to as a Type III error.

In response to these problems, many authors have proposed alternative methods [1, 3, 9, 11–16, 18–22, 25, 26, 28, 33–35, 43–49, 52]. In particular many authors have proposed the application of decision theory and associated expected value of information methods for assessing the evidence from RCTs and for determining optimal sample size for future trials. The application of decision theory to the design and sample size determination is the subject of the remainder of this chapter. In Sect. 2.2 an introduction to the cost-effectiveness analysis of RCTs is given, complete with an illustrative example. The use of decision theory in the design and analysis of RCTs is given in Sect. 2.3 and illustrated with the same example in Sect. 2.4. A summary and discussion are given in Sect. 2.5.

2.2 Cost-Effectiveness Analysis of Randomized Clinical Trials

Consider the cost-effectiveness comparison of a new health care intervention referred to as *Treatment* and labeled T, with an existing health care intervention referred to as *Standard* and labeled S. The health care interventions could be therapeutic, preventive or diagnostic. Let e_j and c_j be the respective mean measure of effectiveness and cost for patients receiving intervention j, where $j = T, S$. The measure of effectiveness is framed in the positive, such as surviving the duration of interest, survival time or quality-adjusted survival time. Cost includes not just cost of the interventions, but all down-stream health care cost over the duration of interest and might, depending on the perspective taken, include non-health care cost, such as time lost from work, etc. Let $\Delta_e = e_T - e_S$ and $\Delta_c = c_T - c_S$.

Initially, cost-effectiveness inference was centred on the parameter $R \equiv \Delta_c/\Delta_e$, which is referred to as the incremental cost-effectiveness ratio (ICER) and is the cost of achieving each additional unit of effectiveness from using *Treatment* rather than *Standard*. For example, suppose the probability of surviving the duration of interest was 0.6 for a patient receiving *Standard* and 0.7 for a patient receiving *Treatment* and the respective mean costs for *Standard* and *Treatment* over the duration of interest were \$14,000 and \$15,000 respectively. The ICER $= (15,000 - 14,000)/(0.7 - 0.6) = \$10,000$ per life saved or death averted. Many authors have discussed inference on the cost-effectiveness ratio [4–6, 8, 27, 30–32, 37, 40, 41, 50].

Due to the concerns regarding ratio statistics, focus has shifted from the incremental cost-effectiveness ratio to the incremental net benefit (INB). Let the net benefit (NB) of intervention j be defined as $\text{NB}_j \equiv e_j\lambda - c_j$ where λ is the threshold value for a unit of effectiveness (e.g., the value of saving a life or the value of a year of life gained). The INB is defined as $b \equiv \text{NB}_T - \text{NB}_S = e_T\lambda - c_T - (e_S\lambda - c_S) = \Delta_e\lambda - \Delta_c$. The term $\Delta_e\lambda$ is the incremental effectiveness (benefits) expressed in monetary terms and the term $-\Delta_c$ subtracts the incremental costs, leaving the incremental *net* benefit. When INB is positive, *Treatment* is considered value-for-money and should be considered for adoption, subject to budgetary constraints and the level of uncertainty. In the simple example above $b \equiv 0.1\lambda - 1,000$ and is positive for values of the threshold greater than \$10,000 (i.e., the ICER). Many authors have discussed inference on the incremental net benefit [2, 7, 24, 29, 36, 38, 39, 53–56].

Suppose $\hat{\Delta}_e$ and $\hat{\Delta}_c$ are the respective estimates of Δ_e and Δ_c from a study, such as a clinical trial or an observational study, where individual patient measures of effectiveness and cost have been recorded. Let $V(\hat{\Delta}_e)$, $V(\hat{\Delta}_c)$ and $C(\hat{\Delta}_e, \hat{\Delta}_c)$ be the relevant variances and covariance. For more on parameter estimation the reader is referred to Willan and Briggs [57]. Assuming no prior information, and invoking the central limit theorem, the posterior *pdf* for the incremental net benefit can be given by $N(b_0, v_0)$, where $b_0 = \hat{\Delta}_e\lambda - \hat{\Delta}_c$ and $v_0 = V(\hat{\Delta}_e)\lambda^2 + V(\hat{\Delta}_c) - 2C(\hat{\Delta}_e, \hat{\Delta}_c)\lambda$. Inference regarding INB, which is an attempt to characterize the cost-effectiveness of *Treatment* compared to *Standard* and the corresponding uncertainty, can best be presented by a plot the cost-effective acceptability curve (CEAC), which is the probability that the INB is positive (i.e., that *Treatment* is cost-effective) as a function of the threshold value for a unit of effectiveness and, invoking the central limit theorem, can be calculated as $\Phi\left(b_0/\sqrt{v_0}\right)$, where $\Phi(\cdot)$ is the *cdf* for the standard normal random variable. The CEAC passes through 0.5 at λ = ICER, crosses the vertical axis at the probability that *Treatment* is cost saving (i.e., $\hat{\Delta}_c < 0$), and is asymptotic to the right to the probability that *Treatment* is more effective (i.e., $\hat{\Delta}_e > 0$). For more on the CEAC the reader is referred to Fenwick et al. [17].

2.2.1 *The CADET-Hp Trial*

The CADET-Hp Trial is a double-blind, placebo-controlled, parallel-group, multi-centre, randomized controlled trial performed in 36 family practitioner centres across Canada. The results are published in Chiba et al. [10] and Willan [51]. Patients 18 years and over with uninvestigated dyspepsia of at least moderate severity presenting to their family physicians were eligible for randomization, provided they did not have any alarm symptoms and were eligible for empiric drug therapy. Patients were randomized between T: Omeprazole 20 mg, metronidazole 500 mg and clarithromycin 250 mg, and S: Omeprazole 20 mg, placebo metronidazole and placebo clarithromycin.

A total of 288 patients were randomized, 142 ($= n_T$) to *Treatment* and 146 ($= n_S$) to *Standard*. Both regimens were given twice daily for 7 days. The binary measure of effectiveness was treatment success and defined as the presence of no or minimal dyspepsia symptoms at 1 year. Total cost was determined from the societal perspective and are given in Canadian dollars. A summary of the parameter estimates is given in Table 2.1. The details regarding parameter estimation are given in the Appendix. Assuming no prior information and invoking the central limit theorem, the posterior *pdf* for the INB is normal with mean $0.1371\lambda + 53.01$

Table 2.1 Parameter estimates for the CADET-Hp trial

Treatment	Standard		
($n_T = 142$)	($n_S = 146$)		
Proportion of successes	0.5070	0.3699	Difference = $\hat{\Delta}_e = 0.1371$
Average cost	476.97	529.98	Difference = $\hat{\Delta}_c = -53.01$
V(Proportion of successes)	0.00176	0.001596	Sum = $\hat{V}\left(\hat{\Delta}_e\right) = 0.003356$
V(Average cost)	2,167	2,625	Sum = $\hat{V}\left(\hat{\Delta}_c\right) = 4,792$
C(Proportion of successes, mean cost)	−0.2963	−0.4166	sum = $\hat{C}\left(\hat{\Delta}_e, \hat{\Delta}_c\right) = -0.7129$

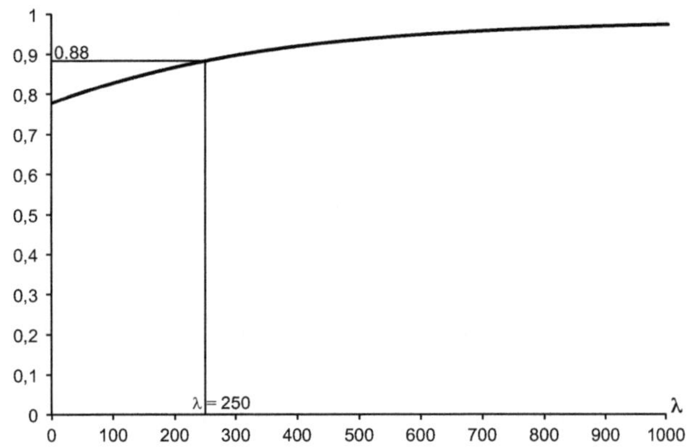

Fig. 2.1 The cost-effectiveness acceptability curve for the CADET-Hp trial

and variance $0.003356\lambda^2 + 4782 + 1.426\lambda$. The cost-effectiveness acceptability curve, given by $\Phi\left(0.1371\lambda + 53.01/\sqrt{0.003356\lambda^2 + 4782 + 1.426\lambda}\right)$, is shown in Fig. 2.1. Because *Treatment* is observed to increase effectiveness (i.e., $\hat{\Delta}_e > 0$) and decrease cost (i.e., $\hat{\Delta}_c < 0$), the INB will be positive and the CEAC will be greater than 0.5 for all positive values of the threshold value (λ).

Because the mean INB is positive regardless of the threshold value, and because *Treatment* is observed to reduce cost and therefore budget constraints may not be an issue, it may seem obvious that *Treatment* should be adopted. But this would ignore the uncertainty regarding the INB (i.e., $v_0 > 0$). Because of this uncertainty, there is a positive probability that the INB is negative. (For $\lambda = 250$, the probability that INB is negative, i.e., $1 -$ CEAC for $\lambda = 250$, is 0.12.) Therefore, there is a positive expected opportunity loss associated with the net benefit maximizing decision (action) to adopt *Treatment* and the optimal action might be to obtain more information (e.g., another trial) to reduce the uncertainty and decrease the expected opportunity loss. Whether or not another trial is optimal, and the optimal size of the trial if it is, will depend on the trade-offs between the additional cost and the reduction in expected opportunity loss. This is covered in the next section.

2.3 Decision Theory and Value of Information in RCT Research

2.3.1 Introduction

In response to the many problems associated with sample size determinations based on tests of hypotheses and power arguments, many authors have proposed alternatives [1, 3, 9, 11–16, 18–22, 25, 26, 28, 33–35, 43–49, 52]. In particular, among others, Willan and Pinto [43], Eckermann and Willan [14–16], Willan [44, 45], Willan and Kowgier [46], and Willan and Eckermann [47–49] propose methods based on decision theory and the expected value of information that determines the sample size for maximizing the difference between the expected cost of the trial and the expected value of the information provided by the results. Fixed, variable and opportunity trial costs are considered. In addition to providing optimal sample sizes, these methods can identify circumstances when the current information is sufficient for decision making, see Willan [44]. Details of the approach are given below.

2.3.2 Opportunity Loss in Decision Making

To recognize the role that decision theory can play in the analysis and design of RCTs one must understand the definition of opportunity loss and how to determine its expected value. To that aim we use an example based on a simple bet on the toss of a (not necessarily fair) coin. The decision to accept the bet on the coin toss has an associated opportunity loss, and one can determine its expected value based on the current information regarding the outcome of a toss of the coin. The more information one has regarding the toss of the coin, the less is the expected opportunity loss. The chance to gather additional information should be accepted only if the cost of doing so is less than the reduction in the expected opportunity cost provided by the additional information. The reduction in the expected opportunity loss provided by additional information is referred to as the expected value of information (EVSI).

Suppose Karl has tossed a particular coin on 12 occasions and noted that it came up heads on 9 of them. He must now decide whether or not to accept the following bet: On a new toss of the coin, if it comes up heads he wins \$1,000 and if it comes up tails hc loses \$1,000. Let the random variable $X = 1$ if the next toss of the coin is a head, and 0 otherwise. A reasonable *pdf* for X to reflect the uncertainty regarding the next toss (i.e., the value of X) is Bernulli(θ), given by $\Pr(X = x) = \theta^x(1-\theta)^{1-x}$, where θ is the probability that the next toss of the coin is a head. The utility of accepting the bet is \$1,000 if the toss is a head and –\$1,000 if it is tail, and as a function of X, equals $1{,}000X - 1{,}000(1 - X) = 1{,}000(2X - 1)$, with expectation $1{,}000(2\theta - 1)$. The utility of refusing the bet is zero, since nothing is gained or lost. Karl's previous experience with the coin has provided him with some knowledge regarding θ. In general, if Karl had observed r heads in n tosses, and assuming he had no other prior knowledge or opinions regarding θ, the posterior distribution for θ is Beta(a_0, b_0), where $a_0 = r + 1$ and $b_0 = n - r + 1$, with mean $a_0/(a_0+b_0)$ and variance $a_0 b_0/\left\{(a_0+b_0)^2(a_0+b_0+1)\right\}$. The *pdf* and *cdf*, denoted by $f_B(\theta; a_0, b_0)$ and $F_B(\theta; a_0, b_0)$ respectively, are given by

$$f_B(\theta; a_0, b_0) = \left[(a_0 + b_0 - 1)!/\left\{(a_0 - 1)!(b_0 - 1)!\right\}\right]\theta^{a_0-1}(1-\theta)^{b_0-1} \quad \text{and}$$

$$F_B(\theta; a_0, b_0) = \sum_{j=a_0}^{a_0+b_0-1} \frac{(a_0 + b_0 - 1)!}{j!(a_0 + b_0 - 1 - j)!}\theta^j(1-\theta)^{a_0+b_0-1-j}$$

Therefore, Karl's current knowledge regarding θ after observing 9 heads in 12 tosses is characterized by a beta distribution with mean $10/14 = 0.7143$ and variance $10 \times 4/\left\{(10+4)^2(10+4+1)\right\} = 0.01361$, and his expected utility for the decision to accept the bet is $1{,}000(2 \times 0.7143 - 1) = 428.6$. Since his expected utility for the decision to refuse the bet is 0, he should accept the bet if he wants to maximize expected utility. Nonetheless, there is an opportunity loss associated with deciding to accept the bet. In general, the opportunity loss associated with a decision is the utility of the best decision *minus* the utility of the decision made. The opportunity loss of accepting the bet depends on whether the coin comes up heads or tails. If it comes up heads there is no opportunity loss because, in that case, accepting the bet is the best decision. If the coin comes up tails, the best decision would have been to refuse the bet. The utility of refusing the bet is zero, but the utility of accepting the bet when it comes up tails is –$1,000. Thus, the opportunity loss of accepting the bet when it comes up tails is the utility of refusing the bet *minus* the utility of accepting the bet, i.e., $0 - (-1{,}000) = \$1{,}000$. Consequently, Karl's opportunity loss function is $1{,}000 \times I(\text{coin comes up tails})$, where $I(\cdot)$ is the indicator function. Therefore, Karl's *expected* opportunity loss based on the current information (EOL_0) is given by

$$\text{EOL}_0 = 1{,}000 \times \Pr(\theta < 0.5) = 1{,}000 \times F_B(0.5; 10, 4). \text{ That is,}$$

$$\text{EOL}_0 = 1{,}000 \sum_{j=10}^{13} \frac{13!}{j!(13-j)!} 0.5^{13} = 46.14.$$

Therefore, based on current information, Karl faces an expected opportunity loss of $46.14 associated with the decision to accept the bet which is his expected utility-maximizing course of action.

Suppose Karl is given the opportunity to pay $20.00 to toss the coin 12 more times. The question is: Is the additional information worth $20.00? In decision theory that question is interpreted as: Will the additional information provided by 12 more tosses reduce the expected opportunity loss by more than $20.00? Suppose Karl tosses the coin 12 more times and observes r heads. The posterior distribution is Beta(a_1, b_1), where $a_1 = a_0 + r = 10 + r$ and $b_1 = b_0 + (12 - r) = 16 - r$. The posterior expected opportunity loss if Karl observes r heads in 12 tosses is

$$\text{EOL}_1 = 1{,}000 \times \Pr(\theta < 0.5) = 1{,}000 \times F_B(0.5; 10 + r, 16 - r).$$

Since the expected opportunity loss is a function of the number of heads observed, the expected value of the expected opportunity loss must be taken with respect to the random variable *number of heads observed*, denoted Y. Thus, the expected opportunity loss including the new information provided by the 12 coin tosses (EOL_1) is given by

$$\begin{aligned}\mathrm{EOL}_1 &= \sum_{r=0}^{12}\{1{,}000\times F_B(0.5;10+r,16-r)\times \Pr(Y=r)\}\\ &= 1{,}000\sum_{r=0}^{12}\left\{\sum_{j=10+r}^{25}\frac{25!}{j!(25-j)!}0.5^{25}\frac{3!}{r!(3-r)!}\theta_0^r(1-\theta_0)^{12-r}\right\}\\ &= 29.87.\end{aligned}$$

where $\theta_0 = a_0/(a_0+b_0) = 0.7143$, Karl's current mean of θ. Therefore, the expected value of sample information provided by the 12 coin tosses is $\mathrm{EOL}_0 - \mathrm{EOL}_1 = 46.14 - 29.87 = \$16.27$, which is less that the offered cost of \$20.00. Therefore, Karl's optimal action is to accept the bet based the information from the initial 12 tosses. To emphasize here, Karl's optimal decision is to accept the bet without paying for the additional information because the cost of the additional information exceeds the amount by which it would reduce the expected opportunity loss. As illustrated in later sections, a similar situation arises in evaluating evidence from a clinical trial. Because of the uncertainty inherent in the evidence, the decision to adopt the utility-maximizing intervention will be associated with an expected opportunity loss. Additional evidence should be sought only if the cost of attaining the evidence is the less than the amount by which it reduces the expected opportunity loss.

Suppose now that Karl was offered the opportunity, prior to deciding whether or not to accept the bet, to make as many tosses as he wished at \$0.50 a toss. If he took 12 tosses the \$6.00 cost would be less than the expected value of information of \$16.27. The question now is: What is the optimal number of tosses? The answer is: It is the number of tosses that maximizes the difference between the expected value of sample information and the cost of making the tosses. The difference between the expected value of information and the cost is referred to as the expected net gain (ENG). If we let m be the number of tosses taken, then the posterior expected opportunity loss is

$$\begin{aligned}\mathrm{EOL}_1(m) &= \sum_{r=0}^{m}\{1{,}000\times F_B(0.5;10+r,4+m-r)\times \Pr(Y=r)\}\\ &= 1{,}000\times 0.5^{16}\times\sum_{r=0}^{m}\left\{\sum_{j=10+r}^{13+m}\frac{(13+m)!m!\theta_0^r(1-\theta_0)^{m-r}}{j!(13+m-j)!r!(m-r)!}\right\},\end{aligned}$$

where $\theta_0 = a_0/(a_0+b_0) = 0.7143$, Karl's current mean of θ. Plots of the EVSI (i.e., $\mathrm{EOL}_0 - \mathrm{EOL}_1(m)$) and total cost (i.e., 0.5 m), as functions of m, are given in Fig. 2.2. By inspection, the difference between the expected value of information and total cost is maximized at 34 tosses, where the expected value of sample

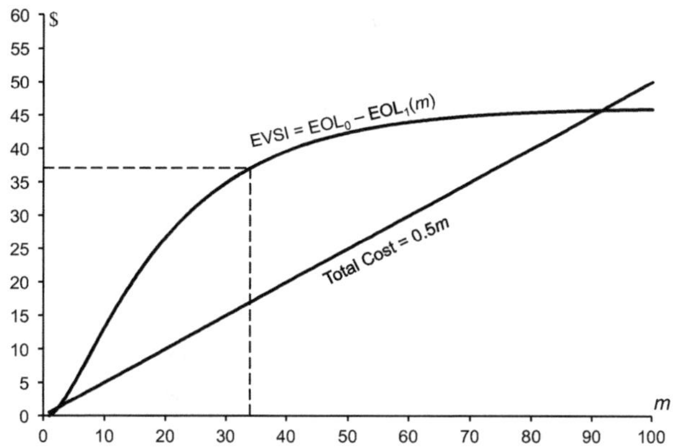

Fig. 2.2 The expected value of information and total cost for the coin toss example

information = \$37.06 and the total cost is \$17.00, yielding an expected net gain of \$20.06. As illustrated in later sections, an analogous situation arises when assessing the evidence from an RCT. One must decide to adopt the utility-maximizing intervention or, if given the chance, perform another trial which should be performed only if the maximum amount by which the expected opportunity loss is reduced (i.e., EVSI) exceeds the cost of the trial. That is, only if the maximum ENG is positive.

It may seem odd that the expected opportunity loss based on the initial 12 tosses is only \$46.14, given that Karl has a $1 - \theta_0 = 0.2957$ probability of losing \$1,000. But the expected opportunity loss relates to the uncertainty regarding θ, not its actual value. That is, if Karl knew for certain that the probability of heads is 0.55, his expected opportunity loss is zero, even though there is a 0.45 probability that he will lose \$1,000. The value of perfect information, when you have it, is zero. Karl accepts the bet if the expected value of the utility (i.e., $1{,}000(2\theta_0 - 1)$) is greater than zero because we have assumed Karl is risk-neutral, that is, a dollar lost has the same value as a dollar won. Being risk-neutral makes most sense if the bet can be accepted or refused numerous times, thus spreading the risk of any single bet over many others. Based on the information Karl has from the initial 12 tosses, the probability that he will lose money on a single toss is $1 - \theta_0 = 0.2857$. However on k tosses he will lose money only if less than a half of them are heads, that is, with probability

$$\sum_{r=0}^{k^*} \frac{k!}{r!(k-r)!} \theta_0^r (1 - \theta_0)^{k-r},$$

where k^* is the largest integer less than $k/2$. So, for 10 tosses the probability of losing money is 0.03764 and for 20 tosses it is 0.01171.

2.3.3 The Expected Value of Sample Information

Consider the problem of determining the sample size for a randomized clinical trial designed to examine the cost-effectiveness of *Treatment* in comparison to *Standard*. The trial is conducted with the purpose of adopting *Treatment* if it is found to be cost-effective. *Treatment* is cost-effective if the INB is greater than zero. Recall from Sect. 2.2 that the INB is defined as $b \equiv \Delta_e\lambda - \Delta_c$, where λ is the threshold value for a unit of health outcome (effectiveness); $\Delta_e = e_T - e_S$, where e_j, for $j = T, S$, is the mean effectiveness for intervention j; and $\Delta_c = c_T - c_S$, where c_j, for $j = T, S$, is the mean cost for intervention j. Recall that $b \equiv e_T\lambda - c_T - (e_S\lambda - c_S)$, so that INB $=$ NB$_T$ $-$ NB$_S$, where NB$_j (\equiv e_j\lambda - c_j)$ is the net benefit for intervention j.

In the following, the threshold value is initially considered fixed for ease of notation but can be allowed to vary, as demonstrated later when examining robustness. Let the current information regarding incremental net benefit be characterized by a normal prior *pdf* with mean b_0 and variance v_0, where $b_0 > 0$ and $v_0 > 0$. Since the prior mean INB (b_0) is positive, adopting *Treatment*, rather than retaining *Standard* maximizes the expected net benefit for future patients. However, since the prior variance of INB (v_0) is positive, adopting *Treatment* is not necessarily the optimum decision facing a decision maker. Consideration must be given to collecting more information, i.e., conducting another trial. Decision uncertainty resulting from a positive v_0 implies that a decision maker faces an opportunity loss when adopting *Treatment*, even though doing so is the decision that maximizes expected net benefit for future patients. The opportunity loss per patient associated with the decision to adopt *Treatment* is defined as the utility of the best decision *minus* the utility of adopting *Treatment*. Since, in this context, utility equals net benefit, the opportunity loss becomes the maximum of (NB$_T$, NB$_S$) *minus* NB$_T$. The maximum of (NB$_T$, NB$_S$) depends on b, the INB. If b is positive, then NB$_T$ $>$ NB$_S$, and NB$_T$ is the maximum. On the other hand, if b is not positive, then NB$_T$ $\leq$ NB$_S$, and NB$_S$ is the maximum. Thus the opportunity loss per patient associated with adopting *Treatment* (OLpp$_T$), as a function of INB, is given by:

$$\text{OLpp}_T(b) = \begin{Bmatrix} \text{Max}(\text{NB}_T, \text{NB}_S) - \text{NB}_T = \text{NB}_S - \text{NB}_T = -b & : b \leq 0 \\ \text{Max}(\text{NB}_T, \text{NB}_S) - \text{NB}_T = \text{NB}_T - \text{NB}_T = 0 & : b > 0 \end{Bmatrix}.$$

When INB is positive there is no opportunity loss associated with adopting *Treatment* since future patients would receive the net benefit-maximizing intervention. However, if *Treatment* is adopted when incremental net benefit is negative, future patients would not receive the net benefit-maximizing intervention and each patient would experience a reduction in net benefit equal to the absolute value of INB. A plot of OLpp$_T(b)$ is given in Fig. 2.3.

Taking the expected value of OLpp$_T(b)$ with respect to the current information regarding incremental net benefit which, as assumed above, is characterized by a normal prior *pdf* with mean b_0 and variance v_0, yields the prior expected opportunity

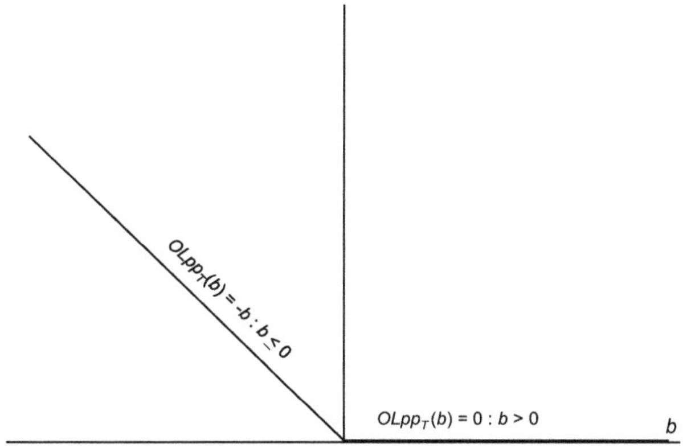

Fig. 2.3 The opportunity loss function per patient of adopting *Treatment*

loss per patient (EOLpp$_{T0}$). Letting $f_N(x;\mu,v)$ be the *pdf* for normal random variable with mean μ and variance v, then

$$\text{EOLpp}_{T0} = \int_{-\infty}^{\infty} \text{OLpp}_T(b) f_N(b;b_0,v_0)\text{d}b = \int_{-\infty}^{0} -b f_N(b;b_0,v_0)\text{d}b = \mathscr{D}(b_0,v_0),$$

where

$$\mathscr{D}(\mu,v) = [v/(2\pi)]^{\frac{1}{2}} \exp\left[-\mu^2/(2v)\right] - \mu\left[\Phi(-\mu/v^{\frac{1}{2}}) - I(\mu \le 0)\right]; \tag{2.1}$$

where $\Phi(\cdot)$ is the *cdf* for the standard normal random variable; and, $I(\cdot)$ is the indicator function, see Willan and Pinto [43] for details. The expected opportunity loss per patient, multiplied by the number of future patients, is the total expected opportunity loss and is also known as the expected value of perfect information, since if the decision maker had perfect information (i.e., $v_0 = 0$), the opportunity loss could be avoided by adopting *Treatment* if b_0 is positive and retaining *Standard*, otherwise. Applying decision theory, as illustrated in Sect. 2.3.2, the expected value of sample information (EVSI) of a new trial is the amount by which the information from the new trial reduces the total expected opportunity loss.

Suppose a new trial of n patients per arm is conducted where $\hat{\Delta}_e$ and $\hat{\Delta}_c$ are the respective estimators of Δ_e and Δ_c from the trial data. Thus, the estimate of INB based on the trial data is $\hat{b} = \hat{\Delta}_e\lambda - \hat{\Delta}_c$ and relying on the central limit theorem regarding the distribution of $\hat{b}$ the posterior mean and variance for incremental net benefit are given by:

$$b_1 = v_1\left(\frac{b_0}{v_0} + \frac{n\hat{b}}{\sigma_+^2}\right) \quad \text{and} \quad v_1 = \left(\frac{1}{v_0} + \frac{n}{\sigma_+^2}\right)^{-1},$$

where σ_+^2 is the sums over treatment arm of the between-patient variances of net benefit, and is assumed known or determinable from prior data. Details for estimating σ_+^2 for the CADET-Hp trial are given in the Appendix. The posterior (i.e., post-trial) expected opportunity cost per patient is given by $\text{EOLpp}_1 = \mathscr{D}(b_1, v_1)$. EOLpp_1 is a function of the random variable $\hat{b}$ and to determine the expected reduction in per-patient opportunity loss, with the purpose of identifying the optimal sample size, the expectation of EOLpp_1 must be taken with respect to $\hat{b}$. Applying the central limit theorem, the predictive distribution for $\hat{b}$ is $N(b_0, v_{\hat{b}})$, where $v_{\hat{b}} = v_0 + \sigma_+^2/n$, and the expected value of EOLpp_1 with respect to $v_{\hat{b}}$ becomes, see Willan and Pinto [43],

$$\text{E}_{\hat{b}}\text{EOLpp}_1 = \text{E}_{\hat{b}}\mathscr{D}(b_1, v_1) = \int_{-\infty}^{\infty} \mathscr{D}(b_1, v_1) f(\hat{b}; b_0, v_{\hat{b}}) \mathrm{d}\hat{b} = I_1 + I_2 + I_3, \text{ where}$$

$$I_1 = \sqrt{v_0/(2\pi)}\sigma_+^2 \exp\left(-b_0^2/2v_0\right)/(nv_{\hat{b}}),$$

$$I_2 = -b_0\Phi\left(-b_0/\sqrt{v_0}\right) + v_0^{3/2} \exp\left(-b_0^2/2v_0\right)/\left(v_{\hat{b}}\sqrt{2\pi}\right), \text{ and}$$

$$I_3 = b_0\Phi\left(-b_0\sqrt{v_{\hat{b}}}/v_0\right) - v_0 \exp\left(-b_0^2 v_{\hat{b}}/(2v_0^2)\right)/\sqrt{2\pi v_{\hat{b}}}.$$

Thus, the expected value of sample information of a trial of n patients per arm is given by

$$\text{EVSI}(n) = B(n)\left\{\mathscr{D}(b_0, v_0) - \text{E}_{\hat{b}}\mathscr{D}(b_1, v_1)\right\},$$

where $B(n)$ refers to the post-trial patient horizon, defined as the number of patients who could potentially receive the new intervention following the trial and therefore can benefit from a reduction in the opportunity loss. For an incidence rate of k patients per year, a time horizon of h years and a trial duration of $t(n)$ years, $B(n) = k\{h - t(n)\}$. The time horizon is the duration for which the decision to either adopt *Treatment* or perform another trial is relevant. Although there is no software packages for determining EVSI, it components can be calculated directly from the formulae using a spreadsheet.

2.3.4 Expected Total Cost

The cost of a trial is assumed to have two components, one financial and the other reflecting opportunity costs. Let C_f be the fixed financial cost of setting up a trial and let C_v be the variable financial cost per patient. Then the total financial cost of a trial with n patients per arm is $C_f + 2nC_v$. The assumption is made that since b_0 is positive, if the trial is not performed, all future patients would receive *Treatment*. This is referred to as the assumption of perfect implementation. It is also assumed that while the trial is performed, all patients outside the trial and half the patient within the trial will receive *Standard*. All patients who receive *Standard* while the trial is performed, denoted as $D(n)$, pay an expected opportunity cost equal to b_0. The decision to perform the trial means that these patients have an expected reduction in net benefit equal to b_0 because they will receive *Standard* rather than *Treatment*. Therefore $D(n) = kt(n) - n$. That is, the number of patients who receive *Standard* because of the trial are all the patient who are incident while the trial is performed, *minus* the n patients who receive *Treatment* in the trial. Therefore, the expected total cost (ETC) of delaying the decision and performing the trial is $\text{ETC}(n) = C_f + 2nC_v + D(n)b_0$.

The function $t(n)$, an important part of the functions of $B(n)$ and $D(n)$, will depend on what assumptions are made regarding the proportion of patients that are recruited into the trial and the duration between when the last patient is randomized and when the trial results are available. These assumptions and their implications are discussed in Sect. 2.4 using the CADET-Hp trial as an example.

2.3.5 The Expected Net Gain and Optimal Sample Size

Given b_0, v_0, σ_+^2, h and k, the EVSI is a function of the sample size n, given as $\text{EVSI}(n) = B(n)\left\{\mathscr{D}(b_0, v_0) - \text{E}_{\hat{b}}\mathscr{D}(b_1, v_1)\right\}$. Likewise, given b_0, C_f and C_v, the expected total cost is a function of the sample size n, given as $\text{ETC}(n) = C_f + 2nC_v + D(n)b_0$. The expected net gain is defined as $\text{ENG}(n) \equiv \text{EVSI}(n) - \text{ETC}(n)$. Considering the trial in isolation, and being free of budget constraints, let n^* be that value of n that maximizes the expected net gain. That is, $\text{ENG}(n^*) \geq \text{ENG}(n)$ for all positive integers n. If $\text{ENG}(n^*) \leq 0$ then optimal sample size is zero and the current information, i.e., b_0 and v_0, is sufficient for decision making. In this case no trial is necessary, since the expected value of the information from the trial is less than the expected total cost, regardless of the sample size. On the other hand, if $\text{ENG}(n^*) > 0$, the decision maker is in a state of equipoise and the optimal decision is to delay adopting *Treatment*, even though $b_0 > 0$, and perform a trial with n^* patients per arm.

2.4 Applying VOI Methods: The CADET-Hp Trial

Suppose, for sake of illustration, the threshold value of a treatment success is \$250, i.e., $\lambda = 250$. Assuming no prior information, the current mean and variance for incremental net benefit is given by:

$$b_0 = 0.1371 \times 250 - (-53.01) = 87.285,$$

$$v_0 = 0.003356 \times 250^2 + 4{,}792 - 2 \times (-0.7129) \times 250 = 5{,}344.7,$$

and, invoking the central limit theorem, the prior *pdf* for INB for the planning of a new trial is $N(87.285, 5{,}344.7)$. The probability that *Treatment* is cost-effective for $\lambda = 250$ is 0.88, see Fig. 2.1. Since $b_0 > 0$ the net benefit maximizing decision, based on current evidence, is to adopt *Treatment* (i.e., add the antibiotics) for future patients. However, since $v_0 > 0$, the decision to adopt *Treatment* is associated with an expected per-patient opportunity loss of $\mathcal{D}(87.285, 5{,}344.6) = 4.1528$ (from Eq. 2.1), and the optimal decision might be to delay the adoption of *Treatment* and perform another trial. Performing another trial would be optimal if the reduction in total expected opportunity loss (i.e., the expected value of sample information) is greater than the expected total cost, that is, if the expected net gain is greater than zero.

2.4.1 Simplifying Assumptions

If we make the simplifying assumptions that all patients in the jurisdiction of interest are recruited into the trial and that the results of the trial are available immediately after the last patient is randomized, then duration of the trial will equal total sample size *divided* by the incidence (i.e., $t(n) = 2n/k$). Under the same assumptions the number of patients that can benefit from the new information (B) will equal the total patient horizon (i.e., kh) minus the $2n$ patients in the trial (i.e., $B(n) = kh - 2n$). The time horizon of a decision is the duration over which the decision is considered relevant. Assuming an incidence of 80,000 per year and a time horizon of 20 years, the plots of EVSI, ETC and ENG as functions of n are given in Fig. 2.4. The fixed (C_f) and variable (C_v) financial cost of the trial were assumed to be \$800,000 and \$2,000, respectively. The optimal sample size is 463 patients per arm, yielding an optimum ENG of \$1,329,020 with an ETC of \$2,692,413 for a return on investment of 49 %.

Willan et al. [51] demonstrates that plotting the combinations of the threshold value (λ) and horizon (h) for which the ENG is zero provides a sensitivity analysis for those variables, see Fig. 2.5. For combinations of λ and h above the curve the

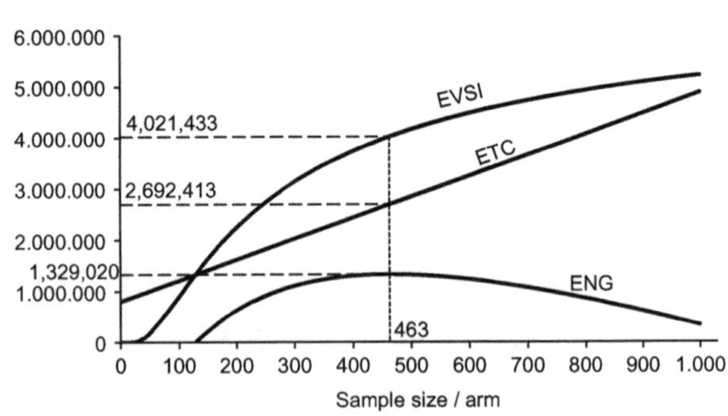

Fig. 2.4 EVSI, ETC and ENG for the CADET-Hp example using the unrealistic assumptions

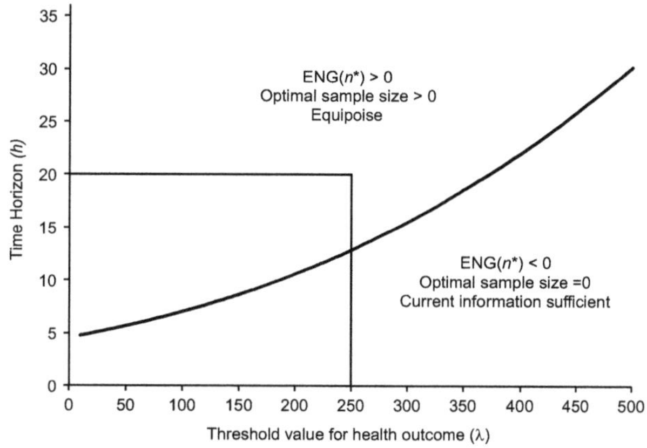

Fig. 2.5 The values of the threshold value for health outcome and horizon for which the ENG is zero for the CADET-Hp example using the unrealistic assumptions

ENG is positive (i.e., a state of equipoise exists) and a new trial is the optimal decision. Whereas, for combinations below the curve the ENG is negative and the current evidence is sufficient for decision making. Note that the combination of $\lambda = 250$ and $h = 20$ lies above the line.

2.4.2 More Realistic Assumptions

In Sect. 2.4.1 it was assumed that all patients in the jurisdiction of interest are recruited into the trial and that the trial results are available immediately after the last patient is randomized. These assumptions almost never hold. Usually only a small fraction of the eligible patients are recruited, and patients need to be followed to observe outcomes. Further, time is required for data entry, cleaning and analysis. If we let the annual accrual rate be denoted by a and the number of years between when the last patient is randomized and the data is analysed be denoted by τ, the trial duration becomes $t(n) = \tau + 2n/a$. Consequently, the number of patients who will benefit from the trial results (B) and the number of patient incurring an opportunity cost (D) are given by:

$$B(n) = k\{h - (2n/a + \tau)\}$$

$$D(n) = k(2n/a + \tau) - n$$

For the CADET-Hp example, if we assume an accrual fraction of 1 % (i.e., $a = 800$ per year), and allow for 1 year of follow-up (necessary to observe the measure of effectiveness) with 3 months for data entry, cleaning and analysis (i.e., $t = 1.25$), the optimal sample size is zero. A plot of the expected total cost and expected value of sample information is given in Fig. 2.6, where it can be seen that costs exceeds value for all sample sizes. The expected total cost have been

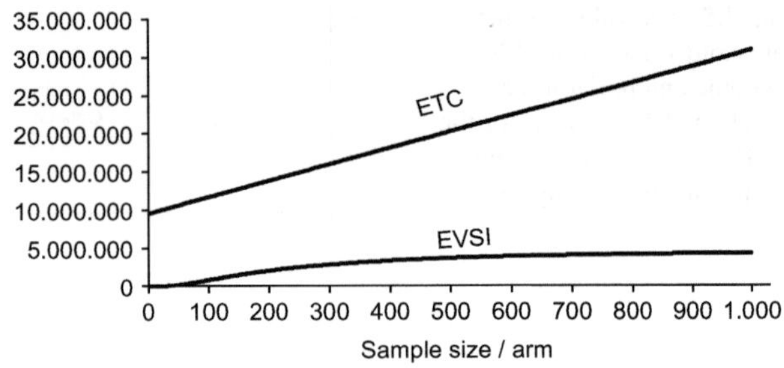

Fig. 2.6 EVSI and ETC for the CADET-Hp example using the realistic assumptions

driven up by the very high expected opportunity cost for the patients who receive *Standard* while the trial is conducted. These patients consist of 99 % of incident cases while the trial is recruiting patients and 100 % during the follow-up period. The expected opportunity cost for the 1.25 years of follow-up alone is $8,728,500. Further, because the trial takes longer to perform the number of patients that can benefit from the new information is reduced, which in turn reduces the EVSI. For details, the reader is referred to Eckermann and Willan [15, 16].

2.4.3 Relaxing the Assumption of Perfect Implementation

For the solution above and in Sect. 2.4.1 the assumption has been made that if the current mean INB is positive (i.e., $b_0 > 0$) that all future patients would receive *Treatment* in the absence of a new trial. Referred to as perfect implementation this assumptions is unlikely to hold. To examine the effect of allowing imperfect implementation Willan and Eckermann [48] assume that the probability that a future patient that would receive *Treatment* if no additional evidence is forthcoming, is a non-decreasing function of the strength of the evidence as measured by the z-statistic, defined as $z_i = b_i / \sqrt{v_i}$, $i = 0, 1$. To demonstrate the dramatic effect that this more realistic assumption has on the solution, the authors use a *sliding step function*, where if $z_i \leq \gamma$, the probability that a future patient would receive *Treatment* is 0, and if $z_i \geq \beta$, the probability that a future patient would receive *Treatment* is 1, where $\gamma \leq \beta$. For $\gamma < z_i < \beta$, a linear function is assumed, where the probability that a future patient receives *Treatment* is $(z_i - \gamma)/(\beta - \gamma)$. For perfect implementation, $\gamma = \beta = 0$.

Relaxing the assumption of perfect implementation can have a dramatic effect on the value of information solution. Firstly, additional information, apart from reducing the expected opportunity cost as before, now has value in increasing the expected proportion of future patients receiving the net benefit maximizing intervention (i.e., $E(z_1) > z_0$). Secondly, the expected opportunity cost of delaying the decision to adopt *Treatment* and performing a future trial is far less since only a portion of the patients would receive *Treatment* in the absence of the future trial and therefore incur an expected opportunity cost.

To demonstrate the effect on the solution using the CADET-Hp example suppose γ and β are chosen to correspond with values of the probability of *Treatment* being cost-effective of 75 and 99 %, respectively, that is, $\gamma = \Phi^{-1}(0.75) = 0.675$ and $\beta = \Phi^{-1}(0.99) = 2.326$. (Recall that for normally distributed incremental net benefit, the CEAC is given by Φ(z-statistic).) Based on the evidence from the existing trial

$$z_0 = b_0 / \sqrt{v_0} = 87.28 / \sqrt{5{,}345} = 1.194,$$

and the probability of a future patient receiving *Treatment* in the absence of a new trial is

$$(z_0 - \gamma)/(\beta - \gamma) = (1.194 - 0.6745)/(2.326 - 0.6745) = 0.3135.$$

Consequently, the expected opportunity cost of performing new trial is less than a third of what it is under the assumption of perfect implementation.

Figure 2.7 contains a plot of the expected value of information and expected total cost as a function of sample size assuming imperfect implementation as characterized by the value of γ and β given above and the same assumption regarding accrual and follow-up given in Sect. 2.4.2. The optimal sample size is 486, with an expected net gain of \$64,299,751. Compared to Fig. 2.6 a dramatic increase in the expected value of information is observed. This is because the new information, apart from reducing the total expected opportunity cost, is expected to increase the proportion of future patients receiving *Treatment*. For the optimal sample size of 486 the post-trial expected probability that a future patient receives Treatment is 0.7. Also observed is a dramatic decrease in the expected total cost, which is a result of the reduction in expected opportunity cost as noted above.

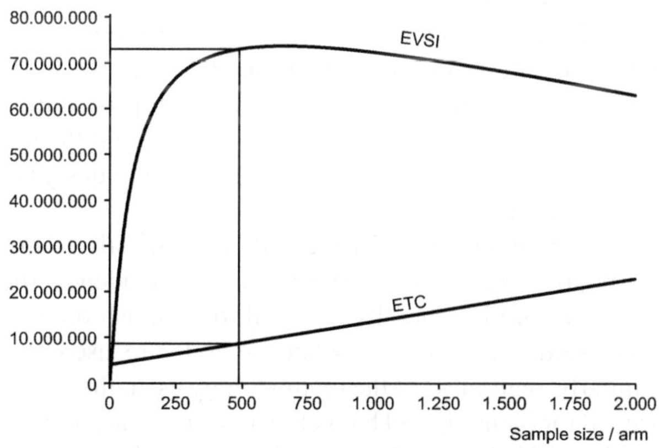

Fig. 2.7 EVSI and ETC for the CADET-Hp example using the realistic assumptions with imperfect implementation

2.5 Discussion

In this chapter the application of decision theory and associated value of information (VOI) methods for the design and analysis of RCTs have been proposed as an alternative to the standard hypothesis testing approach with its reliance on arbitrarily chosen Type I and II error probabilities and smallest clinically important differences. VOI methods allow for the explicit incorporation of important factors, such as the value of health outcomes, incidence and accrual rates, time horizon, current information, follow-up times and trial costs. They also can be used to identify those situations where the evidence is sufficient for decision making and, where evidence is insufficient (equipose), the optimal size of a future trial. Using VOI methods to assess the evidence from a clinical trial or the meta-analysis of several trials provides a more rational alternative to the standard methods since they maximize the expected net benefit for future patients and optimize the allocation of research funding, while providing an operational definition for equipoise. In addition, since the EVSI increases with incidence, interventions for rarer diseases need less evidence for adoption. Thus VOI methods help address the obvious difficulty of patient recruitment in rare diseases.

The use of VOI methods raises a number of issues. Perhaps the most subtle is that of jurisdiction. The assumption is made that trial financial costs are borne by society through government or private donation-based or philanthropic agencies. This raises an issue for research funding agencies. On whose behalf is it acting? The answer to this question has a huge impact on VOI methods since it determines the incidence, which is an important determinant of EVSI. Agencies acting on behalf of small jurisdictions, such as provincial/state governments or health insurers, are less likely to find the funding of additional trials attractive, since the optimal sample size will be zero with greater frequency. However, for federal governments or private donation-based or philanthropic agencies, which may take a more global view, funding additional trials may be more attractive.

Typically VOI methods are based on the assumptions that if a new trial is carried out, the definitive decision regarding the adoption of *Treatment* will be made at the end of the trial. However, the truly optimal procedure would be to repeat the VOI process at the end of the new trial to determine if the updated evidence is sufficient. Relaxing this assumption leads to multi-stage designs as discussed in Willan and Kowgier [46].

The limitations of VOI methods have mostly to do with specifying values for the required parameters. The parameter incidence should be available from the literature, and is generally required to establish the burden of the health condition under study regardless of what methods are used to determine the sample size. Similarly, regardless of the methodology used, the financial cost and accrual rate are needed for planning and budgetary reasons. The parameters that could be considered specific to VOI methods are the threshold value for a unit of health outcome and the time horizon. Various threshold values for a quality-adjusted life-year have been applied in cost-utility analyses, however threshold values for other health outcomes are less well established. The time horizon for a new health care intervention varies depending on the type of intervention (e.g., pharmacological, surgical) and the health condition under study. Time horizons of 20–25 years are often used because they correspond to infinite time horizon with discount rates for future benefits of around 4 or 5 %. It is worth noting that the advantage of VOI methods is that they make the assumptions regarding threshold value of health outcome and time horizon explicit, and although both parameters may be associated with uncertainty, a sensitivity analysis can be performed, as illustrated in the example.

In conclusion, decision theoretic/VOI methods can be used to identify those situations where the evidence is sufficient for decision making, and where evidence is insufficient, the optimal size of a future trial.

Appendix

Let e_{ji} and c_{ji} be the respective observations of effectiveness and cost for patient i receiving intervention j, where $j = T, S$; $i = 1, 2 \ldots n_j$; and n_j is the number of patients on intervention j. For the CADET-Hp trial $e_{ji} = 1$ if the patient is a success, 0 otherwise.

$$\text{Let} \quad \bar{e}_j = \frac{1}{n_j}\sum_{i=1}^{n_j} e_{ji} \quad \text{and} \quad \bar{c}_j = \frac{1}{n_j}\sum_{i=1}^{n_j} c_{ji}$$

Then

$$\hat{\Delta}_e = \bar{e}_T - \bar{e}_S$$

$$\hat{\Delta}_c = \bar{c}_T - \bar{c}_S$$

$$\hat{V}(\hat{\Delta}_e) = \hat{V}(\bar{e}_T) + \hat{V}(\bar{e}_S) = \frac{\bar{e}_T(1-\bar{e}_T)}{n_T} + \frac{\bar{e}_S(1-\bar{e}_S)}{n_S}$$

$$\hat{V}(\hat{\Delta}_c) = \hat{V}(\bar{c}_T) + \hat{V}(\bar{c}_S) = \frac{\sum_{i=1}^{n_T}(c_{Ti} - \bar{c}_T)}{n_T(n_T-1)} + \frac{\sum_{i=1}^{n_S}(c_{Si} - \bar{c}_S)}{n_S(n_S-1)}$$

$$\hat{C}(\hat{\Delta}_e, \hat{\Delta}_c) = \hat{C}(\bar{e}_T, \bar{c}_T) + \hat{C}(\bar{e}_S, \bar{c}_S)$$

$$= \frac{\left(\sum_{i=1}^{n_T} e_{Ti}c_{Ti}\right) - n_T\bar{e}_T\bar{c}_T}{n_T(n_T-1)} + \frac{\left(\sum_{i=1}^{n_S} e_{Si}c_{Si}\right) - n_S\bar{e}_S\bar{c}_S}{n_S(n_S-1)}$$

and

$$\hat{\sigma}_+^2 = \sum_{j=T,S} n_j \left\{\lambda^2 \hat{V}(\bar{e}_j) + \hat{V}(\bar{c}_j) - 2\lambda\hat{C}(\bar{e}_j, \bar{c}_j)\right\},$$

where $\hat{\sigma}_+^2$ is the sum of treatment arms of the between-patient variance of incremental net benefit.

References

1. Adcock, C.J.: Sample size determination: a review. *The Statistician* **46**, 261–283 (1997)
2. Ament, A., Baltussen, R.: The interpretation of results of economic evaluation: Explicating the value of health. *Health Economics* **6**, 625–635 (1997)
3. Berry, D.A., Ho, C-H.: One-sided sequential stopping boundaries for clinical trials: A decision-theoretic approach. *Biometrics* **44**, 219–227 (1988)
4. Briggs, A.H., Wonderling, D.E., Mooney, C.Z.: Pulling cost-effectiveness analysis up by its bootstraps; a non-parametric approach to confidence interval estimation. *Health Economics* **6**, 327–340 (1997)
5. Briggs, A.H., Fenn, P.: Confidence intervals or surfaces? Uncertainty on the cost-effectiveness plane. *Health Economics* **7**, 723–740 (1998)
6. Briggs, A.H., Mooney, C.Z., Wonderling, D.E.: Constructing confidence intervals for cost-effectiveness ratios: an evaluation of parametric and non-parametric techniques using monte carlo simulation. *Statistics in Medicine* **18**, 3245–3262 (1999)
7. Briggs, A.H.: A Bayesian approach to stochastic cost-effectiveness analysis. *Health Economics* **8**, 257–261 (1999)
8. Chaudhary, M.A., Stearns, S.C.: Confidence intervals for cost-effectiveness ratios: An example from a randomized trial. *Statistics in Medicine* **15**, 1447–1458 (1996)
9. Cheng, Y., Su, F., Berry, D.A.: Choosing sample size for a clinical trial using decision analysis. *Biometrika* **90**, 923–936 (2003)
10. Chiba, N., van Zanten, S.J., Sinclair, P., Ferguson, R.A., Escobedo, S., Grace, E.: Treating Helicobacter pylori infection in primary care patients with uninvestigated dyspepsia: the Canadian adult dyspepsia empiric treatment-Helicobacter pylori positive (CADET-Hp) randomised controlled trial. *British Medical Journal* **324**, 1012–1016 (2002)
11. Claxton, K., Posnett, J.: An economic approach to clinical trial design and research priority setting. *Health Economics* **5**, 513–524 (1996)
12. Claxton, K.: The irrelevance of inference: a decision-making approach to the stochastic evaluation of health care technologies. *Journal of Health Economics* **18**, 341–364 (1999)
13. Claxton, K., Thompson, K.M.: A dynamic programming approach to the efficient design of clinical trials. *Journal of Health Economics* **20**, 797–822 (2001)
14. Eckermann, S., Willan, A.R.: Expected value of information and decision making in HTA. *Health Economics***16**, 195–209 (2007)
15. Eckermann, S., Willan, A.R.: Time and EVSI wait for no patient. *Value in Health* **11**, 522–6 (2008)
16. Eckermann, S., Willan, A.R.: The option value of delay in health technology assessment. *Medical Decision Making* **28**, 300–305 (2008)
17. Fenwick, E., O'Brien, B., Briggs, A.: Cost-effectiveness acceptability curves – facts, fallacies and frequently asked questions. *Health Economics* **13**, 405–415 (2004)
18. Gittins, J.: Quantitative methods in the planning of pharmaceutical research. *Drug Information Journal* **30**, 479–487 (1996)
19. Gittins, J., Pezeshk, H.: How large should a trial be? *The Statistician* **49**, 177–197 (2000)
20. Gittins, J., Pezeshk, H.: A behavioral Bayes method for determining the size of a clinical trial. *Drug Information Journal* **34**, 355–363 (2000)
21. Grundy, P.M., Healy, M.J.R., Rees, D.H.: Economic choice of the amount of experimentation. *Journal of the Royal Statistical Society: Series B* **18**, 32–48 (1956)
22. Halpern, J., Brown, Jr B.W., Hornberger, J.: The sample size for a clinical trial: a Bayesian-decision theoretic approach. *Statistics in Medicine* **20**, 841–858 (2001)
23. Hannah, M.E., Hannah, W.J., Hewson, S.H., Hodnett, E.D., Saigal, S., Willan, A.R.: Term Breech Trial: a multicentre international randomised controlled trial of planned caesarean section and planned vaginal birth for breech presentation at term. *The Lancet* **356**, 1375–1383 (2000)
24. Heitjan, D.F.: Fieller's method and net health benefit. *Health Economics* **9**, 327–335 (2000)

25. Hornberger, J.C., Brown, Jr B.W., Halpern, J.: Designing a cost-effective clinical trial. *Statistics in Medicine* **14**, 2249–2259 (1995)
26. Hornberger, J., Eghtesady, P.: The cost-benefit of a randomized trial to a health care organization. *Controlled Clinical Trials* **1p**, 198–211 (1998)
27. Laska, E.M., Meisner, M., Siegel, C.: Statistical inference for cost-effectiveness ratios. *Health Economics* **6**, 229–242 (1997)
28. Lindley, D.V.: The choice of sample size. *The Statistician* **46**, 129–138 (1997)
29. Lothgren, M., Zethraeus, N.: Definition, interpretation and calculation of cost-effectiveness acceptability curves. *Health Economics* **9**, 623–630 (2000)
30. Manning, W.G., Fryback, D.G., Weinstein, M.C. (1996) Reflecting uncertainty in cost effectiveness analysis. In Gold MR, Siegel JE, Russell LB, Weinstein MC (eds) Cost Effectiveness in Health and Medicine, *Oxford University Press*, New York
31. Mullahy, J., Manning, W. (1994) Statistical issues of cost-effectiveness analysis. In Sloan F (ed) Valuing Health Care, *Cambridge University Press*, Cambridge
32. O'Brien, B.J., Drummond, M.F., Labelle, R.J., Willan, A.R.: In search of power and significance: Issues in the design and analysis of stochastic cost-effectiveness studies in health care. *Medical Care* **32**, 150–163 (1994)
33. O'Hagan, A., Stevens, J.W.: Bayesian assessment of sample size for clinical trials of cost effectiveness. *Medical Decision Making* **21**, 219–230 (2001)
34. Pezeshk, H., Gittins, J.: A fully Bayesian approach to calculating sample sizes for clinical trials with binary response. *Drug Information Journal* **36**, 143–150 (2002)
35. Pezeshk, H.: Bayesian techniques for sample size determination in clinical trials: a short review. *Statistical Methodology in Medical Research* **12**, 489–504 (2003)
36. Phelps, C.E. and Mushlin, A.I.: On the (near) equivalence of cost-effectiveness and cost-benefit analysis. *International Journal of Technology Assessment in Health Care* **7**, 12–21 (1991)
37. Polsky, D., Glick, H.A., Willke, R., Schulman, K.: Confidence intervals for cost-effectiveness ratios: A comparison of four methods. *Health Economics* **6**, 243–252 (1997)
38. Stinnett, A.A., Mallahy, J.: Net health benefits: A new framework for the analysis of uncertainty in cost-effectiveness analysis. *Medical Decision Making* **18**(Suppl), S68–S80 (1998)
39. Tambour, M., Zethraeus, N., Johannesson, M.: A note on confidence intervals in cost-effectiveness analysis. *International Journal of Technology Assessment* **14**, 467–471 (1998)
40. van Hout, B.A., Al, M.J., Gordon, G.S., Rutten, F.F.H..: Costs, effects and C/E ratios alongside a clinical trial. *Health Economics* **3**, 309–319 (1994)
41. Wakker, P., Klaassen, M.P.: Confidence intervals for cost/effectiveness ratios. *Health Economics* **4**, 373–381 (1995)
42. Willan, A.R., Cruess, A.F., Ballantyne, M.: Argon green vs krypton red laser photocoagulation of extrafoveal choroidal neovascular lesions: Three-year results in age-related macular degeneration. *Canadian Journal of Ophthalmology* **31**, 11–7 (1996)
43. Willan, A.R., Pinto, E.M.: The expected value of information and optimal clinical trial design. *Statistics in Medicine* **24**, 1791–1806 (2005). Correction: Statistics in Medicine 2006;25:720
44. Willan, A.R.: Clinical decision making and the expected value of information. *Clinical Trials* **4**, 279–285 (2007)
45. Willan, A.R.: Optimal sample size determinations from an industry perspective based on the expected value of information. *Clinical Trials* **5**, 587–594 (2008)
46. Willan, A.R., Kowgier, M.E.: Determining optimal sample sizes for multi-stage randomized clinical trials using value of information methods. *Clinical Trials* **5**, 289–300 (2008)
47. Willan, A.R., Eckermann, S.: Optimal clinical trial design using value of information methods with imperfect implementation. *Health Economics* **19**, 549–561 (2010)
48. Willan, A.R., Eckermann, S.: Accounting for between-study variation in incremental net benefit in value of information methodology. *Health Economics* **21**, 1183–1195 (2012)
49. Willan, A.R., Eckermann, S.: Value of information and pricing new health care interventions. *PharmacoEconomics* **30**, 447–459 (2012)

50. Willan, A.R., O'Brien, B.J.: Confidence intervals for cost-effectiveness ratios: An application of Fieller's theorem. *Health Economics* **5**, 297–305 (1996)
51. Willan, A.R.: Incremental net benefit in the analysis of economic data from clinical trials with application to the CADET-Hp Trial. *European Journal of Gastroenterology and Hepatology* **16**, 543–549 (2004)
52. Willan, A.R., Goeree, R., Boutis, K. (2012) Value of Information Methods for Planning and Analyzing Clinical Studies Optimize Decision Making and Research Planning. *Journal of Clinical Epidemiology* doi: http://dx.doi.org/10.1016/j.jclinepi.2012.01.017
53. Willan, A.R.: Analysis, sample size and power for estimating incremental net health benefit from clinical trial data. *Controlled Clinical Trials* **22**, 228–237 (2001)
54. Willan, A.R., Lin, D.Y.: Incremental net benefit in randomized clinical trials. *Statistics in Medicine* **20**, 1563–1574 (2001)
55. Willan, A.R., Lin, D.Y., Cook, R.J., Chen, E.B.: Using inverse-weighting in cost-effectiveness analysis with censored data. *Statistical Methods in Medical Research* **11**, 539–551 (2002)
56. Willan, A.R., Chen, E.B., Cook, R.J., Lin, D.Y.: Incremental net benefit in clinical trials with quality-adjusted survival. *Statistics in Medicine* **22**, 353–362 (2003)
57. Willan, A.R., Briggs, A.H. (2006) The Statistical Analysis of Cost-effectiveness Data. Wiley, Chichester UK

Chapter 3
Designing Multi-arm Multi-stage Clinical Studies

Thomas Jaki

Abstract In the early stages of drug development there often is uncertainty about the most promising among a set of different treatments, different doses of the same treatment or sets of combinations of treatments. An efficient solution to determine which intervention is most promising are multi-arm multi-stage clinical studies (MAMS). In this chapter we will discuss the general concept to designing MAMS studies within the group sequential framework and provide detailed solutions for multi-arm multi-stage studies with normally distributed endpoints in which all promising treatments are continued at the interim analyses. An approach to find optimal designs is discussed as well as asymptotic solutions for binary, ordinal and time-to event endpoints.

3.1 Background and Motivation

The development of new medicinal products is a time consuming and very expensive process which has been estimated to take 10–15 years and cost several hundred million pounds on average [7]. The reason for the long duration is that, even after a potentially useful compound has been identified, the product needs to undergo pre-clinical animal studies, first in man studies and a series of clinical trials addressing different questions such as safety, dosing and efficacy. Among the largest contributors to both time and cost are confirmatory (Phase III) clinical trials that often involve thousands of patients with follow-up period frequently lasting years [27].

In the early stages of drug development there often is uncertainty about the most promising among a set of different treatments. In order to ensure the best use of resources in such situations it is important to decide which, if any, of the treatments should be taken forward for further testing. This is particularly important as in recent years 45 % of confirmatory clinical trials overall and 59 % of confirmatory trials in oncology have been unsuccessful [15]. An efficient solution to this problem

T. Jaki (✉)
Medical and Pharmaceutical Statistics Research Unit, Department of Mathematics and Statistics, Lancaster University, Lancaster, United Kingdom
e-mail: jaki.thomas@gmail.com

K. van Montfort et al. (eds.), *Developments in Statistical Evaluation of Clinical Trials*, DOI 10.1007/978-3-642-55345-5_3

MAMS studies will be an efficient solution to finding the best experimental treatment if one exists. If there is evidence/believe, however, that one of the potential experimental treatments is markedly better than the others a more traditional design is likely better suited.

A frequently used selling point of MAMS designs and many adaptive designs in general is that one can expect the trial to require fewer subjects than a single stage (fixed sample) study. The cost of this expected advantage is, however, that the maximum sample size will often be larger than the equivalent single stage study. If it is therefore impossible to recruit the number of subjects required for a fixed sample study, adding interim analyses as in MAMS studies will not solve that problem as there will be a chance (although a small one) of needing even more subjects.

The other cost of using an adaptive design such as a MAMS study is that finalizing the design of the study is frequently more time consuming as for fixed sample designs. As a consequence it is possible that, despite a smaller number of subjects being required on average the overall duration of the study (from planning to completion of the analysis) is the same or even longer than a more traditional design. Some of the factors that lead to the increased time to set up a MAMS study are:

- Additional decisions to be made before starting the study, e.g.,
  - How many and which treatments to include,
  - How many analyses time points should be used.
- Thorough simulation studies highly recommended [10] to fully understand the characteristics of the study under many settings.
- Patient information sheets need to cover all possible scenarios. Therefore different information sheets may have to be developed based on the decision at an analysis time point.

There are, however, many reasons to use MAMS studies as well. Not least of all the unique opportunity to compare treatments within the same study which will ensure a fair head-to-head comparison as by construction the same population will be studied, all patients will follow the same protocol and therefore receive the same standard of care due to it being a contemporary comparison. In addition, despite the additional work required up front, in many situations MAMS designs will still reduce the overall duration of the investigation. Furthermore, fewer patients tend to be exposed to ineffective or harmful treatments as these treatments are eliminated quickly from the study. Moreover, if all viable candidates are included within the study, a MAMS design will also lead to a better conclusion at the end due to the possibility of head-to-head comparisons of different active treatments. A further advantage of a MAMS design is that a lower dose than the most promising dose can also easily be included in the study and subsequently selected if safety concerns arise with the higher (often more efficacious) dose. Last but not least MAMS designs tend to be very popular with patients as the increased number of active treatments means that the 'risk' to receive placebo is lower helping recruitment in these studies.

3.3 Methodology

In this section we give some insight into the methodology used to design MAMS studies. For all developments except Sect. 3.3.4 we will assume that the endpoint of interest is normally distributed with known variance which is the same for all treatments. Methods to overcome these assumptions will be discussed in Sects. 3.3.4 and 3.5. Without loss of generality we will assume that an increase in the response is a good outcome and hence desirable. Consequently we will only consider one-sided tests and denote the one-sided type-I-error by α. Generalizations to two-sided tests are, however, straightforward. In addition we will suppose that a J stage design evaluating K experimental treatments against a common control is being planned. We will denote the response of patient $i = 1, 2, \ldots$ on treatment $k = 0, \ldots, K$ by X_{ik} where a subscript of zero represents control. We will denote the number of subjects on control in the first stage by n and define the ratio of the sample size of treatment k at timepoint j over the sample size on control in the first stage as $r_k^{(j)}$, so that the total sample size on treatment k at timepoint j is $r_k^{(j)} n$.

3.3.1 *Constraints on the Design*

We will now discuss the constraints imposed on multi-arm multi-stage designs. The reason why evaluation of multiple experimental treatments within one trial requires specialist statistical methods arises from the fact that more than one hypothesis is tested. In a trial with K active treatments the family of K null hypotheses of interest which is to be tested at each analysis timepoint j is

$$H_{01} : \mu_1 \leq \mu_0, \quad \ldots, \quad H_{0K} : \mu_K \leq \mu_0 ,$$

where μ_k is the mean response of a patient on treatment k. If each of these K hypothesis, that are tested up to J times, were tested at a level of, say, 5 %, the overall error, called the familywise error rate, would be substantially larger than 5 %. The objective is instead to control the familywise error rate at a specific level α, i.e.

$$P(\text{reject at least one true } H_{0k},\ k = 1, \ldots, K) \leq \alpha . \tag{3.1}$$

Furthermore we want to control the above probability under any true $(\mu_0, \ldots, \mu_K)$, i.e. control the familywise error rate in the strong sense. The standard z-statistics for comparing treatment k to control at stage j, defined as

$$Z_k^{(j)} = \frac{\hat{\mu}_k^{(j)} - \hat{\mu}_0^{(j)}}{\sigma \sqrt{\frac{r_k^{(j)} + r_0^{(j)}}{r_k^{(j)} r_0^{(j)} n}}}, \quad k = 1, \ldots K; j = 1, \ldots, J .$$

where $\hat{\mu}_q^{(j)} = \frac{\sum_{i=1}^{r_q^{(j)}n} X_{i,q}}{r_q^{(j)}n}$, $q = 0, \ldots, K$ are the sample means of all observations on treatment q up to analysis time point j, can be used to test the individual null hypotheses H_{0k}.

Just as we wish to control the type I error under the null hypothesis we want to find a sample size that will ensure that we can reject a null hypothesis provided the mean response of the corresponding treatment is large enough. We will utilize the so called 'least favorable configuration' (LFC, [9]) to specify the power requirement. The least favorable configuration is set up to recognize that not all improvements in response are worthwhile, either for practical or financial reasons. It requires two effect sizes, that is improvements of an experimental treatment over control, to be provided by the clinical team: a clinically interesting effect, δ, that, if present, we would like to detect with high probability and an uninteresting effect, δ_0, that, if present, would mean that we would not want to proceed to a further confirmatory study or registration of the treatment. For any effect between δ_0 and δ we are happy with either proceeding or abandoning. Power is then defined as the probability that, without loss of generality, H_{01} is rejected and treatment 1 is recommended given $\mu_1 - \mu_0 = \delta$ and $\mu_k - \mu_0 = \delta_0$ for $k = 2, \ldots, K$. The sample size is then found so that the power under the LFC is large, i.e.

$$P(\text{reject } H_{01} | \mu_1 = \delta_1, \mu_2 = \delta_0, \ldots, \mu_K = \delta_0) \geq 1 - \beta \ . \tag{3.2}$$

This set up is called the LFC since it minimizes the probability of selecting treatment 1 over all choices of μ_k such that $\mu_1 - \mu_0 \geq \delta$ and $\mu_k - \mu_0 \leq \delta_0$, [33].

These two constraints, namely type I error and power constraint are sufficient to design a traditional (two-arm, single stage) study, once the ratio of subjects on control versus experimental is fixed as only two design parameters are left: the critical value for rejecting the null hypothesis and the sample size. In a multi-stage study, however, even after assumptions on the relative sample size, $r_k^{(j)}$, have been made, stopping boundaries are required at each stage, j. Lower boundary values (futility bound), l_j, are used to stop treatments in the trial as soon as the corresponding test statistic, $Z_k^{(j)}$, falls below it. At this point no more patients will be randomized to the corresponding treatment. Once any test statistic exceeds the upper boundary (efficacy bound), u_j, the trial is stopped and the corresponding hypothesis rejected, i.e. it can be concluded that the corresponding treatment is superior to control. If at least one test statistics is between the upper and lower bound while none exceeds the upper bound additional patients are randomized to each of the remaining active treatments plus control. If throughout the trial none of the test statistics exceeds the efficacy bound, then we fail to reject the null hypothesis and conclude that none of the investigated treatments is superior to control. Note that the futility bound and the efficacy bound at the final analysis time point are equal to ensure that a conclusion is reached. As a consequence a multi-stage design has $2 \times J$

parameters – the lower and upper boundaries l_j and u_j at each stage, the boundary at the last stage and the sample size – after the allocation ratio has been specified. In order to obtain a design, there are two principal approaches to overcome the problem of $2 \times J$ unknowns with only 2 constraints:

1. Make additional assumptions about the relationship of critical values and sample size.
2. Use an optimality criterion.

We will defer a discussion of the second approach to Sect. 3.3.3 and focus on the first approach for the moment.

We start by fixing $r_k^{(j)}$, the ratio of sample sizes, a priori in the same way as in a traditional design the ratio of patients on control to experimental treatment is fixed. Commonly equal numbers of patients on each experimental treatment within each stage: $r_1^{(j)} = \ldots = r_K^{(j)} = r^{(j)}$ for $j = 1;\ldots;J$ are assumed. The choice then reduces to setting $r^{(j)}$ and $r_0^{(j)}$ so that setting $r^{(j)} = r_0^{(j)} = jr^{(1)}$ means that the sample size at each stage is equal across all treatments and control. Setting $r^{(j)} = jr^{(1)}$ and $r_0^{(j)} = 2jr^{(1)} = 2r^{(j)}$ allocates twice as many patients to control than each experimental treatment for all stages. The next step is to make all critical values for stages 1 to $J-1$ known functions of the final critical value; $u_j = f(j, u_J)$ and $l_j = g(j, u_J)$, $j = 1, \ldots, J-1$. This approach is identical to the idea frequently used in sequential designs and the same functional relationships can be used. Popular examples are the boundaries due to Pocock [24], O'Brien and Fleming [22] and the triangular test described in [38]. These boundaries are illustrated in Fig. 3.2.

Once a suitable allocation ratio and boundaries have been chosen, two unknown design parameters, the final critical value, u_J, and the sample size, n, remain. These are then obtained by solving the type I error and power equation given in (3.1) and (3.2).

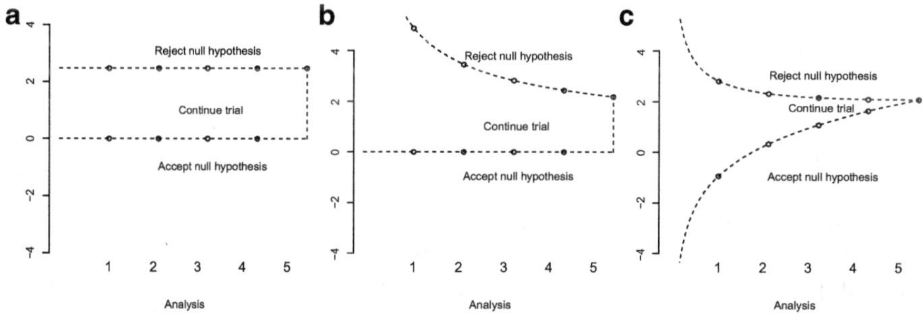

Fig. 3.2 Frequently used boundaries: **(a)** Pocock, **(b)** O'Brien-Fleming, **(c)** Triangular. Note that a futility boundary of zero is used in **(a)** and **(b)**

3.3.2 All-Promising Design

The ideas discussed in the previous subsection give a general framework that can be used to design MAMS studies. Within this framework a variety of different designs that differ mainly in the treatment selection at the interim analyses are available. So called "pick-the-winner" designs select the most promising experimental treatment at the first interim analysis and compare it to control in the subsequent stages (e.g. Stallard and Todd [30]). Stallard and Friede [29] allow more than one treatment to continue beyond the first stage, provided the number of treatment arms within each stage is pre-specified while Kelly et al. [14] advocate using a rule that allows all treatments that are within some margin, ϵ, of the best performing treatment to be selected.

In the discussion below we will consider the select "all-promising" setting where all experimental treatments are selected at each interim analysis, provided that they are promising enough. It is worth noting that the sample size requirement of "pick-the-winner" design tends to be smaller than the "all-promising" design as only two treatment arms remain after the selection while it is possible to recruit patients for all $K + 1$ arms throughout an "all-promising" design. The advantage of the "all-promising" design on the other hand is that it is still possible to recommend any treatment for registration or further testing provided its performance is good enough. Moreover more stringent selection rules such as the ones discussed above can still be used without an inflation of the familywise type I error – the power of the study on the other hand would be compromised.

3.3.2.1 Example

We start by giving an illustrative example of a trial with four novel treatments that are to be compared with control. Suppose that a maximum of three analyses are to be conducted, two interim analyses and one final analysis, and that an equal sample size, n, is used for each treatment at each stage. Moreover, assume that an O'Brien and Fleming boundary shape [22], $u_j = u_J\sqrt{r^{(J)}/r^{(j)}} = u_J\sqrt{3/j}$, and a futility boundary of $(0, 0, u_J)$ is used. Using the results in [17] one can find that, for a familywise error rate of $\alpha = 0.05$ and a power of $1 - \beta = 0.9$, the final critical value, u_J is 2.182 which results in an efficacy boundary of $(3.779, 2.672, 2.182)$ and a futility boundary of $(0, 0, 2.182)$.

Let the interesting effect size, δ, be such that the probability of a randomly selected person on that treatment observing a larger score than a person on control is 0.65 and an uninteresting effect, δ_0 be set such that this probability is 0.55. Note that this, slightly unusual, way to specify an effect size has the advantage that no assumption about the standard deviation, σ, is needed. It is, however, straightforward to obtain the traditional effect size from this formulation. Assuming that the variance is equal to one the interesting effect size, δ, can then be found as $\Phi^{-1}(0.65)\sqrt{2}\sigma = 0.545$ while the uninteresting effect size, δ_0, is $\Phi^{-1}(0.55)\sqrt{2}\sigma = 0.178$. The resulting required sample size per arm and stage is then $n = 31$. As a consequence the maximum sample size for this design is 465 which is necessary if no conclusion is reached before the last analysis, while the smallest sample size is 155 (when either all experimental treatments perform poorly or at least one experimental treatment performs very well). For comparison, the sample size of a single stage design would be 420 under this setting showing that a reduction in sample size by 265 patients is possible. At the same time one can see that in the worst case situation an additional 45 patients would be required. Since the expected sample size, that is the average sample size if the trial were performed numerous times, is around 310 patients if either the null hypothesis or the LFC is true, the benefit does outway the potential risk.

Suppose now that at the first interim analysis test statistics are found to be $(2.330, -1.342, -0.449, 0.621)$. The test statistics corresponding to the second and third experimental treatment fall below the lower boundary at the first analysis so that the corresponding treatments are removed from the further study. Since none of the treatments exceeds the upper boundary a further n patients are randomized to treatments one and four as well as control. Suppose now that at the second interim analysis we find the test statistics for the remaining two treatments based on data from both stages to be $(2.920, 1.662)$. Since the test statistics for the first experimental treatment exceeds the corresponding efficacy boundary, $Z_1^{(2)} = 2.920 > 2.672 = c_2$, the trial is stopped with the conclusion that the first experimental treatment is superior to control. Figure 3.3 provides an illustration of the study.

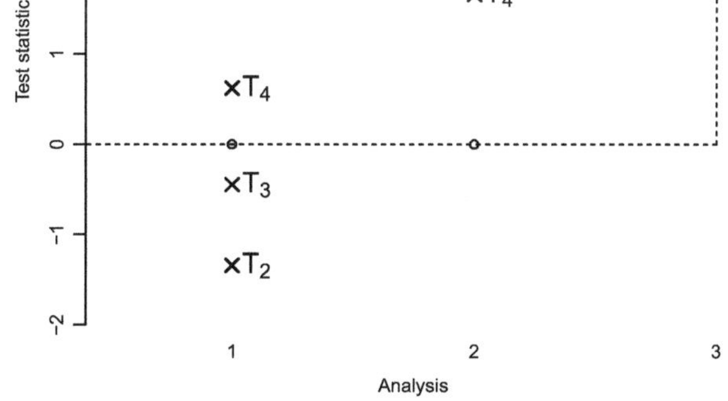

Fig. 3.3 Illustration of an all-promising MAMS design

3.3.2.2 Designing a Two-Stage Trial

We now provide the constraints required to design a 2-stage design with K treatments assuming equal sample size per arm and stage. This special case is discussed here for brevity of notation and because two-stage designs are the most frequently utilized designs in this class. For general results for J stage trials the reader is referred to [17]. A software implementation of the general setting is discussed in Sect. 3.4. To obtain the critical value for a 2-stage design the final critical value u_J can be found by solving

$$\alpha = 1 - \int_{-\infty}^{\infty}\int_{-\infty}^{\infty}\left[\Phi\left(l_1\sqrt{2}+t_2\right) + \Phi_2\left(u_1\sqrt{2}+t_2, u_2\sqrt{2}+\frac{t_1+t_2}{\sqrt{2}}, \frac{1}{\sqrt{2}}\right)\right.$$
$$\left. -\Phi_2\left(l_1\sqrt{2}+t_2, u_2\sqrt{2}+\frac{t_1+t_2}{\sqrt{2}}, \frac{1}{\sqrt{2}}\right)\right]^K d\Phi(t_1)d\Phi(t_2)\ ,$$

where Φ denotes the standard normal distribution function and $\Phi_2(a, b, \sigma_{12})$ denotes the result of integrating the bivariate normal density with mean 0 and covariance matrix with one in the diagonal and σ_{12} in the off diagonal over a region defined by a and b. Note that l_1 and u_1 are chosen to be known functions of the final critical value so that this equation can be solved, and hence the boundaries determined, without knowledge of σ or n. The equation does involve a two-dimensional integral (for the two stages) which in practice needs to be solved numerically. The complexity of the equation does not, however, increase with the number of treatment arms in the study.

To find the required sample size it is worthwhile to point out at this stage that the trial is only stopped with rejection of the null hypothesis at one of the stages. As a consequence the probability of rejecting the null hypothesis can be written as the sum of the rejection probabilities at each of the stages. For a 2-stage design the overall power of the study under the LFC is therefore the sum of two integrals:

$$1-\beta = \int_{-\infty}^{\infty}\Phi\left(u_1\sqrt{2}+t+\frac{\sqrt{n}}{\sigma}\delta\right)\left[\Phi\left(t+\frac{\sqrt{n}}{\sigma}(\delta-\delta_0)\right)\right]^{K-1} d\Phi(t)$$
$$+\int_{-\infty}^{\infty}\int_{-\infty}^{\infty}\left[\Phi\left(l_1\sqrt{2}+t_2-\frac{\sqrt{n}}{\sigma}\delta_0\right)\right.$$
$$+\Phi_2\left(u_1\sqrt{2}+t_2-\frac{\sqrt{n}}{\sigma}\delta_0, t_1+(\delta-\delta_0)\frac{\sqrt{2n}}{\sigma}, \frac{1}{\sqrt{2}}\right)$$
$$\left.-\Phi_2\left(l_1\sqrt{2}+t_2-\frac{\sqrt{n}}{\sigma}\delta_0, t_1+(\delta-\delta_0)\frac{\sqrt{2n}}{\sigma}, \frac{1}{\sqrt{2}}\right)\right]^{K-1}$$

$$\left[\Phi\left(2u_1 + t_2\sqrt{2} - t_1 - \frac{\sqrt{2n}}{\sigma}\delta\right) - \Phi\left(2l_1 + t_2\sqrt{2} - t_1 - \frac{\sqrt{2n}}{\sigma}\delta\right)\right]$$
$$\Phi\left(t_1\sqrt{2} - t_2 - 2u_2\right) d\Phi\left(t_1\right) d\Phi\left(t_2\right) \, .$$

Once again the equation involves a two-dimensional integral which can be solved for n, once the boundaries have been found, for a given δ and δ_0.

3.3.2.3 Deviation from the Planned Design

Having found the design to use, deviations to the plan will invariably occur. The most frequent deviation for the design is that the achieved sample sizes within each arm will deviate slightly from the planned value. To prevent a (typically very small) inflation of the familywise error rate, the boundaries can then be adjusted using the truly observed number of responses. Worked examples how this can be achieved are given in [12].

Besides this, fairly straightforward, departure from the planned design more extreme situations are also likely to occur. It may happen that one of the treatments that is deemed promising based on the efficacy endpoint, i.e. the test statistic exceeds the lower bound at an analysis timepoint, but is found to be unsafe. In this case recruitment to the treatment would naturally be stopped. The implications on the study design are a reduction in the type-I-error rate as well as a reduction in power. Although this behaviour is acceptable in many situations, the conditional error approach due to [21] can be used to, for example, increase the sample size to ensure that the overall power is satisfactory [18].

The design discussed above is set up so that the familywise error is controlled (and exhausted) if the treatment corresponding to the largest test statistic is taken forward to further testing. It is, however, possible that more than one test statistic exceeds the upper boundary for the first time at a given analysis time point. In such a situation it is acceptable to take any treatment forward to further testing provided it exceeds the upper boundary as the familywise error will still be controlled. Note that, due to the binding nature of the futility boundary, the opposite of keeping a treatment that falls below the futility boundary is not acceptable and will inflate the familywise error rate.

A final situation we want to discuss at this stage is when a test statistic does exceed the upper boundary, but the investigators do not want to stop the study at this stage. This may be due to requiring additional information on a secondary or safety endpoint or two treatments performing very similarly. While we stress here that one ought to try to avoid such a situation at the design stage by ensuring that sufficient information on all interesting endpoints is available at the earliest analysis time point already, it is possible to continue with the trial without compromising the familywise error rate. This is provided that any further tests done follow the procedure of the design (i.e. only reject the null hypothesis if a test statistic exceeds the efficacy boundary at the analysis time point).

3.3.3 Optimal Design of MAMS

In the previous section we have discussed an approach where the boundary shape is pre-specified in advance to reduce the number of parameters in the design so that we were able to obtain the final critical value and the required sample size from the traditional type I and power equations. An alternative approach, discussed in [34], is to find boundaries that optimize a certain criterion. Once an appropriate criterion is identified, the parameter space is searched for the solution that optimizes the choosen criterion. In most situations it will not be feasible to evaluate all possible designs so that stochastic search algorithms, such as simulated annealing, are often used.

Besides the computational burden to find an optimal design the biggest challenge is to find a satisfactory optimality criterion. Traditionally the optimality criterion aims to minimize the sample size under a specific parameter configuration. The null-optimal design searches for the minimal expected sample size under the global null hypothesis, $\mu_0 = \ldots = \mu_k = 0$, while the LFC-optimal design optimizes the expected sample size under $\mu_1 - \mu_0 = \delta_1, \mu_2 - \mu_0 = \ldots = \mu_k - \mu_0 = \delta_0$. The drawback of using such an optimality criterion is that, although the expected sample size under the chosen parameter configuration is optimal, their performance under other configurations can be quite poor. To get a design that is better balanced it is therefore advisable to use a more complex criterion such as the δ-minimax criterion. This critcrion minimizes the expected sample size for a parameter configuration of $\mu_1 - \mu_0 = \tilde{\delta}_1, \mu_2 - \mu_0 = \ldots = \mu_k - \mu_0 = \delta_0$ where $\tilde{\delta}_1$ is found to maximize the expected sample size under this configuration. The criterion therefore ensures that under the worst case situation the expected sample size is smallest. As a consequence of this construction the criterion tends to give a good balance of expected sample sizes across many true parameter configurations.

Besides the obvious advantage to use an optimal design (subject to using the appropriate criterion) a further benefit of this approach is that it is not restricted to searching over boundaries. It is, for example, possible to include the allocation ratio between control and experimental treatments in the search as well. The main limitation of using an optimality criterion to design a study is the computational effort required. In their evaluations Wason and Jaki [34], however, find that a triangular boundary shape, defined as $l_j = -u_J(1 - 3\frac{r^{(j)}}{r^{(J)}})/\sqrt{r^{(j)}}$ and $u_j = u_J(1 - \frac{r^{(j)}}{r^{(J)}})/\sqrt{r^{(j)}}$, which has been shown to be asymptotically optimal for traditional group sequential designs, does yield close to optimal solutions in the multi-arm setting in many situations as well. In particular if only few interim analysis are planned the triangular design performs particularly well.

3.3.4 Non-normal Endpoints

The discussion so far has focused on trials with a normally distributed endpoint. In this section we will discuss how asymptotic theory allows the framework developed for normal endpoints to be used with binary, ordinal and time-to-event endpoints as well. Note that binary endpoints are a special case of ordinal endpoints and are hence not discussed separately. For MAMS studies in which only the most promising treatment is selected at an interim analysis full details can be found in [33] for binary endpoints, [39] for ordinal endpoints and [30] for survival endpoints. Asymptotic results for the "all-promising" selection strategy are discussed in [12] for all endpoints. A specific solution for time-to-event endpoints without early stopping for efficacy in which an intermediate endpoint is used for selection of all promising treatments is given in [28].

3.3.4.1 Ordinal Endpoints

Consider an ordinal endpoint with categories C_1 (best), ..., C_a (worst) and denote the probability that a patient on treatment k falls into a category u by $p_{k,u}$. At analysis j, the number of patients on treatment k in category C_u is denoted by $n_{ku}^{(j)}$ and the cumulative probabilities up to category u are denoted by $Q_{k,u} = p_{k,1} + \cdots + p_{k,u}$. Under the proportional odds assumption (e.g. [20]) the parameters of interest, which is the log-odds ratio, can be defined as

$$\beta_k = \log\left\{\frac{Q_{k,u}(1 - Q_{0,u})}{(1 - Q_{k,u})Q_{0,u}}\right\} .$$

The efficient score statistics for β_k evaluated at stage j is

$$S_k^{(j)} = \frac{1}{(r_k^{(j)} + r_0^{(j)})n} \sum_{u=1}^{a} n_{ku}^{(j)}(W_{0,u}^{(j)} - B_{0,u}^{(j)}),$$

where the number of subjects in any of the categories $C_1, \dots, C_{u-1}$ (better than C_u) at stage j is denoted $B_{iu}^{(j)}$, $(B_{i1}^{(j)} = 0)$, and the number falling into a category worse than C_u by $W_{iu}^{(j)}$, $(W_{ia}^{(j)} = 0)$. Whitehead and Jaki [39] show that, under the null hypothesis, the efficient score statistics asymptotically follow a $K \times J$ dimensional multivariate normal distribution with

$$\text{Var}\left(S_k^{(j)}\right) \approx \frac{r_k^{(j)} r_0^{(j)} n}{3(r_k^{(j)} + r_0^{(j)})}\left(1 - \sum_{u=1}^{a} \bar{p}_u^3\right) ,$$

and

$$\text{Cov}\left(S_k^{(j)}, S_{k'}^{(j')}\right) \approx \frac{r_0^{(j)} r_k^{(j)} r_{k'}^{(j')} n}{3(r_0^{(j)} + r_k^{(j)})(r_0^{(j')} + r_{k'}^{(j')})}\left(1 - \sum_{u=1}^{a} \bar{p}_u^3\right), \; (j \leq j'; k \neq k') ,$$

where $\bar{p}_u = p_{0,u} = \ldots = p_{K,u}$ is the anticipated proportion of subjects falling into category u.

The correlations between the standardised test statistics $Z_k^{(j)} = S_k^{(j)} / \sqrt{\text{Var}\left(S_k^{(j)}\right)}$ are now exactly the same as for a normal endpoint so that both the boundaries and sample sizes found for a normal endpoint can be used directly to design a study based on an ordinal endpoint.

3.3.4.2 Time-to-Event Endpoints

To design a MAMS study with a time-to-event endpoint we will use the log-rank test which is a special case of the efficient score statistic under the assumption of proportional hazards and assume that no ties are present. Denote the overall total maximum sample size by N, the calendar time with τ and let $\Delta_p(\tau)$ be the indicator that subject p has had an event by study time τ, $p = 1, \ldots, N$. To denote a subject p being on treatment $q = 1, \ldots, K$ we use the indicator $I_p\{q\}$ and similarly $I_p\{q, q'\}$ indicates that subject p is on either q or q'. Finally let $r_q(\tau)$ denote the number of patients at risk just before calendar time τ on treatment q.

The efficient score statistic for comparing active treatment k $(k = 1, \ldots, K)$ with control at interim j $(j = 1, \ldots, J)$ is

$$S_k^{(j)} = \sum_{p=1}^{N} I_p\{k,0\}\, \Delta_p(\tau_j) \left[-I_p\{k\} + \frac{r_k(\tau_p)}{r_k(\tau_p) + r_0(\tau_p)} \right],$$

where τ_j is calendar time at analysis time point j and τ_p is patient p's calendar event time. Asymptotically, the efficient scores follow a multivariate normal distribution and, assuming an equal allocation of patients to each arm at each stage and that the effect size is small, the variance and covariance of the efficient score statistic for treatment k at stage j can be estimated as

$$\text{Var}\left(S_k^{(j)}\right) \approx \frac{e_{k,0}(\tau_j)}{4},$$

and

$$\text{Cov}(S_k^{(j)}, S_{k'}^{(j')}) \approx \frac{e_{k,k',0}(\tau_j)}{12},$$

where $e_{k,0}(\tau_j)$ is the number of events from patients in the treatment groups k and control up to time τ_j and $e_{k,k',0}(\tau_j)$ is the number of events in treatment groups k, k' and control up to time τ_j. The standardized statistics, $Z_k^{(j)} = S_k^{(j)} / \sqrt{\text{Var}\left(S_k^{(j)}\right)}$ then have again the same correlation structure as for normally distributed endpoints so that the same boundaries can be used. The number of events required is the same as the number of patients required for normal endpoints.

3.4 Software

To date software solutions for MAMS studies are rare and often do require a good understanding of the theory to be able to use the programmes effectively. The add-on package *MAMS* [11] for R [26] provides one of the few existing ready made software solutions and implements the "all-promising" design discussed in the previous sections. In particular the function `mams` allows the user to determine the boundaries of the design for a given familywise error and the required sample size for a given power.

The parameters of the function include `K`, the number of experimental treatments, `J`, the number of stages together with `alpha` and `power` for the one-sided familywise error rate and power. The shape of the boundaries to be used are specified via `u.shape` and `l.shape` for the efficacy and futility boundary, respectively. These parameters can be set to either `"pocock"`, `"obf"` and `"triangular"` for boundaries according to [22, 24] and [38]. Additionally setting the boundaries to `"fixed"` allows a constant boundary to be used. To specify the level of the boundary in this case the parameters `lfix` and `ufix` are used. The interesting effect size is specified via `p` while the uninteresting effect is set via `p0`. Both parameters are parametrized as the probability of a random person on treatment k observing a larger effect than a random person on control. The final set of parameters, `r` and `r0`, specify the allocation ratio of subjects on experimental treatments and control respectively. For both a vector of length J is expected. A choice of `r=1:3` and `r0=c(2,4,6)` corresponds to a trial were twice as many subjects are randomized to control than experimental treatment at each stage.

To illustrate the use of the package we consider the example given in Sect. 3.3.2.1 which seeks a three stage design for four active treatments using an O'Brien and Fleming boundary shape [22] for the efficacy and a zero boundary for futility. Equal sample size per arm and stage is desired, the familywise error rate is set a 5 % and a power of 90 % is used. To obtain the design parameters under this setting the function call

```
mams(K=4, J=3, p=0.65, p0=0.55, r=1:3, r0=1:3, alpha=0.05,
      power=0.9, u.shape="obf", l.shape="fixed", lfix=0)
```

can be used. The output given:

```
Design parameters for a 3 stage trial with 4 treatments

                                        Stage 1 Stage 2 Stage 3
Cumulative sample size per stage (control):
                                             31      62      93
Cumulative sample size per stage (active):
                                             31      62      93
```

```
Maximum total sample size:  465

              Stage 1 Stage 2 Stage 3
Upper bound:     3.78   2.673   2.182
Lower bound:     0.00   0.000   2.182
```

clearly shows the sample size required per stage separately for control and each experimental treatment. The boundaries and maximum total sample size are also provided.

Alternative to this R package there exist a variety of software solutions for the design of multi-arm multi-stage designs using the p-value combination approach. The commercial software AddPlan (www.aptivsolutions.com/adaptive-trials/addplan6/), for example, provides an add-on module for multi-arm trials, while the open-source R package *asd* [23] is a specialized package for multi-arm multi-stage trials.

3.5 Discussion

In this chapter we discussed the statistical concepts of designing multi-arm multi-stage clinical studies within the group sequential framework. Focus has been given to the general ideas useful when designing such studies, to provide some illustrative examples and highlight the advantage of these studies over parallel group and single stage designs. MAMS studies have, however, not only been investigated in the context of a single study but also in the context of the wider drug development process in [35]. The most interesting finding of this work is that inclusion of a large number of treatment arms within one study tends to be optimal when the objective is to find one working treatment at the end of the process. The assumption underlying this result is that the expected effects of all treatments follow the same distribution a priori which essentially means that only treatments that are truly believed to have an effect ought to be included.

As illustrated, MAMS studies offer the opportunity to make an informed decision about the most potent of a number of different treatments in an efficient manner, they do, however, also bring with them a number of practical and more subtle technical challenges. One of the technical details concerns the assumption of a known variance which is being made by virtually all the approaches referenced in this chapter. In practice, however, the best available estimate of the variance would be used to analyse the trial. Using an estimate for the variance instead of a known value does slightly increase the type I error of the study. Taking this added uncertainty into account, however, would make the design of MAMS studies considerably more complex. In studies with a large sample size the effect of estimating the variance will be negligible. For small sample sizes Jennison and Turnbull [13], discuss an adjustment based on transformation to overcome this problem. Note, however, that this approach will only reduce the impact of using an estimate instead of a known variance, but not ensure strict control of the familywise error [36]. The main idea is to transform and back-transform the individual test statistics as, $\Phi^{-1}\left(T_\nu\left(Z_k^{(j)}\right)\right)$, were T_ν is the distribution function of a t-distribution with ν degrees of freedom and the degrees of freedom, ν, are based on the sample size in the trial.

A further challenge arises when estimating the treatment effect at the end of the study. It has been recognised that the maximum likelihood estimate (MLE) of the treatment effect for trials with selection is biased [5]. While the bias of the MLE tends to be small, accurate estimation of a treatments effect is vital for expressing its true worth. Several classes of estimation procedures have been proposed to find unbiased estimates or obtain a reduction in the bias. One general approach is to estimate the bias of the MLE and to use this to iteratively calculate a bias-reduced MLE [31,37].

Besides decisions on the statistical aspects of MAMS studies, there are also a few practicalities to consider when running a MAMS study. One challenge, for example, is ensuring adequate supply of the treatments under investigation which is much more complex due to the stochastic nature of the demand on individual treatments. Also, patient information sheets covering all possible settings after the selection has taken place need to be prepared. An illustrative account of practical and statistical challenges in the context of a specific study is given in [32].

Acknowledgements This work is based on research arising from Dr. Jakis Career Development Fellowship (NIHR-CDF-2010-03-32) supported by the National Institute for Health Research and the MRC grant MR/J004979/1. The views expressed in this publication are those of the author and not necessarily those of the NHS, the National Institute for Health Research or the Department of Health. The author would also like to thank Dr. James Wason and Dominic Magirr for their helpful comments.

References

1. Antonijevic, Z., Pinhero, J., Fardipour, P., Lewis, R.J.: Impact of dose selection strategies used in phase II on the probability of success in phase III. *Statistics in Biopharmaceutical Research* **2**, 469–486 (2010)
2. Barthel, F.M., Parmar, M.K.B., Royston, P.: How do multi-stage, multi-arm trials compare to the traditional two-arm parallel group design–a reanalysis of 4 trials. *Trials* **10:21** (2009)
3. Bauer, P., Kieser, M.: Combining different phases in the development of medical treatments within a single trial. *Statistics in Medicine* **18**, 1833–1848 (1999)
4. Bauer, P., Köhne, K.: Evaluation of experiments with adaptive interim analyses. *Biometrics* **50**, 1029–1041 (1994)
5. Bauer, P., König, F., Brannath, W., Posch, M.: Selection and bias – two hostile brothers. *Statistics in Medicine* **29**, 1–13 (2010)
6. Bretz, F., König, F., Brannath, W., Glimm, E., Posch, M.: Adaptive designs for confirmatory clinical trials. *Statistics in Medicine* **28**, 1181–1217 (2009)
7. DiMasi, J.A., Hansen, R.W., Grabowski, H.G.: The price of innovation: New estimates of drug development costs. *Journal of Health Economics* **22**, 151–185 (2003)
8. Dunnett, C.W.: A multiple comparison procedure for comparing several treatments with a control. *Journal of the American Statistical Association* **50**, 1096–1121 (1955)

9. Dunnett, C.W.: Selection of the best treatment in comparison to a control with an application to a medical trial. In: T.J. Santer, A.C. Tamhane (eds.) *Design of Experiments: Ranking and Selection*, pp. 47–66. Marcel Dekker: New York (1984)
10. Guidance for Industry: *Adaptive Design Clinical Trials for Drugs and Biologics. Draft guidance.* Food and Drug Administration (FDA) (2010). URL www.fda.gov/downloads/Drugs/GuidanceComplianceRegulatoryInformation/Guidances/UCM201790.pdf. Last accessed 11 June 2012.
11. Jaki, T., Magirr, D.: *MAMS: Designing Multi-arm Multi-stage Studies* (2012). URL http://CRAN.R-project.org/package=MAMS. R package version 0.3
12. Jaki, T., Magirr, D.: Considerations on covariates and endpoints in multi-arm multi-stage clinical trials. *Statistics in Medicine* **32**(7), 1150–1163 (2013)
13. Jennison, C., Turnbull, B.W.: *Group Sequential Methods with Applications to Clinical Trials.* Chapman and Hall (2000)
14. Kelly, P.J., Stallard, N., Todd, S.: An adaptive group sequential design for phase II/III clinical trials that involve treatment selection. *Journal of Biopharmaceutical Statistics* **15**, 641–658 (2005)
15. Kola, I., Landis, J.: Can the pharmaceutical industry reduce attrition rates? *Nature Reviews Drug Discovery* **3**, 711–715 (2004)
16. Lehmacher, W., Wassmer, G.: Adaptive sample size calculations in group sequential trials. *Biometrics* **55**, 1286–1290 (1998)
17. Magirr, D., Jaki, T., Whitehead, J.: A generalized dunnett test for multiarm-multistage clinical studies with treatment selection. *Biometrika* **99**(2), 494–501 (2012)
18. Magirr, D., Stallard, N., Jaki, T.: Flexible sequential designs for multi-arm clinical trials. *Statistics in Medicine* **33**(19), 3269–3279 (2014)
19. Marcus, R., Peritz, E., Gabriel, K.R.: On closed testing procedures with special reference to ordered analysis of variance. *Biometrika* **63**, 655–660 (1976)
20. McCullagh, P.: Regression models for ordinal data (with discussion). *Journal of the Royal Statistical Society, Series B* **42**, 109–142 (1980)
21. Müller, H.H., Schäfer, H.: Adaptive group sequential designs for clinical trials: combining the advantages of the adaptive and of classical group sequential approaches. *Biometrics* **57**, 886–891 (2001)
22. O'Brien, P., Fleming, T.: A multiple-testing procedure for clinical trials. *Biometrics* **35**, 549–556 (1979)
23. Parsons, N.: *asd: Simulations for adaptive seamless designs* (2010). URL http://CRAN.R-project.org/package=asd. R package version 1.0
24. Pocock, S.J.: Group sequential methods in the design and analysis of clinical trials. *Biometrika* **64**, 191–199 (1977)
25. Posch, M., König, F., Branson, M., Brannath, W., Dunger-Baldauf, C., Bauer, P.: Testing and estimation in flexible group sequential designs with adaptive treatment selection. *Statistics in Medicine* **24**, 3697–3714 (2005)
26. R Development Core Team: *R: A Language and Environment for Statistical Computing.* R Foundation for Statistical Computing, Vienna, Austria (2012). URL http://www.R-project.org/. ISBN 3-900051-07-0
27. Roberts, T.G., Lynch, T.J., Chabner, B.A.: The phase III trial in the era of targeted therapy: Unraveling the "go or no go" decision. *Journal of Clinical Oncology* **21**, 3683–3695 (2003)
28. Royston, P., Parmar, M.K., Qian, W.: Novel designs for multi-arm clinical trials with survival outcomes with an application in ovarian cancer. *Statistics in Medicine* **22**, 2239–2256 (2003)
29. Stallard, N., Friede, T.: A group-sequential design for clinical trials with treatment selection. *Statistics in Medicine* **27**, 6209–6227 (2008)
30. Stallard, N., Todd, S.: Sequential designs for phase III clinical trials incorporating treatment selection. *Statistics in Medicine* **22**, 689–703 (2003)
31. Stallard, N., Todd, S.: Point estimates and confidence intervals for sequential trials involving selection. *Journal of Planning and Statistical Inference* **135**, 402–419 (2005)

32. Sydes, M.R., Parmar, M.K., James, N.D., Clarke, N.W., Dearnaley, D.P., Mason, M.D., Morgan, R.C., Sanders, K., Royston, P.: Issues in applying multi-arm multi-phase methodology to a clinical trial in prostate cancer: the "MRC STAMPEDE" trial. *Trials* **10**, 39 (2009)
33. Thall, P.F., Simon, R., Ellenberg, S.S.: Two-stage selection and testing designs for comparitive clinical trials. *Biometrika* **75**, 303–310 (1988)
34. Wason, J.M.S., Jaki, T.: Optimal design of multi-arm multi-stage trials. *Statistics in Medicine.* **31**(30), 4269–4279 (2012)
35. Wason, J.M.S., Jaki, T., Stallard, N.: Planning multi-arm screening studies within the context of a drug development programme. *Statistics in Medicine* **32**(20), 3424–3435 (2013)
36. Wason, J.M.S., Magirr, D., Law, M., Jaki, T.: Some recommendations for multi-arm multi-stage trials. Statistical Methods in Medical Research. Published online ahead of print (2014). doi:10.1177/0962280212465498
37. Whitehead, J.: On the bias of maximum likelihood estimation following a sequential trial. *Biometrika* **73**, 573–581 (1986)
38. Whitehead, J.: *The Design and Analysis of Sequential Clinical Trials*. Wiley: Chichester (1997)
39. Whitehead, J., Jaki, T.: One- and two-stage design proposals for a phase II trial comparing three active treatments with a control using an ordered categorical endpoint. *Statistics in Medicine* **28**, 828–847 (2009)

Chapter 4
Statistical Approaches to Improving Trial Efficiency and Conduct

Janice Pogue, P.J. Devereaux, and Salim Yusuf

Abstract Given the trend towards increasing complexity and cost of clinical trials today, future trials may end up being small, complex, and under-powered to detect clinically meaningful treatment effects. In order to continue to perform important research in the future, we need to make clinical trial designs more efficient. Through the retrospective statistical analysis of variation in the design of past trials and the prospective comparisons of clinical trials methods, we can determine which trial procedures truly influence the bias and precision of treatment estimates and where complexity and costs can be reduced. We provide two examples of the retrospective study of clinical trials methods that could change the conduct of future trials. First, an overview of the effect of outcome adjudication on treatment estimates for cardiovascular trials is presented. Second, a prognostic model to detect fraud within multi-centre trials is developed as part of a system of central statistical monitoring. There are many more unanswered questions about efficiencies in clinical trials methodology that need to be examined by statisticians and researchers.

4.1 Background

Researchers have always been interested in studying and improving clinical trials methodology. It is only natural that one who lives by the scientific method may apply this to the process of science itself. Trials today have benefited from many statistical advances including: proper random assignment to treatment groups, intention-to-treat analysis populations, conservative statistical monitoring boundaries, statistical adjustments for multiplicity, proportional hazards models, and many others. To date research has primarily focused on improving trial design and statistical approaches to analyses, but given the changes that are happening, it is the area of trial conduct that now needs our collective thoughts and innovation.

The environment of clinical trials is increasing in complexity and bureaucracy. The current climate for clinical trials is now much more complex, fuelled by

J. Pogue (✉) • P.J. Devereaux • S. Yusuf
Department of Clinical Epidemiology and Biostatistics and Faculty of Health Sciences, McMaster University, Hamilton, ON, Canada
e-mail: poguej@mcmaster.ca; philipj@mcmaster.ca; yusufs@mcmaster.ca

K. van Montfort et al. (eds.), *Developments in Statistical Evaluation of Clinical Trials*, DOI 10.1007/978-3-642-55345-5_4

layers of regulation and a misplaced hypersensitivity and fear of litigation. Take for example, the typical trial informed consent form. This was once a page or two containing an explanation of the purpose of the trial, a summary of the efforts asked of the participant, and a listing any foreseeable risks involved. It is now an equivalent to a corporate contract: over 20 pages is now common, incomprehensible to anyone except lawyers, and full of protection clauses in case of class-action lawsuits, without distinguishing between usual risks seen in clinical practice and any significant risks due to the experimental design or interventions. Thus, due to increasing complexity and bureaucracy today's "informed consent" form no longer fulfils its purpose.

What has happened to the informed consent is mirrored in many parts of clinical trials conduct today where there is a disproportionate focus on minor deviations or inaccuracies in use of inclusion criteria, inappropriate over emphasis on the precision of individual data points and reporting every minor "adverse event" (even those which are part of the natural history of the disease or conditions common in a particular age group). These procedures have gotten in the way of ensuring the precision of the outcomes that matter because so much cost is going into collecting unnecessary data and monitoring aspects of trial conduct that ultimately do not matter.

The danger is that if we cannot determine how to perform well-designed trials in a more efficient manner, we may be left pursuing only expensive small complex trials that have little hope of finding effective treatments for the world's burden of disease. Statisticians and scientists must determine what trial methods, rules, and regulations are necessary, as they really affect trial results and participant safety, and which ones are wasting scientific and monetary resources. Rather than face validity, personal experiences, or legal opinion, what we need now is objective evidence obtained from past and future trials that evaluate whether detailed procedures materially influence trial results and validity.

4.2 Growing Complexity in Modern Trials and its Effect

Prentice [32] pointed out the paradox that the randomized controlled trial, the research design most insulated from confounding, is subject to the most effort and expense to record and control confounders. Compared to cohort studies, trials typically have more complex and detailed inclusion and exclusion criteria, extensive baseline characteristics and follow-up data collection, multiple quality control procedures, standardized outcome monitoring, definitions and reporting, and outcome adjudication. This complexity is ever-present in trials and continues to grow. Getz [18] and Wampler [40] have documented a steady increase in protocol complexities since 1990.

Getz [18] conducted a retrospective analysis of 10,038 phase one to four clinical trials protocols, from pharmaceutical and biotechnology, hosted in the Fast Track System [26]. They estimated a growth in the average total procedures in trials of 6.5 % per year from 1999 to 2005, with procedures in phase four trial protocols increasing by 9.1 % per year. The total burden of work required of the site investigator by trial protocols increased by 10.5 % per year. The number of eligibility criteria increased three-fold, the median number of reported adverse events grew by a 122 % increase, and the median number of serious adverse events reported per participant in the year 2005 was 12.5 times that of 1990. The average number of case report forms per trial increased from 55 pages to a staggering 180 pages. The length of consent forms has more than doubled, and the work load on REBs has increased substantially yet the number of trials reviewed per month has declined. Performance by sites within trials has been declining within the context of increasing demands. Getz [18] found a 16 % absolute drop in site enrolment between 1999 and 2005, while the rates of retention of participants in trials fell by 21 %.

Others have also documented the growing demands on trial site investigators and participants, the associated increased cost of trials, and an associated decline in site performance. Eisenstein et al. [16] have documented a doubling in the cost of clinical trials over the past decade within the United States from 37 to 64 % of total expenditures of the pharmaceutical industry and the National Institutes of Health from 1994 to 2003. Yet there was a reduction in the number of Food and Drug Administration approvals from 35.5 to 23.3 entities per year over the same period. Clearly this increase in cost has not translated into greater availability of effective disease therapies. Yet the cost of individual trials is substantial, with estimates ranging from 83 to 142 million US dollars for multicenter cardiovascular trials [16]. Trials that require such large investment will by their very nature be limited in number, leading to fewer clinical trials in many areas.

While these "perverse" trends are of great concern, many have suggested that trial methodology can be made more efficient. At the heart of the large simple trial design is the tenant that simple efficient designs will produce the clearest results, and focusing efforts on those methods can influence trial results [12, 43, 44]. Since then others have made further suggestions for increased efficiencies. Thornquist [38] predicted up to a 12 % decrease in total trial budget if non-compliance could be reduced by 50 %. Eisenstein [16] suggested that trials in congestive heart failure and acute coronary syndrome that followed a simplified design could reduce their costs by up to 43 %, without reducing sample size. Simplifications could include: reduced data collection, less use of on-site monitoring, lower site payments, and more efficient site management and data management. Eisenstein et al. [15] suggested that 59 % of the cost of running trials could be saved through reducing planning time for trials, time to recruit the full sample size, reductions in the number of case report forms, smaller numbers of sites, use of electronic data capture systems, and efficient site management practices.

Given that cost and complexity can (and must) be reduced, the question we must now answer is what aspects of trial methodology need to be reduced, improved, increased, or maintained in order to produce reliable, precise, unbiased trial results. We now need researchers experienced in trial methodology to focus their efforts on determining the value of clinical trials practices as a guide to finding needed simplifications and cost reductions. In particular statisticians can assist in this endeavour by quantifying or measuring the effect of these practices on the treatment effects in terms of bias and precision for both efficacy and safety outcomes. One good source for such methodology research would involve examining the data from past clinical trials, both individually and over multiple datasets to compare and estimate the effect of various clinical trials practices.

All clinical trial practices need to be evaluated in terms of their ability to serve three important functions. First, they may help to minimize the difference between the trial treatment estimate φ and the true effect $g(\theta)$, known as bias in the estimation of the treatment effect:

$$b_\varphi(\theta) = E_\theta\big(\varphi(X)\big) - g(\theta) .$$

While the true effect $g(\theta)$, is never known, it may be expected that a valuable clinical trial practice which reduces bias would bring the estimated treatment effect closer to the true treatment effect. Second, a clinical trial methodology may increase the precision τ of the trial treatment estimate θ or decrease the variance, such that

$$\tau = \frac{1}{\sigma_\theta^2} .$$

Lastly, means of performing trial functions efficiently but with less resources, thus reducing the cost of trials, are worthy of study. Means of minimizing bias and increasing precision directly affect trials results, but cost (c) is also indirectly related to precision:

$$\tau \propto \frac{1}{c} .$$

If the cost of enrolling and following an individual subject in a trial is high, trialists may reduce sample size or select less clinically important outcomes in order to make the trial feasible, and this will decrease the precision of the treatment estimate. Having recognised this we now use our units of measurements to examine useful and wasteful procedures in trials. We would like to present two examples where parts of typical trial methodologies were examined to determine their value and suggest possible improvements in efficiency.

4.3 Example 1: Estimating the Effect of Outcome Adjudication

To evaluate the effectiveness of a treatment on a series of protocol-defined outcomes, we require an unbiased collection and validation of these outcomes [17]. One of the key clinical trials methods to ensure this is to have central adjudication of all important outcomes, to determine that they truly meet the protocol definitions.

Outcome adjudication commonly involves a group of experts who examine the supporting documentation for each outcome event and either except it as a valid outcome or reject it from the trial database. The goals of outcome adjudication are clear. Evaluating non-fatal or complex outcomes that may have a subjective component or variability in their ascertainment, in theory could decrease precision due to added "noise" in any trial, regardless of treatment blinding. Adjudication seeks to minimize this potential noise and bias by enforcing standardized outcome definitions through the review of source documents and tests [19,34,39]. It should be able to correct any systematic misclassification based on investigators a priori beliefs about the effectiveness of the treatments being compared. Trial credibility is thought to increase if outcome adjudication is used for the trial's important outcomes [19,34]. This process is thought to provide a check on the quality and consistency for the trial's outcomes [14, 19]. This method has appeal to many trialists, as it may increase trial precision by eliminating outcomes that may not be affected by the treatment being studied [14, 19]. It may also help eliminate biased reporting in trials where the therapy is not masked to the participant and/or the site physicians [5, 19, 21, 23]. It is important to recognize, however, that adjudicators typically only evaluate events that were submitted and if biased reporting happens in a trial—whereby there is under-reporting of borderline events in one of the treatment group—this problem is usually not overcome by central adjudication. The process of outcome adjudication does increase the cost of trials [11,19,39]. Source documents need to be collected centrally, after redacting all direct participant identifiers, masking and sham masking of treatment information needs to be complete for open trials, translations to the language of the adjudicator may be required for international trials, costs for document shipment, tracking software, and adjudicator remuneration are required.

It remains to be demonstrated if the potential value of outcome adjudication is worth its cost. There have been few systematic efforts to estimate the value of outcome adjudication. Individual trials have occasionally commented that the treatment effect based on investigator-reported outcomes differed from that based on adjudicated outcomes. For example, CHARM-Preserved trial found that candesartan was superior to placebo in reducing the composite of cardiovascular deaths or hospitalization of heart failure with a hazard ratio of 0.85 ($p = 0.028$) based on reported outcomes, but after adjudication this decreased to 0.89 ($p = 0.12$) [45]. The EPIC trial found the opposite pattern that a reported hazard ratio of 0.73 ($p = 0.12$) changed to 0.65 ($p = 0.008$) after the outcomes were adjudicated [36]. Yet other trials, such as the CDP [7] and PARAGON-B [25], have documented consistency between reported and adjudicated trial results. Generally comments about the effect of outcome adjudication on the estimated treatment effect in specific trials are relatively rare and potentially influenced by publication bias.

To address this problem, we conducted a systematic comparison to evaluate and estimate the effect of outcome adjudication within the large cardiovascular trials conducted at the Population Health Research Institute of McMaster University, between 1993 and 2006 [30]. This involved 10 trials with >95,000 trial participants randomized and >9,000 outcomes. It included trials with and without blinding

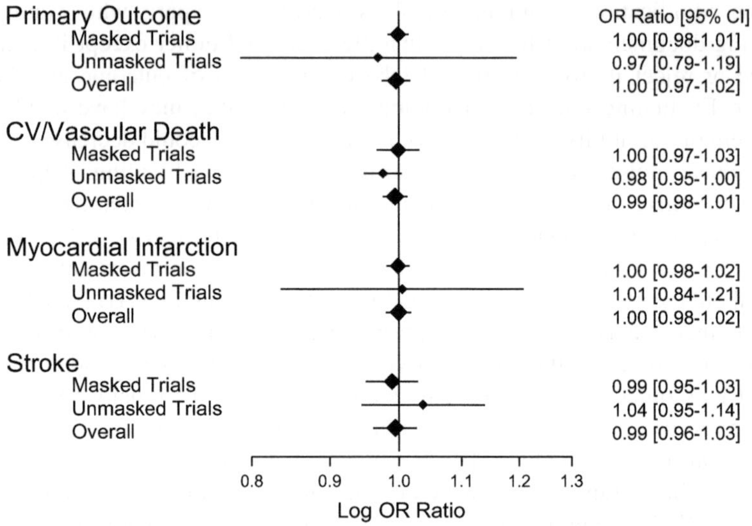

Fig. 4.1 Masked vs. unmasked trials: ratio of odds ratios for adjudicated vs. reported outcomes (Reprinted from [30] with permission)

or masking of treatments. For each trial, we determined the odds ratios for treatment effects using investigator reported events and the treatment odds ratio based on events after adjudication, and pooled these odds ratios across trials with trial as a random effect. The paired difference of the natural logarithm of the odds ratio for adjudicated outcomes minus the natural log odds ratio for reported outcomes was regressed over all trials. Exponentiating this mean difference produced a ratio of odd ratios, where 1.0 indicates no evidence of a treatment difference due to outcome adjudication. This analysis was performed overall and then separately for trials with blinding of treatment group and for trials without blinding. All analyses were weighted by trial size. Figure 4.1 displays the effect of outcome adjudication on the primary outcome for each trial and overall, showing a ratio of odd ratios of 1.00 with 95 % confidence interval 0.97–1.02, implying that we cannot reject the null hypothesis that

$$b_{\varphi}(\theta) = 0 \, .$$

This estimate was similar for trials with and without blinding [OR ratio = 1.00 (0.98–1.01) and OR ratio = 0.97 (0.79–1.19), respectively]. Similar comparisons were done for individual outcomes included cause specific cardiovascular death, myocardial infarction, and stroke. No significant effect of outcome adjudication was found.

These analyses suggest that the quality monitoring part of systematic and complete outcome adjudication could be eliminated or replaced by a random sampling approach for major cardiovascular mortality and morbidity. Similar analyses need to be conducted on trials from other coordinating centres and in other research areas (i.e., based on trials with different types of outcomes) so that we may understand when we do and do not need outcome adjudication to minimize bias and maximize precision in trials.

4.4 Example 2: Central Statistical Monitoring as an Alternative to Onsite Monitoring

The gold standard of site monitoring for clinical trials is thought to be frequent on-site visits where all trial data are verified against local source documents. ICH E6 states that, "In general there is a need for on-site monitoring, before, during, and after the trial" [20]. It then goes on to state that central monitoring accompanied by appropriate investigator training and guidance may replace regular on-site monitoring in "exceptional circumstances". As a result of this guidance document, the use of on-site monitoring is wide spread within industry or clinical research organization trials (84 %), although less commonly used in less well funded academic or government sponsored trials (31 %), based on a survey of 65 research organization that conduct clinical trials in 2009 [27].

On-site monitoring is a costly component of trial methodology, often consuming 20–30 % of the entire cost of a trial, representing tens of millions of dollars for large multi-site trials. Yet there have been surprisingly few evaluations of the effectiveness of on-site monitoring to detect either problems in implementing the trial protocol or possible data fabrication. Published summaries of FDA audits [33] indicate that serious deficiencies are sometime detected (4 % of data audit conducted), but the definition of a serious deficiency is not provided. This means that the reader cannot determine if any of these would have altered trial results. This summary does give examples where data fabrication was detected, but fails to quantify the number of times this misconduct was identified directly by on-site auditors. In contrast to this, others have found that on-site monitoring did not find important problems at sites and did not alter important trial results. The National Institute of Cancer's on-site monitoring program did not change the agreement rate for treatment failures or the number of protocol deviations [41]. A program of on-site audits started near the end of the GUSTO trial found no errors that would have changed the trial results [17]. The National Surgical Adjuvant Breast and Bowel Project on-site monitoring program found no unknown treatment failures or deaths, and only a very small number of not previously known ineligible participants [9].

In contrast to this dearth of evidence for the effectiveness of on-site monitoring, there have been some limited successes reported with the use of statistical methods and central statistical monitoring to confirm or identify fabricated data. Several authors have used statistical methods to illustrate the implausibility of data sets that were suspected to contain fabricated data. When fraud was suspected in a diet trial submitted for publication to the British Medical Journal, a comparison of these data with that from another diet trial found that their suspicions may have been warranted. In comparing the intervention to control group within each trial, Al-Marzouki et al. [1] found many more statistically significant differences within the data set thought to be fabricated. Kranke [24] and Carlisle [8] separately used probability models to calculate the chances of observing the group of summary statistics presented in multiple publications ($n = 47$ and $n = 169$) published by a single researcher. Carlisle [8] compared summary binary patient characteristics (e.g., sex or use of antihypertensive medications) to the expected binomial distribution, allowing for a separate population rate per trial across this one researcher's published trials. The discrepancy between these reported and expected distribution was quantified using a Fisher's exact test. For each trial's mean continuous variables ($\overline{m}$) (e.g., weight or blood pressure) a similar comparison was done using the central limit theorem.

$$\frac{\overline{m} - \mu}{\text{SEM}} \times \left(1 + \frac{SD_{\text{SEM}}}{\sqrt{\text{SEM}}}\right).$$

Here μ is the grand mean over all trials and SEM is the standard error of the mean from each individual trial. They each concluded that these trials collectively resulted in implausible published data. Central statistical monitoring, in various forms, has been used successfully to identify sites within a multi-center trial that have fabricated data. These trials include the AMPIM [3], MRFIT [28], NSABP-06 [9], Second European Stroke Prevention Study [37], COMMIT-1 [10], POISE [13], and other trials. In many of these cases central statistical monitoring identified the problem, while on-site existing monitoring had failed to find the problem.

While the above case studies demonstrate promise for the use of central statistical monitoring in trials, further work in this area is needed. Just as we commonly develop risk models to predict disease in patients, central statistical monitoring could use risk models to identify sites at high-risk for fabricating data, within a multi-centre trial. If a statistical model with sufficient predictive ability could be developed, then their use within central statistical monitoring could replace the function of on-site monitors in fraud detection. We used data from the POISE Trial to retrospectively build a series of prognostic logistic regression models conducted on site-level data to identify the sites that had fabricated their data [29]. Let y take on the value 1 if the site committed fraud and 0 otherwise and suppose there are k independent variables (x_1 to x_k) that predict fraud. Then the probability of fraud having occurred at the jth centre is:

$$p = P\langle y = 1 \mid X = x\rangle ,$$

$$\ln\left(\frac{p_j}{1 - p_j}\right) = \beta_0 = \beta_1 x_1 + \cdots + \beta_k x_k .$$

POISE was a multi-centre, multi-national randomized controlled trial testing the effectiveness of a peri-operative beta-blocker in preventing cardiovascular outcome in high-risk patients undergoing non-cardiac surgery. Of the 196 participating clinical sites from 24 countries, 9 were found to have fabricated data, representing 947 patients out of the total 9,298 randomized within this trial.

For the purpose of building a prognostic model, we used data from all sites that had randomized at least 20 trial patients ($N = 109$ sites). An analytic strategy was followed to develop these prognostic models. First, a wide variety of statistical tests were included since authors have suggested that many types of data and statistics may be useful to identify fabricated data [1–4, 6, 10, 22, 31, 35, 37, 42]. Variables were included from baseline characteristics (binary or continuous), combinations of baseline variables, compliance, site performance, concomitant medications, physical measurements in follow-up, and efficacy and safety outcomes (see Pogue et al. [29] for the complete description). Second, for these models data were summarized at the site level, as opposed to the patient level, since the goal was to identify sites at high-risk for data fabrication. We focused on comparing data across sites and determining how different each site was from the others. We required that these summaries be unit-less, and not dependent on the exact variables collected in the POISE Trial.

These risk models will only be useful for future trials if their prognostic variables may be replaced by the different variables collected in each trial. Making the independent variables unit-less is likely to assist in this goal. Primarily, this involved using probability values (p-values) to quantify how different a site was from all other sites combined for a particular variable. We made no assumptions about direction of effect for these summaries, but instead analysed p-values as continuous possible predictors, rather than using pre-defined cut-offs.

Seven different types of statistical summaries were used. We tested whether each site was different from the rest using a two-by-two frequency comparison for binary variables, such as history of diabetes, and summarized as that site's Pearson chi-square test p-value. We tested how different each site was from the rest for continuous variables (e.g., systolic blood pressure) using two-sample t-tests and calculated a p-value for each site. We compared digit preference for variables such as day of week for randomization for each site versus all others using Pearson's chi-squared test p-values. The variances of continuous variables were compared for each site versus all others using Folded F-test p-values. Distance measures (d_j) were derived for each site for continuous variables indicating how far away one site's data are from the overall mean ($\overline{y}$) across all centers, standardized by the overall standard deviation (s), using data from the ith trial participant at the jth center. The natural logarithm of distance was used as a possible predictor.

$$d_j = \sum_i \left(\frac{y_{ij} - \overline{y}}{s} \right)^2 .$$

For the comparison of outcomes and compliance, we calculated the probability of observing an outcome rate as extreme as that observed at a site, assuming a Poisson distribution, adjusting for country variation. Instead of testing each center against all the others, we directly calculated each site's cumulative probability distribution (CDF) value from these models. Lastly, for repeated physical measurements over

Table 4.1 Risk scores predicting fabricated data (Reprinted from [29] with permission)

Model terms	Categories	Category	Score
Predictor 1:	SBP over time Intra-class correlations	1	+0
		2	+1
		3	+2
		4	+3
		5	+4
Predictor 2:	DBP Mean Comparison – t-test p-values	1	+0
		2	+1
		3	+2
		4	+3
		5	+4
Predictor 3:	Model 1: SBP digit preference χ^2 p-values	1	+0 (+4)
	Model 2: Surgery: Intrathoracic or intraperitoneal – frequency χ^2 p-value	2	+1 (+3)
	Model 3: Anaesthesia: General – frequency χ^2 p-value	3	+2 (+2)
	Model 4: ACE-I/ARB χ^2 p-value	4	+3 (+1)
	Model 5: Compliance Outcome Probability – CDF	5	+4 (+0)

Note: point reversed for model 5 only and provided in brackets
Categories: ICC and p-values: 1 $=\leq$ 0.20, 2 = 0.21–0.40, 3 = 0.41-0.60, 4 = 0.61–0.80, 5 = 0.81+

time, the intra-class correlation coefficient (ICC) itself was used as a unit-less summary for a site's data.

This lead to a long list of potential predictors for fabricated data, and we then eliminated redundancy among these using factor analysis with varimax rotation. Out of 52 possible predictors, 18 independent factors were identified and the predictor with the highest loading for each of these factors was selected for inclusion into a series of logistic regression with fraud at each site as the outcome. We used the best subsets of models using the branch and bound algorithm of Furnival and Wilson to find models with the largest score statistic for including different numbers of variables. The final series of models was selected based on no significant increase in the score test for increasing the number of variables in the model. These models were checked for lack of fit using the Hosmer and Lemeshow goodness of fit test. Out of these, the five best predictive models were selected. We then converted these into risk scores using a points system. These are summarized in Table 4.1.

These risk scores were tested in an independent data set in a trial that had on-site monitoring and contained no known data fabrication and produced low-risk score for almost all sites (see Fig. 4.2). These risk scores appear to distinguish well between sites with and without fabricated data, but will require further validation across different types of trials. Where the specific variables used in these score are not collected within a trial, the focus should be on substituting other similar repeated physical measurements or baseline characteristics into these risk scores. The goal is to look for the combination of a site with both greater than normal correlation

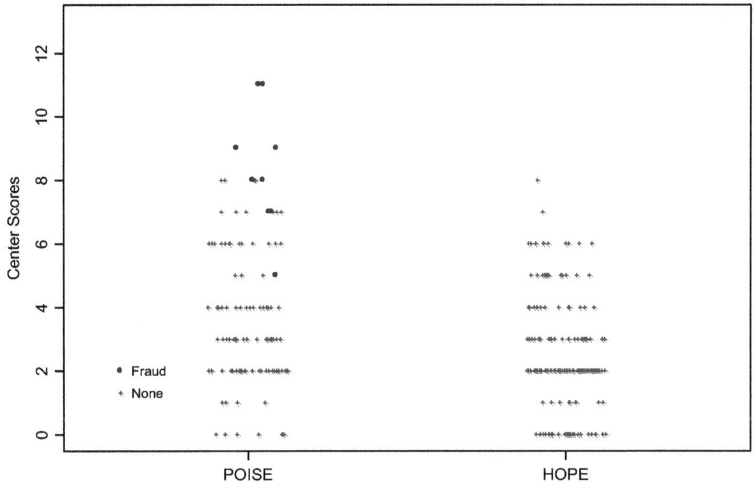

Fig. 4.2 External validation of Model 1 on a trial without fabricated data: a comparison of the distribution of Center Risk Score in POISE (with nine fraudulent centers) and HOPE (no fraudulent centers) (Reprinted from [29] with permission)

over time in physical measurement (high ICCs), and baseline characteristics that look extremely similar to all the other sites (high χ^2 p-values). More research into this area is needed potentially leading to a toolbox of statistical risk scores that can more effectively guide monitoring within trials and lead to greater efficiencies for trials.

4.5 Improving Future Trials

We have illustrated only two investigations into determining what efficient trial conduct should involve. There are many other trial methodologies that need to be studied. It would be useful to estimate the effect of conducting a pilot study prior to launching a full-scale trial, and what are the characteristics of a good pilot study. The effect of complex inclusion/exclusion criteria on speed of recruitment and study power could be estimated. The effect of a run-in period on compliance in the main trial should be studied. These are just a few important unanswered questions that we could examine, using retrospective trial database analyses or overviews of prior trials.

In the future, we may be able to build in tests of differing trial methodology prospectively within given trials. The only way to argue against increasing complexity and bureaucracy is through scientific evidence. We need to separate the elements that matter in conducting a trial that leads to an unbiased, precise answer, from those methodologies that represent a waste of resources. The quality and quantity of future trials may depend on us doing so.

References

1. Al-Marzouki, S., Evans, S., Marshall, T., Roberts, I.: Are these data real? Statistical methods for the detection of data fabrication in clinical trials. *British Medical Journal* **331**, 267–270 (2005)
2. Baigent, C., Harrell, F., Buyse, M., Emberson, J., Altman, D.: Ensuring trial validity by data quality assurance and diversification of monitoring methods. *Clinical Trials* **5**, 49–55 (2008)
3. Bailey, K.: Detecting fabrication of data in a multicenter collaborative animal study. *Controlled Clinical Trials* **12**, 741–752 (1991)
4. Benford, F.: The law of anomalous numbers. *Proceedings of the American Philosophical Society* **78**, 551–572 (1938)
5. Boutron, I., Estellat, C., Guittet, L., Dechartres, A., Sackett, D., Hrobjartsson, A., Ravaud, P.: Methods of blinding in reports of randomized controlled trials assessing pharmacologic treatments: A systematic review. *PLOS Medicine* **3**, 1931–1939 (2006)
6. Buyse, M., George, S., Evans, S., Geller, N., Ranstam, J., Scherrer, B., Lesaffre, E., Murray, G., Edler, L., Hutton, J., Colton, T., Lachenbruch, P., Verma, B., for the ISCB Subcommittee on Fraud: The role of biostatistics in the prevention, detection and treatment of fraud in clinical trials. *Statistics in Medicine* **18**, 3435–3451 (1999)
7. Canner, P., Krol, W., Forman, S.: External quality control programs. *Controlled Clinical Trials* **4**, 441–466 (1983)
8. Carlisle, J.: The analysis of 169 controlled trials to test data integrity. *Anaesthesia* pp. 1–17 (2012)
9. Christian, M., McCabe, M., Korn, E., Abrams, J., Kaplan, R., Friedman, M.: The national cancer institute audit of the national surgical adjuvant breast and bowel project B-06. *New England Journal of Medicine* **333**, 1469–1474 (1995)
10. COMMIT (CLOpidogrel and Metroprolol in Myocardial Infarction Trial) collaborative group: Addition of clopidogrel to aspirin in 45852 patients with acute myocardial infarction: randomized placebo-controlled trial. *The Lancet* **366**, 1607–1621 (2005)
11. Cook, D., Walter, S., Cook, R., Freitag, A., Guyatt, G., Devitt, H., Meade, M., Griffith, L., Sarabia, A., Fuller, H., Turner, M., Gough, K.: Incidence of and risk factors for ventilator-associated pneumonia in critically ill patients. *Annuals of Internal Medicine* **129**, 433–440 (1998)
12. DeMets, D., Califf, R.: Lessons learned from recent cardiovascular clinical trials: Part II. *Circulation* **106**, 880–886 (2002)
13. Devereaux, P., Yang, H., Yusuf, S., Guyatt, G., Leslie, K., Villar, J., Xavier, D., Greenspan, L., Lisheng, L., Xu, S., Malaga, G., Avezum, A., Jacka, M., Choi, P.: Effects of extended-release metoprolol succinate in patients undergoing non-cardiac surgery (POISE trial): a randomised controlled trial. *The Lancet* **371**, 1839–1847 (2008)
14. Eisenbud, R., Assman, S., Kalish, L., van der Horst, C., Collier, A., for the Viral Activation Transfusion Study (VATS) Group: Differences in difficulty adjudicating clinical events in patients with advanced HIV disease. *Journal of Immune Deficiency Syndrome* **28**, 43–46 (2001)
15. Eisenstein, E., Collins, R., Cracknell, B., Posesta, O., Reid, E., Sandercock, P., Shakhov, Y., Terrin, M., Sellers, M., Califf, R.: Sensible approaches for reducing clinical trials costs. *Clinical Trials* **5**, 75–84 (2008)
16. Eisenstein, E., Lemons II, P., Tardiff, B., Schulman, K.A., Jolly, M., Califf, R.: Reducing the costs of phase III cardiovascular clinical trials. *American Heart Journal* **149**, 482–488 (2005)
17. Friedman, L.M., Furberg, C.D., DeMets, D.L.: *Fundamentals of Clinical Trials*. Springer, New York (2010)
18. Getz, K., Wenger, J., Campo, R., Seguine, E., Kaitin, K.: Assessing the impact of protocol design changes on clinical trial performance. *American Journal of Therapeutics* **15**, 450–457 (2008)

19. Granger, C., Vogel, V., Cummings, S., Held, P., Fiedorek, F., Lawrence, M., Neal, B., Reidies, H., Santarelli, L., Schroyer, R., Stockbridge, N., Zhao, F.: Do we need to adjudicate major clinical events. *Clinical Trials* **5**, 56–60 (2008)
20. International Conference on Harmonization of Technical Requirements for Registration of Pharmaceuticals for Human Use: *Guidelines for good clinical practice E6. Step 4* pp. 1–53 (1996)
21. Kestle, J., Milner, R., Drake, D.: An assessment of observer bias in the Shunt Design Trial. *Pediatric Neurosurgery* **30**, 57–61 (1999)
22. Knatterud, G., Rockhold, F., George, S., Barton, F., Davis, C., Fairweather, W., Honohan, T., Mowery, R., O'Neill, R.: Guidelines for quality assurance in multicenter trials: A position paper. *Controlled Clinical Trials* **19**, 477–493 (1998)
23. Kradjian, S., Gutheil, J., Baratelle, A., Einstein, S., Kaslo, D.: Development of a charter for an endpoint assessment and adjudication committee. *Drug Information Journal* **39**, 53–61 (2005)
24. Kranke, P., Apfell, C., Roewer, N.: Reported data on granisetron and postoperative nausea and vomitting by Fujii et al. are incredibly nice! *Anesthesia and Analgesia* **90**, 1004–1007 (2000)
25. Mahaffey, K., Roe, M., Dyke, C., Newby, L., Kleiman, N., Connolly, P., Berdan, L., Sparapani, R., Lee, K., Armstrong, P., Topol, E., Califf, R., Harrington, R., for the PARAGON-B Investigators: Misreporting of myocardial infarction end points: results of adjudication by a central clinical events committee in the PARAGON-B trial. *American Heart Journal* **143**, 242–248 (2002)
26. Mathieu, M.P.: Fast track systems. Index of clinical study complexity by year, phases I-IV. In: M.P. Mathieu (ed.) *PAREXEL's Pharmaceutical R&D Statistical Sourcebook* (2007)
27. Morrison, B., Cochran, C., White, J., Harley, J., Kleppinger, C., Liu, A., Mitchel, J., Nickerson, D., Zacharias, C., Kramer, J., Neaton, J.: Monitoring the quality of conduct of clinical trials: a survey of current practices. *Trials* **8**, 342–349 (2011)
28. Neaton, J., Bartsch, G., Broste, S., Cohen, J., Simon, N., for the MRFIT Research Group: A case of data alteration in the Multiple Risk Factor Intervention Trial (MRFIT). *Controlled Clinical Trials* **12**, 731–740 (1991)
29. Pogue, J., Devereaux, P., Yusuf, S.: Central statistical monitoring: A model to predict fraud in clinical trials. *Clinical Trials* **10**, 225–235 (2013)
30. Pogue, J., Walter, S., Yusuf, S.: Evaluating the benefits of event adjudication of cardiovascular outcomes in large simple RCTs. *Clinical Trials* **6**, 239–251 (2009)
31. Preece, D.: Distribution of final digits in data. *Statistician* **30**, 31–60 (1981)
32. Prentice, R.: Opportunities for enhancing efficiency and reducing cost in large scale disease prevention trials: A statistical perspective. *Statistics in Medicine* **9**, 161–172 (1990)
33. Shapiro, M., Charrow, R.: The role of data audits in detecting scientific misconduct: Results of the FDA program. *Journal of the American Medical Association* **261**, 2505–2511 (1989)
34. Sicurella, J., Roberts, R., Gent, M.: The operation of a central adjudication committee and its effect on the validity of the assessment of treatment benefit. *Controlled Clinical Trials* **11**, 283 (1990)
35. Taylor, R., McEntegart, D., Stillman, E.: Statistical techniques to detect fraud and other data irregularities in clinical questionnaire data. *Drug Information Journal* **36**, 115–125 (2002)
36. The EPIC Investigators: Use of a monoclonal antibody directed against the platelet glycoprotein IIb/IIIa receptor in high-risk coronary angioplasty. *New England Journal of Medicine* **330**, 956–961 (1994)
37. The EsPS2 Group: European stroke prevention study 2. Efficacy and safety data. *Journal of Neurological Sciences* **151**, S1–S77 (1997)
38. Thornquist, M., Urban, N., Tseng, A., Edelstein, C., Lund, B., Omenn, G.: Research cost analysis to aid in decision making in the conduct of a large prevention trial, CARET. *Controlled Clinical Trials* **14**, 325–339 (1993)
39. Walter, S., Cook, D., Guyatt, G., King, D., Troyan, S., for the Canadian Lung Oncology Group: Outcome assessment for clinical trials: how many adjudicators do we need. *Controlled Clinical Trials* **18**, 27–42 (1997)

40. Wampler, S.: Tackling protocol complexity. *Good Clinical Practice Journal* **7**, 6–8 (2000)
41. Weiss, R., Vogelzang, N., Peterson, B., Panasci, L., Carpenter, J., Gavigan, M., Sartell, K., Frei III, E., Mitchel, J., McIntyre, O.: A successful system of scientific data audits for clinical trials: A report from the Cancer and Leukemia Group B. *Journal of the American Medical Association* **270**, 459–464 (1993)
42. White, C.: Suspected research fraud: difficulties of getting the truth. *British Medical Journal* **331**, 281–288 (2005)
43. Yusuf, S., Bosch, J., Devereaux, P., Collins, R., Baigent, C., Granger, C., Califf, R., Temple, R.: Sensible guidelines for the conduct of large randomized trials. *Clinical Trials* **5**, 38–39 (2008)
44. Yusuf, S., Collins, R., Peto, R.: Why do we need some large, simple randomized trials. *Statistics in Medicine* **3**, 409–420 (1984)
45. Yusuf, S., Pfeffer, M., Swedberg, K., Granger, C., Held, P., McMurray, J., Michelson, E., Olofsson, B., Ostergren, J.: Effects of candesartan in patients with chronic heart failure and preserved left-ventricular ejection fraction: the CHARM-Preserved Trial. *Lancet* **362**, 777–781 (2003)

Chapter 5
Competing Risks and Survival Analysis

Kees van Montfort, Peter Fennema, and Wendimagegn Ghidey

Abstract The analysis of time-to-event data in the presence of competing risks is part of many studies today. However, the impact of the interrelationship between the competing risks on the interpretation of the results seems to be unclear to many researchers, however. We try to provide a guide to researchers interested in analysing competing risks data. Estimation of the cause-specific hazard function, the cumulative incidence function, the Gray test statistic, and the multi-stage models for analysing competing risks data are explained. Furthermore, we apply the theoretical methodology and illustrate the fundamental problems of interpreting the results of competing risk analyses by using empirical data in the field of outcome research in orthopaedics.

5.1 Introduction

In clinical trials, time to an event is frequently studied as endpoint. Competing risks data are encountered when subjects under study are at risk of more than one mutually exclusive events, like death from different causes. If after removal of one cause of failure the risks of failure of the remaining causes are unchanged, we may use classical statistical methodology. However, this situation is rather rare. Although the statistical methodology for analysing such competing risks data has been known for decades (see Kalbfleisch and Prentice [16]; Prentice and Kalbfleisch [22]), there is still great uncertainty in the medical research about how to approach this type of data. This is reflected by many of the recent publications in the medical and biostatistical literature reconsidering (see Tai et al. [27]; Tai, Peregoudov and

K. van Montfort (✉)
Department of Biostatistics, Erasmus Medical Center, Rotterdam, The Netherlands
e-mail: c.vanmontfort@erasmusmc.nl

P. Fennema
Advanced Medical Research, Männedorf, Switzerland
e-mail: peter.fennema@amr-cro.com

W. Ghidey
Department of Hematology, Erasmus Medical Center, Rotterdam, The Netherlands
e-mail: w.ghideyalemayehu@erasmusmc.nl

K. van Montfort et al. (eds.), *Developments in Statistical Evaluation of Clinical Trials*, DOI 10.1007/978-3-642-55345-5_5

Machin [28]; Satagopan et al. [26]; Friedlin and Korn [9]; Fennema and Lubsen [6]; Crowder [5]).

The Kaplan-Meier curves are commonly used for assessing survival curves in the presence of competing risks data (see Kaplan and Meier [17]). The Kaplan-Meier method describes the time to any one single event, such as revision for any reason. It assumes the independence of the event of interest from any competing event that precludes it (see Alberti et al. [1]; Fennema and Lubsen [6]). However, when the competing event is defined as death without any event of interest, censoring these patients will affect the incidence of the event of interest by modifying the number of exposed patients. Classical survival methods then assume that patients who are censored and are no longer part of the risk set owing to an unrelated competing event (for instance, death) have a similar probability of the event occurring to those who are not censored. As we have noted, this patient group consists of at least two categories, namely those who died without the occurrence of the event of interest, and those who are still alive. In this setting, the Kaplan-Meier curve estimates how the survival curve would look if all censored patients were allowed to continue until they had the event of interest. Part of the curve is therefore attributable to patients who died. This is a hypothetical setting that cannot be tested statistically (see Alberti et al. [1]; Grunkemeier, Anderson and Starr [15]; Pepe and Mori [21]). Consequently, the estimated failure probabilities of the classical survival methods cannot be interpreted as the probabilities of failure of the cause of interest in the presence of competing risks because, under these circumstances, it leads to biased results and conclusions. The bias mainly depends on the magnitude of the correlation between the events of interest and the competing events (see Gooley et al. [13]).

The basic theoretical approaches used for analysing competing risks data will be presented in the following section. In particular, the concept of the cause-specific hazard function, the cumulative incidence function, and the Gray test will be explained. Next, we will deal with multi-state models—which are an extension of competing risks models—and the available software to run a competing risks model. Furthermore, the analysis of competing risk models will be illustrated using two examples of outcome research data in the field of orthopaedics. The results of the classical survival methodology, which do not account for competing risks, will be compared to the results of methodology correcting for these competing risks. Finally, an example with generated data will be discussed. The dependency between the occurrences of the primary risk and competing risk will be modeled by using the Clayton copula.

5.2 Theoretical Framework

5.2.1 Estimation

The competing risks data are represented by the failure time T (i.e. continuous and positive), the failure cause D (i.e. values in the finite set $\{1, \ldots, m\}$) and a matrix of covariates **Z**. An option to model these data is by using multivariate failure time models. In such models each subject is assumed to have a potential failure time for each type of event. The earliest event is actually observed and the others are latent. This approach focuses on the joint distribution of the failure times of the m different failure types, described by the joint survival function

$$S(t_1, \dots, t_m) = P(T_1 > t_1, \dots, T_m > t_m).$$

The marginal hazard function of cause j

$$h_j(t) = \lim_{\Delta t \to 0} P(t \leq T_j < t + \Delta t \mid T_j \geq t)/\Delta t$$

is defined by the marginal survival

$$S_j(t) = P(T_j > t) = S(0, \dots, 0, t, \dots, 0).$$

Without any additional assumptions, no joint survival function is identifiable from the observed data, nor are there any marginal distributions identifiable (Fürstovà and Valenta [10]). Therefore, in the presence of competing risk(s) these multivariate failure time models have little practical use.

Other concepts in competing risks models use the cause-specific hazard function or the cumulative incidence function. These two functions specify the joint distribution of the failure time T and the failure cause D. The cause-specific hazard function for cause j (with $j = 1, 2, \dots, m$) is defined as follows:

$$\lambda_j(t) = \lim_{\Delta t \to 0} P(t \leq T_j < t + \Delta t, D = j \mid T_j \geq t)/\Delta t.$$

This cause-specific hazard for cause j corresponds to the instantaneous failure rate from this cause in the presence of all other possible causes of failure. The probability of failure from cause j until time t in the presence of all other possible causes is known as cause-specific cumulative distribution function and depends on the cause-specific hazards for all other causes:

$$F_j(t) = P(T \leq t, D = j).$$

Parameter estimates for the cause-specific hazard of a cause j may be obtained by maximizing the likelihood function involving cause j. It should be noted that these parameters are the same with the likelihood function that would be obtained by treating failures from all other causes, except for j, as censored observations. For the cause-specific hazard functions we may assume the well-known semi-parametric proportional hazard model (with vector of covariates **z**):

$$\lambda_j(t, \mathbf{z}) = \lambda_{j_0}(t) \exp(\boldsymbol{\beta}_j^T \mathbf{z}).$$

The function $\lambda_j(t, \mathbf{z})$ should not be interpreted as a marginal survival function, unless the competing event time distributions and the censoring distributions are independent. The corresponding partial likelihood function is (with $j = 1, \ldots, m$):

$$L(\boldsymbol{\beta}_1, \ldots, \boldsymbol{\beta}_m) = L(\boldsymbol{\beta}_1) \cdots L(\boldsymbol{\beta}_m),$$

with

$$L(\boldsymbol{\beta}_j) = \exp(\boldsymbol{\beta}_j^T \mathbf{z}_1) \cdots \exp(\boldsymbol{\beta}_j^T \mathbf{z}_n) / \sum_{i=1}^{n} \exp(\boldsymbol{\beta}_j^T \mathbf{z}_i),$$

where $j = 1, \ldots, m$; $i = 1, \ldots, n$; and $\mathbf{z}_i$ the realization of the vector of covariates corresponding to observation i.

Parameter estimates for the cause-specific hazard for cause j may be obtained by maximizing the likelihood for the vector of parameters $\boldsymbol{\beta}_j$. This may be applied by treating observations with failure from all other causes except j as censored and fitting a Cox proportional hazards model on these data. It is possible to simultaneously fit cause-specific hazard models for all causes and to test the equality of effects of specific covariates on different failure types. We need to multiply the records m times, one for each failure type, and to generate a failure type identifier, so that each record of a subject corresponds to one cause of failure. The failure indicator takes the value 1 in the record of the subject that corresponds to the actual failure cause and 0 in the remaining records of this subject. Thus, the failure indicator takes the value 0 in all corresponding records for subjects who have not (yet) failed.

The Kaplan-Meier estimate (based on cause-specific hazard) does not always provide an appropriate estimate for cumulative incidence of cause j, as it generally overestimates that quantity (Gichangi and Vach [12]). The cumulative incidence of the competing risks method, first described by Kalbfleisch and Prentice [16], overcomes the shortcomings of Kaplan-Meier and provides a valid estimate of cumulative incidence in the presence of competing risks. In medical literature, cumulative incidence of competing risks is also known as "cause-specific failure probability" (Gaynor et al. [11]), "crude incidence curve" (Korn and Deroy [19]), and "cause-specific risk" (Benichou and Gail [4]). It takes into account the informative nature of censoring due to competing risks by assessing the risk of failure of the cause of interest and that of the competing risk.

As mentioned above, a second concept in competing risks models is the use of the cumulative incidence function. The cumulative incidence for a particular cause of failure is defined as the probability of experiencing the cause of failure until a specific time point t, in the presence of all other causes. The cumulative incidence of competing risks acknowledges that a competing risk influences the risk of occurrence of the event of interest. A competing risk is handled as another type of event (see Fürstovà and Valenta [10]).

The cumulative incidence function of a specific cause does not only depend on the hazard of that specific cause, but also on the hazards of all other causes. Therefore, the relation of the cumulative incidence function of a specific cause for two different covariate values does not only depend on the effect of the covariate of the specific cause, but also on the effects of the covariate on all other causes and on the baseline hazards of all other causes.

A well-known form of cumulative incidence is the proportional hazard model for the subdistribution of a competing risk (see Fine and Gray [7]). This method uses the hazard of a subdistribution, which is a function of the cumulative incidence for the corresponding cause of failure and may be defined as (with vector of covariates $\mathbf{z}$):

$$\begin{aligned}\lambda_j^{\text{sub}}(t;\mathbf{z}) &= \lim_{\Delta t \to 0} P(t \leq T < t + \Delta t, D = j \mid T \geq t \vee (T \leq t \wedge D \neq j), \mathbf{z})/\Delta t \\ &= \partial \log(1 - F_j(t;\mathbf{z}))/\partial t,\end{aligned}$$

where T is the failure time (≥ 0) and D the cause of failure (for example, $D = 1$ or $D = 2$).

The above-mentioned function includes at time t subjects who did not fail yet as well as subjects who failed from other causes before t who are not really at risk at that time. The semiparametric proportional hazard form of the hazard of the sub distribution (of event j) is defined as follows:

$$\lambda_j^{\text{sub}}(t;\mathbf{z}) = \lambda_{j0}^{\text{sub}}(t) \exp(\boldsymbol{\beta}_j^T \mathbf{z}).$$

We may define the indicator function $I_i(t)$, which is equal to value 1 if it is known that subject i has not been censored or failed until time point t. Based on this indicator function and the estimated survival distribution (i.e. Kaplan-Meier estimator) of the factor of interest, we can estimate the vector of coefficients $\boldsymbol{\beta}$ (see Bakoyannis and Touloumi [3]).

It should be noted that the Kaplan-Meier estimate (based on the cause-specific hazards) does not guarantee to provide an appropriate estimate for cumulative incidence for cause j, as it generally overestimates this quantity. It may be applied to estimate the cumulative incidence for cause j in the ideal situation in which failures from other causes were eliminated and the cause-specific hazard of interest does not change after this elimination (i.e. failure times for the different events are uncorrelated), which is seldom.

5.2.2 Comparing the Survival of Different Groups

The cause-specific hazard and cumulative incidence functions are the most important approaches to analyze competing risks data in biomedical research. The effect of a covariate on the cause-specific hazard function for a particular cause may differ from its effect on the cumulative incidence of the corresponding cause. Therefore the estimation of the cause-specific hazard function on the one hand and the cumulative incidence function on the other hand may yield different results. The choice of the best approach depends on the research question of interest and the assumptions with respect to the independence of the competing risks. For instance, if we are interested in how many subjects failed by a specific cause, then we should use cause-specific cumulative incidence functions. If we are interested in how the risk of failure due to different causes changes over time, we should use cause-specific hazard functions. Nevertheless, in many situations the research question is specified in terms of investigating the difference between two (or more) patient groups with respect to the occurrence of the various causes.

In standard survival analysis the comparison of the cause-specific cumulative incidence functions among different groups is done by using the nonparametric tests comparing curves generated by the Kaplan-Meier method (i.e. the log-rank test). In the presence of competing risks, these tests are not appropriate. Gray [14] proposed a class of linear rank statistics for testing equality of the cumulative incidence functions. The tests are based on comparing weighted averages of the hazards of the cumulative incidence function for the failure type of interest (see Fürstovà and Valenta [10]). A study of Williamson, Kolamunnage-Dona and Tudur Smith [29] provided simulation results for the log-rank test comparing cause-specific hazard rates and Gray's test comparing cause-specific cumulative incidence curves. In settings where there are effects in opposite directions for the two event types, Gray's test has greater power to detect treatment differences than the log-rank analysis has.

5.3 Multi-state Models and Competing Risks

In fact, multi-state models are an extension of the competing risks models. Competing risks models deal with one initial state and one or more mutually failure/success states. For instance, a patient's disease or recovery process may consist of intermediate events that can neither be classified as initial states nor as final states. Multi-state models also deal with different states in the course of the time (see Klein, Keiding and Copelan [18]).

A property that is often assumed in practice is that the multi-state model is a Markov model, i.e., given the present state and the event history of a patient, the next visited state and the time at which this will occur will only depend on the present state. Competing risk models are always Markov models, since there is no event history (see Putter, Fiocco and Geskus [23]).

If one is especially interested in the event of interest as a first event, the other events are competing risks. The intermediate event types, not equal to the initial state, nor equal to the event of interest, provide more detailed information on the failure/recovery process and allow for more precision in predicting the prognosis of patients. The occurrence of an intermediate event may be considered a transition from one state to another. Multi-state models provide a framework that allows for the analysis of such transitions. They are an extension of competing risk models, since they extend the analysis to what happens after the first event. The estimation of the transition probabilities is discussed in Andersen, Abildstrøm and Rosthøj [2]).

5.4 Available Software

Cumulative incidence curves in a competing risk setting may be estimated by using the software packages R, Stata, SAS and S-Plus/R (`cmprsk` library). The R package `cmprsk` allows the user to calculate and plot cumulative incidence functions. This package has functions for performing cumulative incidence regression described by Fine and Gray [7]. The R package `mstate`, which deals with multi-state models, may also be used for competing risks. It implements the reduced rank approach of Fiocco et al. [8].

In the Stata function `stcompet.ado`, we observe an event of interest and one or more competing events, the occurrence of which precludes or alters the probability of occurrence of the first one. `stcompet.ado` creates variables containing cumulative incidence, a function that, in this case, appropriately estimates the probability of the occurrence of each endpoint, the corresponding standard error, and confidence bounds. Among others, the cumulative incidence for failure type j will be estimated as

$$I_j(t) = \sum_{i\,:\,t_i \le t} S(t_{i-1}) d_{j,i} / n_i,$$

where $S(t_{i-1})$ is the Kaplan-Meier estimate of the overall survival function, that is, considering failures of any kind, and the second factor is an estimate of the hazard of failure type j (see Marubini and Valsecchi [20]). The sum of these incidences equals $(1 - S(T))$, the complement of the overall Kaplan-Meier estimate of survival considering failures of any kind.

Furthermore, in Stata version 12 and higher the function `stcrreg.ado` fits, via maximum likelihood, competing-risks regression models according to the method of Fine and Gray [7]. The function `stcrreg.ado` posits a model for the subhazard function of a failure event of primary interest. In the presence of competing failure events that impede the event of interest, a standard analysis using Cox regression is able to produce incidence-rate curves that either (1) are appropriate only for a hypothetical universe where competing events do not occur or (2) are appropriate for the data at hand, yet the effects of covariates on these curves are not easily quantified. Competing-risks regression, as performed using `stcrreg.ado`, provides an alternative model that can produce incidence curves that represent the observed data and for which describing covariate effects is straightforward.

To our knowledge, no procedure is available for computing cumulative incidence functions in SAS (see http://support.sas.com/resources/papers/proceedings12/344-2012.pdf). However, you can find SAS macros on the internet for performing the calculations (see http://mayoresearch.mayo.edu/mayo/research/biostat/sasmacros or http://www.biostat.mcw.edu/software/softmenu). In addition, Rostøj et al. [25] wrote a set of SAS macros that allows you to translate results on cause-specific hazards into cumulative incidence curves (see website http://www.pubhealt.ku.dk/).

5.5 Empirical Examples

Total hip arthroplasty (implantation of an artificial hip) generally yields 10-year survivorship of around 95 %. As hip arthroplasty is typically performed in the elderly population; there is a high probability that the prosthesis outlives the patient. Follow-up data of an established total hip system were evaluated and assessed for the effect of the presence of competing risk (Fennema and Lubsen [6]). A slightly different subset of this cohort of 406 patients has been published previously by Zweymüller et al. [30]. The patients were operated on between January 1993 and May 1994. A retrospective study was initiated in 1995. At that time, 91 patients (22.4 %) had died from unrelated causes, and 7 patients (1.7 %) had undergone revision for various reasons. The remaining 299 patients had their prosthesis in situ.

All risks were measured on a daily basis, and expressed as the daily probability of an event occurring. Death was unrelated to the implantation of the artificial joint. Competing events were not included if they followed revision surgery. Stata 12.1 (Stata Corp LP, College Station, TX, USA) was used to perform the analysis, and the `stcompet.ado` command was used to calculate the cumulative incidence function, which was compared to standard Kaplan-Meier analysis (1 − K-M).

The Kaplan-Meier and cumulative incidence approach are displayed in Fig. 5.1. Taking revision for any cause as an endpoint, the 1 − K-M was 4.3 %, whereas the cumulative incidence was 3.3 % at 15 years. The 1 − K-M thus overestimated the incidence of revision by 1.0 %. The relative difference was 31.3 %.

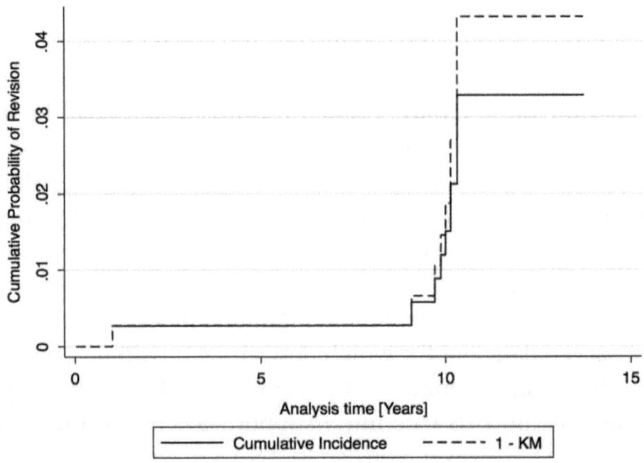

Fig. 5.1 Comparison of the incidence of revision for the cumulative incidence of competing risk and 1 − K-M methods for dataset 1

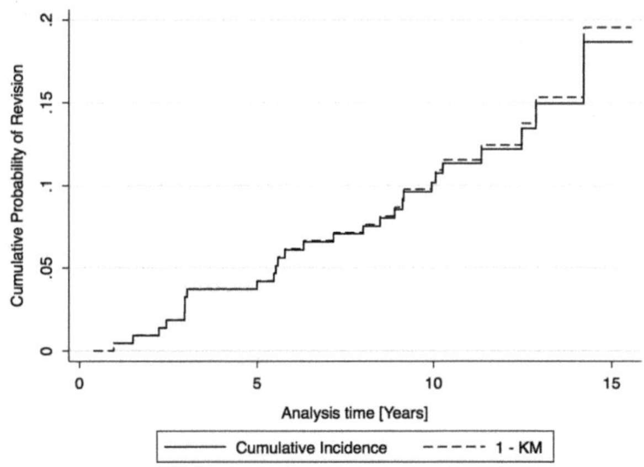

Fig. 5.2 Comparison of the incidence of revision for the cumulative incidence of competing risk and 1 − K-M methods for dataset 2

In a second example, we analysed 15 year follow-up of a total hip system, which has recently been published by Repantis et al. [24]. The authors report a high probability of aseptic loosening that is stated to have been caused by the low carbide metal-on-metal articulation. Repantis et al. [24], however, do not take into account the presence of competing risks. In the current study, revision for other causes, as well as (unrelated) death can be considered as competing events. Similar methodology as described above was applied to these data.

Between 1994 and 1999, 217 patients were operated on, of which 27 (12.4 %) had experienced the event of interest, 10 patients (4.6 %) had a competing event and 21 (9.7 %) patients died from unrelated causes. The 1 − K-M was 19.5 % at 15 years, while the cumulative incidence was 18.7 %, at a relative difference of 4.7 % (Fig. 5.2).

5.6 Simulation Study

In this example we generate data with respect to a primary risk, a competing risk and a censoring event. The durations of the risk events are drawn from exponential distributions with the shape parameter value corresponding to a hazard rate lambda of 1/0.0125, 1/0.2 and 1/0.4 for the primary risk and 1/0.1 for the competing risk. These parameters settings correspond to hazard ratios of 0.125, 2.0 and 4.0. The values of the lambda's of the censoring duration are chosen equal to the values of the lambda's of the primary risk.

The dependency between the occurrences of the two risks is modeled by using the Clayton copula, with the chosen parameter values of 0 (i.e. independency between both risks), 1 and 2. So, in this Monte Carlo simulation we will vary the hazard ratios and the dependencies of the two risks.

Table 5.1 Simulation results: cumulative incidence (first line) and 1-KM percentage (second line) of the primary risk and the competing risk after 1 year (i.e. means and standard deviations between parentheses)

Clayton cupola parameter	0	1	2
Hazard ratio			
0.125	0.46 (0.07)	0.43 (0.07)	0.42 (0.07)
	0.70 (0.11)	0.62 (0.10)	0.60 (0.10)
2	0.64 (0.07)	0.70 (0.07)	0.77 (0.08)
	0.88 (0.10)	0.88 (0.10)	0.88 (0.09)
4	0.77 (0.08)	0.86 (0.08)	0.89 (0.08)
	0.92 (0.09)	0.93 (0.08)	0.93 (0.09)

The cumulative incidence and the 1-KM percentage (after 1 year) of the primary risk and competing risk will be calculated for each value of the hazard ratio (i.e. 0.125, 2.0 and 4.0) and the dependency measure (i.e. Clayton copula parameter value of 0, 1 and 2). Each parameter setting consists of 200 replications of a clinical trial with 100 included patients. The results of the simulation study are presented in Table 5.1.

The differences between the cumulative incidences and the 1-KM percentages depend on the hazard ratios and the dependency of the primary and competing risk. From our simulation results it follows that the impact of competing risk on KM increases with decreasing hazard ratio and with decreasing clayton copula parameter (i.e. independency of the event of interest and the competing event(s)).

5.7 Conclusions

The limitations of the Kaplan-Meier method in the presence of competing risks have been well described in the literature. Alternative approaches have been discussed, including the cumulative incidence function and Gray's linear rank statistics for testing the equality of the cumulative incidence functions (Gray [14]), as well as the proportional hazard model for the sub-distribution of a competing risk (Fine and Gray [7]). Despite the availability of these approaches in many modern software packages, researchers frequently fall back on standard statistical approaches that do not take into account the presence of competing risks for the analysis of time-to-event data.

Empirically, we have shown that the use of Kaplan-Meier leads to statistical bias in the presence of competing risk, although the magnitude of this bias varied considerably. Comparing Kaplan-Meier curves in different cohorts of patients (with different rates of competing risk) obviously affects the results of statistical significance testing and leads to biased conclusions.

In view of the frequent occurrence of competing risks in biomedical research, we encourage the use of methods that account for the presence of competing risk approaches in order to improve the interpretability of time-to-event analyses.

References

1. Alberti, C., Métivier, F., Landais, P., et al.: Improving estimates of event incidence over time in populations exposed to other events: application to three large databases. *Journal of Clinical Epidemiology* **56**, 536–545 (2003)
2. Andersen, P.K., Abildstrøm, Rosthøj: Competing risks as a multi-state model. *Statistical Methods in Medical Research* **11**, 203–215 (2002)
3. Bakoyannis, G., Touloumi, G.: A practical guide on modelling competing risk. *Statistical Methods in Medical research* **21**(3), 257–272 (2012)
4. Benichou, J., Gail, M.: Estimates of absolute cause-specific risk in cohort studies. *Biometrics* **46**, 813–826 (1990)
5. Crowder, M.J.: *Multivariate survival analysis and competing risks*. Chapman and Hall CRC Texts in Statistical Science, Boca Raton, USA (2012)
6. Fennema, P., Lubsen, J.: Survival analysis in total joint replacement: an alternative method of accounting for the presence of competing risk. *Journal of Bone and Joint Surgery* **92**(5), 701–706 (2010)
7. Fine, J., Gray, R.: A proportional hazards model for the subdistribution of a competing risk. *Journal of the American Statistical Association* **94**, 496–509 (1999)
8. Fiocco, M., Putter, H., Van de Velde, C.J.H.: Reduced proportional hazard models for competing risks: an application. *Journal of Statistical Planning and Inference* **136**, 1655–1668 (2006)
9. Friedlin, B., Korn, E.: Testing treatment effects in the presence of competing risks. *Statistics in Medicine* **24**, 1703–1712 (2005)
10. Fürstovà, J., Valenta, Z.: Statistical analysis of competing risks: overall survival in a group of Chronic Myeloid Leukaemia patients. *European Journal for Biomedical Informatics* **7**(1), 2–10 (2011)
11. Gaynor, J., Feuer, E., Tan, C., et al.: On the use of cause-specific failure and conditional failure probabilities: examples from clinical oncology data. *Journal of the American Statistical Association* **88**, 400–409 (1993)
12. Gichangi, A., Vach, W.: *The analysis of competing risk data*. Research report, University of Southern Denmark, Denmark (2005)
13. Gooley, T.A., Leisenring, W., Crowley, J., Storer, B.E.: Estimation of failure probabilities in the presence of competing risks: new representations of old estimators. *Statistics in Medicine* **18**, 695–706 (1999)
14. Gray, R.J.: A class of k-sample tests for comparing the cumulative incidence of competing risk. *The Annals of Statistics* **16**, 1141–1154 (1988)
15. Grunkemeier, G., Anderson, R., Starr, A.: Actuarial and actual analysis of surgical results: empirical validation. *Annals of Thorac Surgery* **71**, 1885–1887 (2001)
16. Kalbfleisch, J., Prentice, R.: *The statistical analysis of failure time data*. Wiley, New York (1980)
17. Kaplan, E., Meier, P.: Nonparametric estimation from incomplete observations. *Journal of the American Statistical Society* **53**, 457–481 (1958)
18. Klein, J.P., Keiding, N., Copelan, E.A.: Plotting summary predictions in multistate survival models: probabilities of relapse and death in remission for bone-marrow transplantation patients. *Statistics in Medicine* **12**, 2315–2332 (1994)

19. Korn, E., Derey, F.: Applications of crude incidence curves. *Statistics in Medicine* **11**, 813–829 (1992)
20. Marubini, E., Valsecchi, M.G.: *Analysing survival data from clinical trials and observational studies*. John Wiley and Sons, Chichester, UK (1995)
21. Pepe, M., Mori, M.: Kaplan-Meier, marginal and conditional probability curves in summarizing competing risks failure time data. *Statistics in Medicine* **12**, 737–751 (1993)
22. Prentice, R., Kalbfleisch, J.: The analysis of failure times in the presence of competing risks. *Biometrics* **34**, 541–554 (1978)
23. Putter, H., Fiocco, M., Geskus, R.B.: Tutorial in biostatistics: competing risks and multi-state models. *Statistics in Medicine* **26**, 2389–2430 (2007)
24. Repantis, T., Vitsas, V., Korovessis, P.: Poor mid-term survival of the low carbide metal-on-metal Zweymüller total hip arthroplasty: a concise follow-up, at a minimum of 10 years, of a previous report. *The Journal of Bone & Joint Surgery* **95**, 331–334 (2013)
25. Rosthøj, S., Andersen, P.K., Abildstrøm, S.Z.: SAS macrofor estimation of the cumulative incidence functions based on a Cox regression model for competing risk survival data. *Computer Methods and Programs in Biomedicine* **74**, 69–75 (2004)
26. Satagopan, J., Porat-Ben, L., Robson, M., et al.: A note on computing risks in survival analysis. *British journal of Cancer* **91**, 1229–1235 (2004)
27. Tai, B., Machin, D., White, I., et al.: Competing risk analysis of patients with osteosarcoma: A comparison of four different approaches. *Statistics in Medicine* **20**, 661–684 (2001)
28. Tai, B., Peregoudov, A., Machin, D.: The competing risk approach of the analysis of alternative intra-uterine devices (IUDS) for fertility regulations. *Statistics in Medicine* **20**, 3589–3600 (2001)
29. Williamson, P., Kolamunnage-Dona, R., Tudur Smith, C.: The influence of competing-risks setting on the choice of hypothesis test for treatment effects. *Biostatistics* **8**(4), 689–694 (2007)
30. Zweymüller, K.: Good results with an uncoated grit-blasted tapered straight stem at ten years. *Interactive Surgery* **2**, 197–205 (2007)

Chapter 6
Recent Developments in Group-Sequential Designs

James M.S. Wason

Abstract In a group-sequential trial, patients are recruited in groups, and their response to treatment is assessed. After each group is assessed, an interim analysis is conducted. At each interim analysis, the trial can stop for futility, stop for efficacy, or continue. The main advantage of group-sequential designs is that the expected number of patients is reduced compared to a design without interim analyses. There are infinitely many possible group-sequential designs to use, and the choice strongly affects the operating characteristics of the trial. This chapter discusses optimal and admissible group-sequential designs. Optimal designs minimise the expected sample size at some specified treatment effect; admissible designs optimise a weighted sum of trial properties of interest, such as expected sample size and maximum sample size. Methods for finding such designs are discussed, including a detailed description of an R package that implements a quick search procedure. Recent applications of group-sequential methodology to trials with multiple experimental treatments being tested against a single control treatment are also described.

6.1 Group-Sequential Designs Background

The traditional approach to analysing a randomised controlled trial is to conduct a statistical test of some null-hypothesis after a planned number of patients are recruited. In most disease areas, the number of patients is limited and so recruitment is generally time-consuming. Thus, data on the effect of treatment on early patients are available before recruitment is finished. A group-sequential design allows for multiple tests of the null-hypothesis as data is accrued. These earlier tests are referred to as interim analyses. The trial design may allow for early stopping if results from an interim analysis suggest the experimental treatment is significantly better than the control treatment. This is referred to as stopping for *efficacy*. The design may also allow for early stopping for *futility* if the results at an interim analysis suggest the trial is unlikely to end in success. A third reason for stopping

---

J.M.S. Wason (✉)
MRC Biostatistics Unit Hub for Trials Methodology Research,
Institute of Public Health, Cambridge, United Kingdom
e-mail: james.wason@mrc-bsu.cam.ac.uk

K. van Montfort et al. (eds.), *Developments in Statistical Evaluation of Clinical Trials*, DOI 10.1007/978-3-642-55345-5_6

is for safety – for example if the new treatment causes unacceptable side-effects. We just consider designs that allow stopping on the basis of whether or not the new treatment is effective, but one can also incorporate safety monitoring [5] into group-sequential designs.

The main advantages of group-sequential designs over designs that have no interim analyses (referred to as fixed sample-size trials) are:

1. Due to the possibility of early stopping, the expected sample size used in a trial will be lower than a fixed sample-size trial with the same significance level and power;
2. If the experimental treatment is less effective than the control treatment, the trial may stop early, meaning fewer patients are subjected to an ineffective experimental treatment;
3. In the long run, due to lower expected sample sizes, a limited set of patients can support more trials.

Group-sequential designs also have disadvantages:

1. More analyses means more statistical and data-management support is required;
2. Interim analyses require data to be unblinded before the end of the trial, meaning more potential for bias;
3. Since the null hypothesis is tested multiple times, the significance level of each analysis must be lower than that of the fixed sample-size trial in order to control the overall significance level; thus, if the trial continues to the end without stopping, the sample size used in the group-sequential trial will be larger than the fixed sample size trial.

Group-sequential designs are less useful when the outcome of interest takes a long time to observe, since recruitment will often be completed before the data on the effect of treatment on early patients are available. In settings where the treatment outcome is observed relatively quickly, the efficiency and ethical advantages of group-sequential designs are generally thought to outweigh the disadvantages.

In this chapter we will restrict attention to one-sided group-sequential designs. These are used when the null-hypothesis is tested against a one-sided alternative hypothesis. One-sided group-sequential tests are more relevant in clinical trials, as the experimental treatment is generally not of interest if it is worse than the control treatment.

A one-sided group-sequential design is parametrised by: (1) the number of patients to be recruited at each stage; (2) the futility boundaries, determining the threshold for futility stopping at each analysis; and (3) the efficacy boundaries, determining the thresholds for efficacy stopping at each analysis. The constraints on the design are the overall type-I error rate and power of the design. Since there are more parameters than constraints, there are an infinite number of possible designs to choose from. The choice of design is extremely important as it affects the statistical properties of the design, such as expected sample size.

There are three main approaches to choosing a design. The first is to constrain the stopping boundaries using some shape function. Commonly used functions are those of Pocock [22], O'Brien and Flemming [20], and Whitehead [33]. The main advantage of this approach is that it is quick to find a design; the main disadvantage is that the properties of the design, such as expected sample size, may not be desirable for the investigator. A second approach is to use a more flexible family of stopping boundary functions. For example, the power-family of group-sequential tests, proposed by Pampallona and Tsiatis [21], is a single-parameter family of stopping boundary shapes. By varying the parameter, the properties of the resulting design differ. A third approach, is to search over the full set of parameters in order to choose the design that best matches the desired properties of the investigator.

This chapter provides an overview of some recent methodological developments in group-sequential designs, and is organised as follows: in Sect. 6.2, notation for the rest of the chapter is given; in Sect. 6.3 some background on optimal designs is provided; in Sect. 6.4 the δ-minimax design is motivated, and a simulated annealing technique to find optimal designs is discussed; in Sect. 6.5 the concept of admissible designs is motivated and discussed; in Sect. 6.6 the problem of not knowing the variance of the treatment response at the design stage is addressed; in Sect. 6.7 an R package which allows quick finding of admissible designs is described; in Sect. 6.8, extensions of group-sequential methods to multi-arm multi-stage designs are discussed; finally in Sect. 6.9, some limitations and possible extensions of the methods in the chapter are discussed.

6.2 Notation

Consider a randomised two-arm group-sequential design with up to J analyses. The jth analysis takes place after n_j patients have been randomised to each arm, and their treatment response measured. The response of patient i on the control arm, X_{0i}, is assumed to be distributed as $N(\mu_0, \sigma^2)$, with the response of patient i on the experimental arm, X_{1i}, is assumed to be distributed as $N(\mu_1, \sigma^2)$. Here, the value of σ^2 is assumed to be known, although unknown variance will be addressed in Sect. 6.6. The parameter of interest is the difference in mean response between the experimental and control arms, $\mu_1 - \mu_0$, and is labelled δ. The null-hypothesis tested is $H_0 : \delta \leq \delta_0$. A design is required such that the probability of rejecting the null is at most α when H_0 is true, and at least $1 - \beta$ when $\delta \geq \delta_1$, where δ_1 is the clinically relevant difference (CRD). The value of δ_0 will generally be set to 0, indicating that any improvement is of interest. These two constraints are referred to as the type-I error and power constraints respectively. A design which meets both constraints is called *feasible*.

At a given interim analysis j, the z-statistic for testing H_0, Z_j, is calculated:

$$Z_j = \sqrt{\frac{n_j}{2\sigma^2}} \frac{\sum_{i=1}^{n_j} X_{i1} - \sum_{i=1}^{n_j} X_{i0}}{n_j}. \qquad (6.1)$$

If $Z_j > e_j$, the trial stops for efficacy; if $Z_j \leq f_j$, the trial stops for futility. If it is between the two thresholds, the trial continues to stage $j+1$. The value of e_J is set to f_J to ensure that a decision is made at the last interim analysis.

The number of design parameters is $3J-1$: J parameters for the sample size at each stage, J efficacy parameters $e = (e_1, \ldots, e_J)$, and J futility parameters $f = (f_1, \ldots, f_{J-1})$ (actually $J-1$ free parameters as $f_J = e_J$). Generally the number of parameters is reduced by assuming a constant number of patients recruited per stage to each treatment arm, n, called the *group-size*. With this assumption, the value of n_j will be equal to jn. This reduces the number of parameters to $2J$.

The vector of random variables $(Z_1, Z_2, \ldots, Z_J)$ has a multivariate normal distribution with mean vector $\left(\sqrt{\frac{n}{2\sigma^2}}\delta, \sqrt{\frac{2n}{2\sigma^2}}\delta, \ldots, \sqrt{\frac{Jn}{2\sigma^2}}\delta\right)$, and covariance matrix Σ, where the (i,j)th entry of Σ, Σ_{ij}, is equal to $\sqrt{\frac{\min(i,j)}{\max(i,j)}}$, [31]. Finding the probability of stopping for efficacy at stage j, Π_j, involves multivariate integration. Stopping for efficacy at the jth stage happens if and only if $(Z_1, \ldots, Z_{j-1})$ were all between the futility and efficacy stopping boundaries, and Z_j is above e_j. The probability of this is:

$$\Pi_J(\delta) = \int_{f_1}^{e_1} \int_{f_2}^{e_2} \cdots \int_{f_{j-1}}^{e_{j-1}} \int_{e_j}^{\infty} f_{Z_{(j)}}(z_1, \ldots, z_j) dz_j \ldots dz_1 \, , \tag{6.2}$$

where $f_{Z_{(j)}}$ is the pdf of $(Z_1, \ldots, Z_j)$. Note that the mean of $Z_{(j)}$ depends on δ, but the covariance does not. Equation (6.2) can be evaluated using the technique of Genz and Bretz [10], or the technique of Armitage [2, 18], described further in Chap. 19 of Jennison and Turnbull [13]. Note that the normality of the test statistics is the main assumption used and not the normality of the treatment endpoint – therefore other types of endpoints such as binary and time-to-event for which there are asymptotically normally distributed test statistics can be considered within this framework [13].

The probabilities $\Pi_1(\delta), \ldots, \Pi_J(\delta)$ can be summed to give the total probability of stopping for efficacy. Setting $\delta = \delta_0$ will give the type-I error rate, and setting $\delta = \delta_1$ will give the power. A similar formula as (6.2) can be used to find the probability of stopping for futility at each stage. From the probabilities of stopping for futility and efficacy at each stage, the expected sample size can be straightforwardly found.

6.3 Optimal Group-Sequential Designs

Within the context of group-sequential designs, an optimal design is one that satisfies the required type-I error rate and power (i.e. it is *feasible*), and out of all possible feasible designs, it optimises some criterion of interest. Criteria considered tend to be some function of the sample size, for example the expected sample size at some value of δ.

Finding an optimal design involves searching over the stopping boundary parameters as well as the sample size parameters. With the constraints described in Sect. 6.2, searching for an optimal J-stage group-sequential design involves searching over $2 \times J$ parameters, as the final futility and efficacy threshold are set to be the same. There are just two constraints: the type-I error and the power. This is a computationally challenging problem when $J > 2$, as the number of parameters is large and there are many local optima in the set of designs to be searched.

The method of dynamic programming was proposed for finding symmetric (i.e. the type-I error rate, α, is equal to the type-II error rate, β) optimal one-sided group-sequential designs [7] and optimal two-sided designs [8]. This was extended to non-symmetric designs by Barber and Eales [3]. The method works by defining a Bayes decision theory problem for which the optimal group-sequential design is the solution. The decision theory problem is to decide between $D_0 : \delta = 0$ and $D_\delta : \delta = \delta^*$, with the cost of making decision D with true treatment difference δ equal to $C(D_0, \delta^*) = d_\delta$ for $D = D_0$, and $C(D_\delta, 0) = d_0$ for $D = D_\delta$. For any other value of δ, $C(D, \delta)$ is set to be 0. Backwards induction can be used to find the design that minimises a given objective function, such as expected sample size at the null hypothesis. A numerical search over (d_0, d_δ) is conducted in order to find the design giving the correct type-I error rate and power. This final design will then be the optimal one.

Generally this method can be used to find an optimal design when the optimality criterion is the expected sample size at a specific value of δ (or sums of expected sample sizes at different values of δ). However, in the next section an optimality criteria is proposed that is of potential interest and that cannot be optimised using dynamic programming.

6.4 δ-Minimax Design and Simulated Annealing

The expected sample size of a group-sequential design depends on the true treatment effect. If an optimal design is chosen for a particular treatment effect, then the design may perform poorly when the true treatment effect varies from the design value. For designs allowing stopping for both futility and efficacy, the expected sample size increases in δ monotonically to a maximum and then decreases monotonically. Intuitively this is because as δ increases, the probability of the trial stopping early for futility decreases monotonically, but the probability of the trial stopping early for efficacy increases monotonically. A slightly more formal explanation is given in Wason, Mander and Thompson [31].

Thus each design has a treatment effect, $\tilde{\delta}$, that leads to the design having the maximum expected sample size over all possible values of δ. This is called the worst-case-scenario treatment effect. The optimality criterion of choosing the feasible design with the lowest maximum expected sample size was proposed for two-stage trials with binary outcomes by Shuster [25]. The design showed some good properties such as low expected sample sizes at the null treatment effect and CRD. The design was extended to two-stage trials with normally distributed outcomes by Wason and Mander [30] and named the δ-minimax design, as it has the lowest maximum expected sample size over δ. To find the δ-minimax design for two-stages, it is feasible to use a grid-search technique, as the number of parameters (i.e. futility and efficacy boundary parameters, and group-size) is low. For more than two-stages, there are too many parameters to perform a grid-search. The dynamic programming algorithm proposed by Barber and Jennison [3] works when the optimality criterion is independent of the design; however the value of $\tilde{\delta}$ depends on the design, thus a different method must be used for $J > 2$. In Wason et al. [31], use of a stochastic search technique called simulated annealing was proposed to find the δ-minimax design.

The simulated annealing algorithm is described in detail in the supplementary material of Wason et al. [31], and C code is available on the author's website (http://sites.google.com/site/jmswason). Each iteration of the simulated annealing process consists of two steps. The first step is to generate a new candidate design from the current design (i.e. the design which the process is currently at). The second step is to decide whether the process should move from the current design to the candidate design. Both steps rely on so-called 'temperature' parameters. At the end of each iteration, the temperature parameters are reduced. As the temperature parameters fall: (1) the candidate design generated at each iteration will, on average, be closer to the current design; and (2) the process is less likely to move to a design that is worse. In this way, the process is more likely to explore the space of designs towards the beginning, with the aim of avoiding getting stuck at a local optimum.

For two and three-stage designs, simulated annealing is quick and reliable, with results not varying considerably between independent runs. However, for four or more stages, the process takes longer and becomes less reliable. The reason that it takes longer is that evaluating the operating characteristics of a design is more time-consuming when there are more stages. The process is less reliable because the number of parameters is greater and there are more local optima in the space of possible designs. One can run the simulated annealing process for longer in order to improve reliability, but of course this takes longer. With four or five stages, it is recommended that a number of independent simulated annealing processes with different random number seeds are run. The best resulting design can then be chosen.

The δ-minimax design is comparable to the triangular design proposed by Whitehead and Stratton [33]. In the case of a *symmetric* ($\alpha = \beta$) and fully sequential (i.e. interim analyses after each patient), as the type-I error rate converges to 0, the resulting triangular stopping boundaries minimise the maximum expected sample size. It is thus of interest to see whether the δ-minimax design adds anything over the use of the triangular stopping boundaries. Table 6.1 shows, for $\delta_0 = 0$, $\delta_1 = 1$, $\sigma = 3$ and various values of J, the expected sample size of the null-optimal design (optimal at $\delta = \delta_0$), the CRD-optimal design (optimal at $\delta = \delta_1$), the δ-minimax design, and the triangular test at: (1) the null treatment effect, i.e. 0; (2) the CRD, i.e. 1; (3) the worst-case-scenario treatment effect. Also shown is the maximum sample size used by the design if early stopping does not take place.

Table 6.1 Expected and maximum sample sizes per arm of investigated designs for different numbers of stages. The random variable N denotes the sample size per arm used with a specified design

		Null-optimal	CRD-optimal	δ-minimax	Triangular design
J=2	$\mathbb{E}(N\|\delta = \delta_0)$	107.6	118.0	110.9	111.2
	$\mathbb{E}(N\|\delta = \delta_1)$	130.5	117.1	119.4	117.6
	$\mathbb{E}(N\|\delta = \tilde{\delta})$	138.9	136.8	133.3	132.2
	Maximum sample size	170	172	180	180
J=3	$\mathbb{E}(N\|\delta = \delta_0)$	94.9	105.7	98.0	100.4
	$\mathbb{E}(N\|\delta = \delta_1)$	128.9	107.0	109.2	108.4
	$\mathbb{E}(N\|\delta = \tilde{\delta})$	137.3	130.0	125.9	125.5
	Maximum sample size	183	186	189	192
J=4	$\mathbb{E}(N\|\delta = \delta_0)$	88.7	98.0	92.7	98.3
	$\mathbb{E}(N\|\delta = \delta_1)$	119.1	102.2	105.0	106.1
	$\mathbb{E}(N\|\delta = \tilde{\delta})$	130.6	125.5	122.0	124.9
	Maximum sample size	192	196	196	204
J=5	$\mathbb{E}(N\|\delta = \delta_0)$	85.4	92.1	89.2	96.0
	$\mathbb{E}(N\|\delta = \delta_1)$	113.1	99.3	102.8	103.9
	$\mathbb{E}(N\|\delta = \tilde{\delta})$	126.8	122.5	119.6	123.0
	Maximum sample size	200	210	205	210

Table 6.2 Group-size, futility stopping boundaries, and efficacy stopping boundaries of five-stage optimal and triangular designs

Design	n	f	e
Null-optimal	40	$(-0.24, 0.37, 0.76, 1.09, 1.56)$	$(3.01, 2.47, 2.23, 2.03, 1.56)$
CRD-optimal	42	$(-0.51, 0.29, 0.83, 1.33, 2.05)$	$(2.14, 2.05, 2.09, 2.15, 2.05)$
δ-minimax	41	$(-0.52, 0.34, 0.92, 1.38, 1.83)$	$(2.54, 2.09, 2.03, 1.96, 1.83)$
Triangular	42	$(-0.85, 0.30, 0.98, 1.49, 1.90)$	$(2.55, 2.10, 1.96, 1.91, 1.90)$

When $J = 2$ or 3, the δ-minimax and triangular designs have very similar expected sample size properties. The δ-minimax design in fact has a higher maximum expected sample size for $J = 2, 3$, but this is because the equations determining the triangular design, given in Jennison and Turnbull [13], for given $\alpha \neq \beta$ do not result in the feasibility constraints being met exactly (the triangular design has $\alpha = 0.0517$ and 0.0512 for $J = 2$ and $J = 3$ respectively). For $J = 4$ and 5 the δ-minimax design is more distinct, having a 7.1 % reduction in expected sample size under the null, and a 5.7 % reduction for $J = 4$, compared to the triangular design.

Table 6.2 shows the design parameters for each five-stage design, and Fig. 6.1 shows the stopping boundaries of the three optimal designs graphically. Although the expected sample size patterns are similar, the stopping boundaries of the δ-minimax and triangular designs are somewhat different. Generally the δ-minimax design is marginally more likely to stop at the first stage, although this is balanced

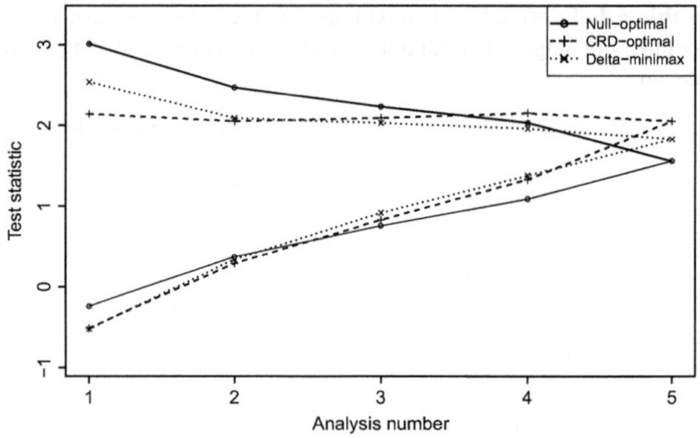

Fig. 6.1 Futility and efficacy stopping boundaries, in terms of test statistics, of the null-optimal, CRD-optimal, and δ-minimax design for $\alpha = 0.05, \beta = 0.1, \sigma = 3, \delta_0 = 0, \delta_1 = 1$

by it being slightly less likely to stop once the trial is at a later stage. The maximum sample sizes are similar, but differ between the designs for some values of J (see Table 6.1).

The δ-minimax design has desirable properties in comparison to the other two optimal designs. By definition it has the lowest maximum expected sample size of the three designs, but it also has low expected sample sizes across the range of treatment effects considered. When the treatment effect is close to δ_0, its expected sample size is only slightly higher than that of the null-optimal design; similarly its expected sample size is only slightly higher than that of the CRD-optimal design when δ is close to δ_1. The optimal designs perform well when δ is close to the treatment effect for which they are optimal, but poorly when δ is different. As one would expect, the expected sample size curves shifts downwards as J increases, indicating that including more stages results in lower expected sample sizes at each value of δ. The relative shapes of the curves change slightly, especially as δ increases past δ_1.

Minimising the expected sample size is an important objective in trials, but it is also of interest to control the maximum potential sample size. A design which yields a small improvement in expected sample size at a cost of a large increase in maximum sample size is unlikely to be preferred in practice. Table 6.1 shows that the δ-minimax and triangular designs generally have larger maximum sample sizes compared to null-optimal and CRD-optimal designs. All the optimal designs have maximum sample sizes noticeably larger than the sample size required for the one-stage design (155).

6.5 Admissible Designs

Optimal designs tend to have large maximum sample sizes, which can be problematic for planning individual trials. In addition, they may perform poorly with respect to other criteria of interest. For example, the null-optimal design has a relatively high maximum expected sample size. Admissible designs have been proposed in order to balance over more than one criteria of interest.

The first work on admissible designs was in the context of two-stage trial designs with binary outcomes and only futility stopping allowed. These designs have been well studied in the literature due to their relative simplicity and the fact that all possible designs can be enumerated (as sample size and stopping boundary parameters are all integers). Simon [26] discussed and recommended two designs for this type of trial. The first was the 'optimal' design (in the terminology of Sect. 6.4, the null-optimal design). The second was the 'minimax' design, which chooses the design with the lowest expected sample size at the null out of all designs that have the lowest maximum sample size. Jung et al. [14] noted that the optimal design has a relatively large maximum sample size, and the minimax design has a relatively large expected sample size. These observations motivated investigation of 'admissible' designs, which would balance the two criteria.

To do this, the authors specified a loss function as the weighted sum of the expected sample size under the null treatment effect and the maximum sample size: $\omega\mathbb{E}(N|H_0) + (1-\omega)\max(N)$, for $\omega \in [0,1]$. Admissible designs are feasible designs that minimise the loss function for some value of w. Additional information is available in Jung et al. [14] about how this corresponds to admissible decision rules in Bayesian decision theory. The optimal and minimax designs are admissible (for $\omega = 1$ and 0 respectively), but other admissible designs also exist which balance the two quantities in different ways. Admissible designs exist that show very small increases in expected sample size compared with the optimal design, but large decreases in the maximum sample size. In practice, such a design may be preferable to the optimal design, as a small maximum sample size is desirable.

Mander et al. [17] extend the ideas in Jung et al. to phase II trials with binary outcomes allowing early stopping for efficacy. When stopping for efficacy is allowed, the expected sample sizes at treatment effects other than the null are also of interest. Designs that are admissible with respect to the expected sample size at the null, the expected sample size at the CRD, and the maximum sample size are evaluated.

When considering normally distributed endpoints, finding admissible designs is more challenging. This is because the stopping boundary parameters are non-integer and so infinitely many feasible designs exist. This is as opposed to the binary outcome case where the stopping boundary parameters are integers, and so all designs can be enumerated. Instead, in Wason et al. [31], it was argued that the maximum expected sample size could be used as a surrogate for all expected sample sizes of interest. The loss function in this case is:

$$\omega\mathbb{E}(N|\tilde{\delta}) + (1-\omega)\max(N)\ . \tag{6.3}$$

The advantage of just considering the two criteria in (6.3) is that it is computationally feasible to find all admissible designs. For each possible maximum sample size, the futility and efficacy parameters can be chosen so that the maximum expected sample size is minimised. Any other design with that maximum sample

Table 6.3 Properties of admissible designs for $J = 3$; $\max(N) =$ maximum sample size per arm, ω interval gives the values of ω that would lead to that design being the admissible design of choice

$\max(N)$	$\mathbb{E}(N\|\delta_0)$	$\mathbb{E}(N\|\delta_1)$	$\mathbb{E}(N\|\tilde{\delta})$	ω interval
156	117.29	124.73	139.02	[0,0.426)
159	107.34	121.43	134.97	[0.426,0.539)
165	102.05	114.78	129.83	[0.539,0.713)
168	101.44	112.55	128.62	[0.713,0.820)
171	100.21	111.23	127.96	[0.820,0.843)
177	98.19	111.14	126.84	[0.843,0.921)
186	98.74	109.19	126.07	[0.921,0.981)
189	98.20	109.88	126.01	[0.981,1]

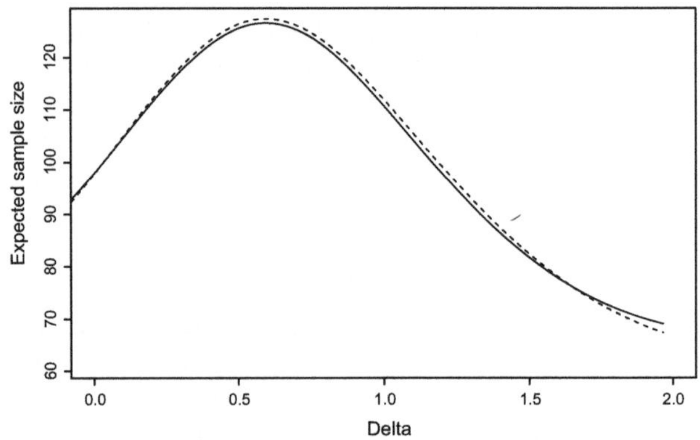

Fig. 6.2 Expected sample sizes of δ-minimax design (*solid line*) and admissible design with $N = 171$ in Table 6.3 (*dashed line*)

size cannot be admissible because loss function (6.3) will always be higher (unless $\omega = 0$). No design with maximum sample size greater than that of the δ-minimax design can be admissible, as such a design would have both a higher maximum sample size and a higher maximum expected sample size.

As an illustration, Table 6.3 displays the properties of the possible admissible designs for $\delta_0 = 0, \delta_1 = 1, \sigma = 3, \alpha = 0.05, 1 - \beta = 0.9$.

From Table 6.3, using the value of $\max(\mathbb{E}(N))$ as an admissibility criterion is a good surrogate for jointly considering $\mathbb{E}(N|\delta_0)$ and $\mathbb{E}(N|\delta_1)$, since the two latter quantities generally decrease as the former does. The table includes the range of ω's (i.e. the weighting put on the maximum expected sample size) for which each design is best. For instance, if the two quantities are each given equal weight ($\omega = 0.5$), the second design in the table is the best one to pick. The choice of ω may depend on several factors. For instance, if the trial is being carried out in an area with limited patient numbers, ω might be chosen to be low, since it would be desirable to reduce the maximum sample size. In other situations, a higher value of ω may be preferred, since on average the number of patients required is reduced.

Figure 6.2 shows the expected sample size curve of the δ-minimax design for a range of values of δ. Also included is the expected sample size curve for the admissible design from Table 6.3 with $\max(N) = 171$. The difference in the expected sample size curves is very small, but there is a 9.5 % reduction in the maximum sample size. This indicates that by relaxing the requirement for optimality very slightly, a big improvement in other characteristics of interest is possible.

6.6 Unknown Variance

A common assumption made in the design of group-sequential trials is that the variance of the treatment response, σ^2 is known for each arm. In practice this is unlikely to be the case, and if the postulated value is incorrect, then the operating characteristics of the trial can be strongly affected.

Various techniques to allow for unknown variance have been proposed in the literature. Shao and Feng [24] suggest using Monte-Carlo simulation to choose an appropriate critical value. Although this technique would be too computationally intensive to be used in conjunction with a search for optimal designs, it could be used to modify the final design's stopping boundaries. Jennison and Turnbull [12] show how one can convert boundaries for the known variance case to the unknown variance case using a recursive algorithm.

Jennison and Turnbull [13] propose a method for converting the stopping boundaries that is simpler than the recursive algorithm and less computationally intensive than simulation. Recall that f_j and e_j are the stopping boundaries for analysis j, and n_j is the number of patients per arm that are randomised by the time of the analysis. Then the thresholds for stopping in terms of p-values are attained from the quantile of the normal distribution, i.e. $1 - \Phi(e_j)$ and $1 - \Phi(f_j)$ respectively. With unknown variance, when $\delta = 0$, the test-statistics would be marginally distributed as a Student's t-distribution with $2n_j - 2$ degrees of freedom. Therefore by substituting in new stopping boundaries $f_j' = T_{2n_j-2}(1 - \Phi(f_j))$ and $e_j' = T_{2n_j-2}(1 - \Phi(e_j))$, where T_p is the cumulative distribution function of Student's t-distribution with p degrees of freedom, the design will marginally have the correct stopping characteristics (under the null) at each stage. The overall type-I error rate of the trial will still differ from its nominal value because the assumed correlation between test-statistics when the variance is known will differ from the actual correlation when it is unknown (the size of the difference is investigated later on in this section).

Table 6.4, taken from Wason et al. [31], shows the type-I error rate and power for the five-stage δ-minimax design for $\delta_1 = 1, \sigma = 3, \alpha = 0.05, \beta = 0.1$ as the true value of σ differs from 3. Three scenarios are considered: (1) no modification is made, (2) t-tests are used with the known-variance stopping boundaries, (3) t-tests are used with the stopping boundaries modified using the quantile-substitution method. The type-I error rate and power are estimated from 250, 000 independent replicates each.

The simulated type-I error rates show that methods (2) and (3) both work well. The type-I error rates are very close to the required level of 0.05, with quantile-substitution working slightly better. The power is not controlled as the value of

Table 6.4 Type-I error rate and power estimates as the true standard deviation varies from the assumed value of 3

	Type I error			Power		
σ	Z-test	T-test	T-test with modified boundaries	Z-test	T-test	T-test with modified boundaries
1	0.000	0.051	0.050	1.000	1.000	1.000
1.5	0.000	0.052	0.050	0.998	1.000	1.000
2	0.000	0.051	0.050	0.984	0.995	0.995
2.5	0.021	0.052	0.050	0.95	0.965	0.965
3	0.050	0.051	0.050	0.900	0.900	0.899
3.5	0.086	0.052	0.050	0.851	0.810	0.809
4	0.124	0.052	0.051	0.807	0.714	0.712
4.5	0.158	0.052	0.051	0.768	0.626	0.623
5	0.189	0.051	0.050	0.737	0.550	0.547

Table 6.5 Type-I error rate and power estimates as the true standard deviation varies from the assumed value of 1

	Type I error			Power		
σ	Z-test	T-test	T-test with modified boundaries	Z-test	T-test	T-test with modified boundaries
0.25	0.000	0.070	0.054	1.000	1.000	1.000
0.5	0.000	0.069	0.052	0.997	1.000	1.000
0.75	0.011	0.069	0.053	0.964	0.986	0.985
1	0.050	0.069	0.052	0.900	0.902	0.893
1.25	0.102	0.068	0.052	0.832	0.768	0.750
1.5	0.154	0.069	0.052	0.775	0.64	0.613
1.75	0.201	0.069	0.052	0.726	0.533	0.503
2	0.236	0.069	0.052	0.691	0.455	0.424

σ increases however. To overcome this, an adaptive design would be required in which the sample size of the rest of the trial is chosen depending on the estimated variance; an example of this is given in Whitehead et al. [34]. The good performance of both methods (2) and (3) could be due to the large group-size resulting in the degrees of freedom of the t-distribution being sufficiently high to allow the standard normal to be a good approximation. To see what happens when the group-size is lower, results are shown for the five-stage δ-minimax design with $\sigma = 1$. This results in a group-size of 4, $f = (-0.914, -0.026, 0.698, 1.177, 1.761)$, and $e = (2.980, 2.308, 2.048, 1.976, 1.761)$. It is clear that the type-I error rate is less well controlled in this case (Table 6.5), although the T-test in conjunction with the quantile-substitution method controls the type-I error rate fairly well.

Thus it seems that quantile substitution is a straightforward, but effective method to control the type-I error rate when the variance is unknown.

6.7 OptGS: An R Package for Optimal and Admissible Group-Sequential Designs

The consideration of optimal and admissible group-sequential designs has been motivated in the previous sections. All the theory to implement finding such designs is available in the literature, but it takes a lot of work to implement from scratch. There are some existing software packages that implement group-sequential designs, summarised by Wassmer and Vandemeulebroecke [32]. The IML module in SAS® contains routines that allow calculation of stopping boundaries that give a specified type-I error rate. In R, the package gsDesign [1] allows the user to find boundaries and group-size required for several group-sequential designs, including O'Brien-Flemming and Pocock. Commercially available stand-alone programs that implement group-sequential designs include ADDPLAN, East, PASS, and PEST. However, none of these software packages include a function that searches for optimal or admissible designs.

FORTRAN code that implements searching for optimal designs using dynamic programming, as described in Barber and Jennison [3] is available from Stuart Barber's webpage (http://www1.maths.leeds.ac.uk/~stuart/Research/Software/0118.tar). Compilation of the code requires some technical computing knowledge, as it requires installation of the GNU Scientific Library. The code would also not be extendable to all optimality criterion, for example the maximum expected sample size. In this section, the R package OptGS [28] is described, which is freely available from the author's website (http://sites.google.com/site/jmswason). The package allows quick searching for designs that are near-optimal, or admissible with respect to four optimality criteria. Instead of simulated annealing, an extension of the Power-family is used. This extension allows a wide range of stopping boundary shapes, but considerably reduces the time taken to search. A quick method for searching is desirable so that investigators may explore many possible admissible designs in a short time.

6.7.1 Two-Parameter Power Family

The power family of group-sequential tests was first proposed by Emerson and Flemming [9] for symmetric designs (i.e. $\alpha = \beta$). Pampallona and Tsiatis [21] extended the family to allow non-symmetric designs ($\alpha \neq \beta$). In this section we consider the formulation of Pampallona and Tsiatis. The family is indexed by a parameter Δ, which determines the shape of the stopping boundaries. The power-family stopping boundaries are:

$$e_j = C_e(J, \alpha, \beta, \Delta)(j/J)^{\Delta - 0.5}$$

$$f_j = \delta_1 \sqrt{\mathscr{I}_j} - C_f(J, \alpha, \beta, \Delta)(j/J)^{\Delta - 0.5} \,,$$

where $\mathscr{I}_j = 2n_j/\sigma^2$.

To meet the required constraint $e_J = f_J$, the value of $\mathscr{I}_J$ is set to:

$$\mathscr{I}_J = 2n_J/\sigma^2 = \frac{\{C_e(J,\alpha,\beta,\Delta) + C_f(J,\alpha,\beta,\Delta)\}^2}{\delta^2}. \tag{6.4}$$

For a specific value of Δ, $C_f(J,\alpha,\beta,\Delta)$ and $C_e(J,\alpha,\beta,\Delta)$ take values such that the design has correct type-I error rate and power. Varying Δ changes the shape of the boundaries, and thus the operating characteristics of the design, with higher values generally giving designs with lower expected sample sizes, but higher maximum sample sizes.

Although the power-family provides a flexible range of stopping boundary shapes, it does not provide enough flexibility to include optimal designs. For optimal designs, the shape of the efficacy stopping boundaries will differ from the shape of the futility stopping boundaries.

OptGS uses a straightforward extension to the power family: introducing two shape parameters Δ_f and Δ_e, allowing the shape of the futility and efficacy boundaries to differ, and thus allowing greater flexibility in shape. The stopping boundaries are:

$$\begin{aligned} e_j &= C_e(J,\alpha,\beta,\Delta)(j/J)^{\Delta_e-0.5} \\ f_j &= \delta_1\sqrt{\mathscr{I}_j} - C_f(J,\alpha,\beta,\Delta)(j/J)^{\Delta_f-0.5}. \end{aligned} \tag{6.5}$$

Note that Eq. (6.4) still ensures $e_J = f_J$.

Given values of $(J,\Delta_f,\Delta_e,C_f,C_e)$, the group-size and stopping boundaries are determined from (6.4) and (6.5). As in Pampallona and Tsiatis [21], for each value of (Δ_f,Δ_e), values of C_f and C_e exist so that the design has desired type-I error rate, α, and power, $1-\beta$. These values can be found by searching for the values of (C_f,C_e) that minimise the following function:

$$(\alpha^*(J,\Delta_f,\Delta_e,C_f,C_e) - \alpha)^2 + (\beta^*(J,\Delta_f,\Delta_e,C_f,C_e,\delta) - \beta)^2, \tag{6.6}$$

where $\alpha^*(\cdot)$ and $\beta^*(\cdot)$ are the type-I and type-II error rate of the design given by $(J,\Delta_f,\Delta_e,C_f,C_e)$. The value of (6.6) is 0 if and only if the type-I error rate and power of the design are equal to the required values. In OptGS, this minimisation is performed using the Nelder-Mead algorithm [19].

The Nelder-Mead algorithm is also used to search over values of (Δ_f,Δ_e) in order to find an optimal design. Almost surely, the optimal value of (Δ_f,Δ_e) will imply a non-integer group size. To get a final design with integer group-size, two additional optimisations are run. The first with the constraint that the final group-size is equal to the group-size implied by the optimal (Δ_f,Δ_e) rounded up. The second instead rounding down. Of the designs found, the one that is closer to optimal is picked as the final design. Additional details are provided in Wason [28].

OptGS allows the user to find a design that balances the three optimality criteria discussed in Sect. 6.4 as well as the maximum sample size. A vector of weights, $(\omega_1, \omega_2, \omega_3, \omega_4)$, is specified by the user such that all are non-negative. Then the feasible design is found that minimises the following function:

$$\omega_1 \mathbb{E}(N|\delta = \delta_0) + \omega_2 \mathbb{E}(N|\delta = \delta_1) + \omega_3 \max \mathbb{E}(N) + \omega_4 J n_1 \ . \tag{6.7}$$

This design balances the three optimality criteria together with the maximum sample size. Note that one of ω_1, ω_2, and ω_3 must be strictly positive, because an infinite number of designs will exist with the lowest maximum sample size.

6.7.2 Comparison of OptGS and Simulated Annealing

Table 6.6, taken from Wason [28], shows the time taken to find J-stage null-optimal designs using SA and using OptGS. A single M5000 SPARC 2.4 GHz processor was used to carry out all computation. Ten independent simulated annealing (SA) searches were carried out for each value of J because SA is a stochastic process, and results may vary between runs. The average and minimum expected sample size under the null over the ten processes are shown in the table.

For several values of J, the optimal design found from OptGS is actually better than that found from the best of 10 runs of SA. This is despite the shape constraint imposed by use of the extended power-family. Only for $J = 5$ does SA show some improvement over OptGS. OptGS is substantially faster than even one SA run. Clearly, using OptGS has substantial advantages over using simulated annealing.

Table 6.7, also taken from Wason [28], shows the optimal values of $\Delta_f, \Delta_e, C_f, C_e$ for the three types of optimal design implemented in OptGS as well as the $(1, 1, 1, 1)$-admissible design, i.e. the admissible design that puts equal weight on all four operating characteristics. The results show that allowing Δ_f to differ from Δ_e is necessary to allow optimal designs to be found – the null-optimal and CRD-optimal designs have Δ_f and Δ_e designs with opposite signs. Interestingly,

Table 6.6 Comparison of run-time and expected sample size at $\delta = \delta_0$ of designs found from simulated annealing (SA) and OptGS

	$\mathbb{E}(N\delta_0)$			Time taken	
J	Average from 10 SA runs	Minimum from 10 SA runs	OptGS	Average SA run (s)	OptGS (s)
2	108.2	107.9	107.5	18.2	0.27
3	95.2	94.8	94.8	193.5	9.91
4	89.9	89.6	89.0	373.7	13.7
5	85.6	85.7	85.8	573.6	25.5

Table 6.7 Optimal design parameters $(\Delta_f, \Delta_e, C_f, C_e)$ for various optimality criteria and number of stages. The rows labelled $(1, 1, 1, 1)$ correspond to the $(1, 1, 1, 1)$-admissible design. Note that the expected and maximum sample sizes shown are for both treatment arms

Design	J	Δ_f	Δ_e	C_f	C_e	$\mathbb{E}(2N\delta_0)$	$\mathbb{E}(2N\delta_1)$	$\max \mathbb{E}(2N)$	$\max(2N)$
Null-optimal	2	0.45	−0.34	1.50	1.57	215.0	285.1	293.4	340
	3	0.52	−0.55	1.66	1.52	189.6	276.0	283.2	366
	4	0.52	−0.41	1.74	1.53	178.0	261.4	272.1	384
	5	0.53	−0.37	1.81	1.52	171.6	256.2	267.8	400
CRD-optimal	2	−0.18	0.46	1.25	1.84	241.0	234.6	276.9	344
	3	−0.15	0.48	1.26	1.96	231.1	214.8	265.6	372
	4	−0.13	0.49	1.27	2.03	222.7	205.5	259.0	392
	5	−0.01	0.48	1.31	2.06	207.3	200.0	250.6	410
δ-minimax	2	0.30	0.33	1.40	1.74	221.4	238.7	266.5	356
	3	0.33	0.33	1.48	1.79	196.8	219.4	251.9	384
	4	0.32	0.32	1.51	1.82	185.8	210.0	244.2	400
	5	0.32	0.32	1.53	1.84	179.7	204.2	239.4	410
$(1, 1, 1, 1)$	2	−0.01	0.08	1.32	1.68	226.3	245.7	272.9	324
	3	0.06	0.05	1.37	1.68	206.1	233.1	259.0	336
	4	0.12	0.12	1.41	1.71	194.3	220.2	248.8	352
	5	0.08	0.04	1.42	1.70	191.3	219.2	246.4	350

the δ-minimax and $(1, 1, 1, 1)$-admissible designs would be well approximated by the original one-parameter power-family, as Δ_f and Δ_e are very close in value.

6.7.3 Tutorial on Use of OptGS

OptGS contains a single function `optgs()`. The arguments taken by `optgs` are documented in the help file. The default arguments will produce a two-stage design with $\delta_0 = 0, \delta_1 = 1, \sigma = 3, \alpha = 0.05, 1 - \beta = 0.9$, and $(\omega_1, \omega_2, \omega_3, \omega_4) = (0.95, 0, 0, 0.05)$. The entries of ω imply that the design of interest is the admissible design that puts 0.95 weight on the expected sample size at the δ_0, and 0.05 weight on the maximum sample size. The output is as follows:

```
> optgs()

Groupsize:   84
Futility boundaries   0.5781 1.5776
Efficacy boundaries   2.9559 1.5776
ESS at null:        107.522
ESS at CRD:         145.325
Maximum ESS:        148.302
Max sample-size: 168
```

The output shows the required group-size (i.e. patients to be recruited per arm per stage); the futility and efficacy boundaries; and the operating characteristics of the design. Note that the expected sample sizes and maximum sample size are per arm. If the user wanted a design with three stages, then they could change the `J` argument:

```
> optgs(J=3)

Groupsize:   60
Futility boundaries   0.1388 0.9458 1.5551
Efficacy boundaries   3.9195 2.1874 1.5551
ESS at null:        94.935
ESS at CRD:         132.496
Maximum ESS:        137.018
Max sample-size: 180
```

Note that the futility and efficacy boundaries now have three entries. The expected sample sizes have all fallen, and the maximum sample size has risen, as one would expect. The above designs put weight on the expected sample size at the null, so will tend to have high expected sample sizes at the CRD, and also high maximum sample sizes. If the user wanted to put some of the weight on the expected sample size at the CRD, they could change the `weights` argument as follows:

```
> optgs(J=3,weights=c(0.5,0.45,0,0.05))

Groupsize:   62
Futility boundaries   -0.0062 1.0382 1.77
Efficacy boundaries   2.2247 1.9258 1.77
ESS at null:        98.945
ESS at CRD:         110.062
Maximum ESS:        126.107
Max sample-size: 186
```

Note that the resulting design has a somewhat higher expected sample size at the null, but considerably reduced expected sample size at the CRD (and also a reduced maximum expected sample size and an increased maximum sample size despite the respective weights not having changed).

As discussed in Sect. 6.6, in practice the assumption of known variance is not reasonable. OptGS uses the quantile-substitution method to convert the known-variance stopping boundaries to unknown-variance stopping boundaries. Setting the `sd.known` 0 argument to F will return unknown-variance stopping boundaries:

```
> optgs(J=3,weights=c(0.5,0.45,0,0.05),sd.known=F)

Groupsize:   62
Futility boundaries   -0.0062 1.0404 1.7749
Efficacy boundaries   2.2522 1.9351 1.7749
```

```
ESS at null:      98.945
ESS at CRD:       110.062
Maximum ESS:      126.107
Max sample-size: 186
```

Notice that in this case the stopping boundaries do not differ considerably to previously. This is because the group-size is fairly large. If the group-size was smaller, there would be a more noticeable difference between the two.

6.8 Multi-arm Multi-stage Clinical Trials

In this section, we briefly discuss recent work that extends group-sequential design methodology to allow testing of multiple experimental treatments against a control treatment. If more than one experimental treatment is available for testing, then testing all within a multi-arm trial is more efficient than separate randomised trials of each. That is because only one control group is needed instead of one control group per treatment. Applying group-sequential methodology to a multi-arm trial gives a multi-arm multi-stage (MAMS) clinical trial. At each interim analysis, treatments may be dropped for futility, or the whole trial may be stopped if an effective treatment is found.

6.8.1 *Notation*

Consider a MAMS trial with J stages and K experimental treatments and one control treatment. At each stage n patients are allocated to each remaining treatment. The treatment response of patient i on treatment k ($k = 0$ represents the control group), X_{ik}, is assumed to be distributed as $N(\mu_k, \sigma_k^2)$. The parameters of interest are $(\delta^{(1)}, \ldots, \delta^{(K)})$, where $\delta^{(k)} = \mu_k - \mu_0$. There are K null hypotheses being tested in the trial; the kth is $H_0^{(k)} : \delta^{(k)} \leq 0$.

At a given interim analysis j, the z-statistic for testing $H_0^{(k)}$, $Z_j^{(k)}$, is calculated:

$$Z_j^{(k)} = \sqrt{\frac{jn}{\sigma_k^2 + \sigma_0^2}} \frac{\sum_{i=1}^{jn} X_{ik} - \sum_{i=1}^{jn} X_{i0}}{jn}. \tag{6.8}$$

If $Z_j^{(k)} \leq f_j$, arm k is dropped for futility. If $Z_j^{(k)} > e_j$, then the trial stops for efficacy, and $H_0^{(k)}$ is rejected.

6.8.2 Designing a MAMS Trial

As in the group-sequential case, designing a MAMS trial involves choosing the group-size, futility boundaries and efficacy boundaries so that the type-I error and power are as required. The type-I error is more complicated than previously as there are multiple hypotheses. Magirr et al. [16] explain that it is sufficient to consider the probability of rejecting any null hypothesis when $\delta^{(1)} = \delta^{(2)} = \ldots = \delta^{(K)} = 0$, because this strongly controls the family-wise error rate. In other words, the probability of rejecting any true null hypothesis is maximised when $\delta^{(1)} = \delta^{(2)} = \ldots = \delta^{(K)} = 0$. The authors derive an analytic formula for this probability.

The power is also more complicated. Magirr et al. recommend powering the trial at the least favourable configuration (LFC) of Dunnett [6]. This is the probability of rejecting $H_0^{(1)}$ when $\delta^{(1)} = \delta_1$ and $\delta^{(2)} = \delta^{(3)} = \ldots = \delta^{(K)} = \delta_0$. Here, δ_1 is the clinically relevant difference, and δ_0 is the threshold such that if $\delta^{(k)}$ is below δ_0, treatment k is considered uninteresting. A suitable value of δ_0 could be 0, with higher values requiring a larger sample size but making it more likely that the best treatment will be picked.

Magirr et al. show how to apply traditional stopping boundaries to MAMS trials, for example those of Pocock. However, the same ideas of optimal and admissible designs discussed previously can be applied. Wason and Jaki [29] discuss considerations for searching for optimal designs in the case of a MAMS trial.

6.8.3 Future Work for Design of MAMS

MAMS trials are a very broad class of designs, with the ones considered above being relatively straightforward. In practice, MAMS trials have been used when the endpoints considered differ at each interim analysis, such as in the MRC STAMPEDE trial [27]. The methodology for this is described in Royston et al. [23], and consists of powering each individual stage separately. Efficiency could be gained by considering the whole trial at once, as Magirr et al. do, but this becomes difficult when the endpoint differs at each stage. Currently this area is an important priority for research.

6.9 Discussion

There are strong ethical and efficiency arguments for the use of group-sequential designs in practice. They reduce the average number of patients used in a trial, and therefore allow more trials to be run using the same limited population of patients. Statistical research in group-sequential designs has been ongoing since the 1970s, and shows no sign of slowing down. Greater computational power has allowed considerable progress in areas such as searching for optimal group-sequential designs and group-sequential multi-arm multi-stage trial designs. This chapter has provided a summary of some of the recent research on group-sequential designs.

We have just considered normally distributed endpoints with known variance. Although this may at first seem highly restrictive, in fact asymptotically normally distributed test statistics are used for binary and survival endpoints. Thus, with some modification, methods discussed in this chapter can be used for other types of endpoints. The known variance assumption can be overcome with methods discussed in Sect. 6.6.

In practice, analyses may not take place when the planned number of patients have been assessed. Some patients may have dropped out of the trial, or practical considerations may have determined that the interim analysis must be at a certain time. In a time-to-event trial, it is particularly hard to ensure the planned number of events have taken place. As long as the total number of analyses is not varied this does not cause a problem as the stopping boundaries can be modified. Jennison and Turnbull [13] describe a method to adapt stopping boundaries from the one-parameter power family to allow different numbers of patients at each analysis. Additionally, fixed stopping boundaries from an optimal or admissible group-sequential design can be interpolated using an error spending function, as described by Kittelson and Emerson [15]. Both of these approaches control the overall type-I error, but not necessarily the power.

Group-sequential designs are less useful when the endpoint takes a long time to observe, such as in a time-to-event trial. In this case, one cannot pause recruitment until a group of patients have had the effect of treatment fully observed. Although group-sequential designs will not be able to reduce the expected number of patients recruited, they can still be useful in order to determine if a trial should be stopped early. Hampson and Jennison [11] propose group-sequential methods for when treatment responses are delayed. A Bayesian approach could also be used to incorporate early information to improve decision making at interim analyses, as discussed in chapter 5 of Berry et al. [4].

References

1. Anderson, K.: *gsDesign: Group Sequential Design* (2012). URL http://CRAN.R-project.org/package=gsDesign. R package version 2.6-04
2. Armitage, P., McPherson, C.K., Rowe, B.C.: Repeated significance tests on accumulating data. *Journal of the Royal Statistical Society A* **132**, 235–244 (1969)
3. Barber, S., Jennison, C.: Optimal asymmetric one-sided group sequential tests. *Biometrika* **89**, 49–60 (2002)
4. Berry, S.M., Carlin, B.P., Lee, J.J., Muller, P.: *Bayesian adaptive methods for clinical trials.* CRC Press (2010)
5. Cook, R.J., Farewell, V.T.: Guidelines for monitoring efficacy and toxicity responses in clinical trials. *Biometrics* **50**, 1146–1152 (1994)

6. Dunnett, C.W.: A multiple comparison procedure for comparing several treatments with a control. *Journal of the American Statistical Association* **50**, 1096–1121 (1955)
7. Eales, J.D., Jennison, C.: An improved method for deriving optimal one-sided group sequential tests. *Biometrika* **79**, 13–24 (1992)
8. Eales, J.D., Jennison, C.: Optimal two-sided group sequential tests. *Sequential Analysis* **14**, 273–286 (1995)
9. Emerson, S.S., Flemming, T.R.: Symmetric group sequential designs. *Biometrics* **45**, 905–923 (1989)
10. Genz, A., Bretz, F.: Methods for the computation of multivariate t-probabilities. *Journal of Computational and Graphical Statistics* **11**, 950–971 (2002)
11. Hampson, L.V., Jennison, C.: Group sequential tests for delayed responses. *Journal of the Royal Statistical Society B* **75**, 1–37 (2013)
12. Jennison, C., Turnbull, B.W.: Exact calculations for sequential t,χ^2 and f tests. *Biometrika* **78**, 133–141 (1991)
13. Jennison, C., Turnbull, B.W.: *Group sequential methods with applications to clinical trials.* Chapman and Hall (2000)
14. Jung, S.H., Lee, T., Kim, K., George, S.L.: Admissible two-stage designs for phase II cancer clinical trials. *Statistics in Medicine* **23**, 561–569 (2004)
15. Kittelson, J.M., Emerson, S.: A unifying family of group sequential test designs. *Biometrics* **55**, 874–882 (1999)
16. Magirr, D., Jaki, T., Whitehead, J.: A generalized Dunnett test for multiarm-multistage clincial studies with treatment selection. *Accepted by Biometrika* **99**, 494–501 (2012)
17. Mander, A.P., Wason, J.M.S., Sweeting, M.J., Thompson, S.G.: Admissible two-stage designs for phase ii cancer clinical trials that incorporate the expected sample size under the alternative hypothesis. *Pharmaceutical Statistics (in press)* **11**, 91–96 (2012)
18. McPherson, K., Armitage, P.: Repeated significance tests on accumulating data when the null hypothesis is not true. *Journal of the Royal Statistical Society A* **134**, 15–25 (1971)
19. Nelder, J.A., Mead, R.: A simplex method for function minimization. *The Computer Journal* **7**, 308–313 (1965)
20. O'Brien, P.C., Flemming, T.R.: A multiple-testing procedure for clinical trials. *Biometrics* **35**, 549–556 (1979)
21. Pampallona, S., Tsiatis, A.A.: Group sequential designs for one-sided and two-sided hypothesis testing with provision for early stopping in favor of the null hypothesis. *J. Statist. Planning and Inference* **42**, 19–35 (1994)
22. Pocock, S.J.: Group sequential methods in the design and analysis of clinical trials. *Biometrika* **64**, 191–199 (1977)
23. Royston, P., Parmar, M.K.B., Qian, W.: Novel designs for multi-arm clinical trials with survival outcomes with an application in ovarian cancer. *Statistics in Medicine* **22**, 2239–2256 (2003)
24. Shao, J., Feng, H.: Group sequential t-tests for clinical trials with small sample sizes across stages. *Contemporary Clinical Trials* **28**, 563–571 (2007)
25. Shuster, J.: Optimal two-stage designs for single-arm phase II cancer trials. *Journal of Biopharmaceutical Statistics* **22**, 39–51 (2002)
26. Simon, R.: Optimal two-stage designs for phase II clinical trials. *Controlled Clinical Trials* **10**, 1–10 (1989)
27. Sydes, M.R., Parmar, M.K.B., James, N., et al.: Issues in applying multi-arm multi-stage methodology to a clinical trial in prostate cancer: the MRC STAMPEDE trial. *Trials* **10** (2009)
28. Wason, J.M.S.: OptGS: An R package for finding near-optimal group-sequential designs. *Journal of Statistical Software* **Accepted** (2013)
29. Wason, J.M.S., Jaki, T.: Optimal design of multi-arm multi-stage trials. *Statistics in Medicine* **31**, 4269–4279 (2012)
30. Wason, J.M.S., Mander, A.P.: Minimising the maximum expected sample size in two-stage phase II clinical trials with continuous outcomes. *Journal of Biopharmaceutical Statistics, in press* **22**, 836–852 (2012)

31. Wason, J.M.S., Mander, A.P., Thompson, S.: Optimal multi-stage designs for randomised clinical trials with continuous outcomes. *Statistics in Medicine* **31**, 301–312 (2012)
32. Wassmer, G., Vandemeulebroecke, M.: A brief review on software developments for group sequential and adaptive designs. *Biometrical Journal* **48**, 732–737 (2006)
33. Whitehead, J., Stratton, I.: Group sequential clinical trials with triangular continuation regions. *Biometrics* **39**, 227–236 (1983)
34. Whitehead, J., Valdes-Marquez, E., Lissmats, A.: A simple two-stage design for quantitative responses with application to a study in diabetic neuropathic pain. *Pharmaceutical statistics* **8**, 125–135 (2009)

Chapter 7
Statistical Inference for Non-inferiority of a Diagnostic Procedure Compared to an Alternative Procedure, Based on the Difference in Correlated Proportions from Multiple Raters

Hiroyuki Saeki and Toshiro Tango

Abstract In a clinical trial of diagnostic procedures to indicate non-inferiority, the efficacy is generally evaluated on the basis of the results from multiple raters who interpret and report their findings independently. Although we can handle the multiple results from the multiple raters as if there were a single rater by considering consensus evaluations or majority votes, this handling is not recommended for the primary evaluation. Therefore, all results from the multiple independent raters should be used in the analysis. This chapter addresses a non-inferiority test, confidence interval and sample size formula, for inference of the difference in correlated proportions between the two diagnostic procedures based on the multiple raters. Moreover, we illustrate the methods with data from studies of diagnostic procedures for the diagnosis of oesophageal carcinoma infiltrating the tracheobronchial tree and for the diagnosis of aneurysm in patients with acute subarachnoid hemorrhage.

7.1 Introduction

In situations where an accepted standard diagnostic procedure exists, it is possible to plan a clinical trial to confirm that a new diagnostic procedure is superior to the standard diagnostic procedure. However, if it will be expected that the efficacy of the new diagnostic procedure is not lower than that of the standard diagnostic procedure and the new diagnostic procedure is less or non-invasive, less or non-toxic, inexpensive or easy to operate in comparison with the standard

H. Saeki (✉)
FUJIFILM RI Pharma Co. LTD., Chuo-ku, Tokyo, Japan
e-mail: sahiroyuki@ffri.co.jp

T. Tango
Center for Medical Statistics, Minato-ku, Tokyo, Japan
e-mail: tango@medstat.jp

K. van Montfort et al. (eds.), *Developments in Statistical Evaluation of Clinical Trials*, DOI 10.1007/978-3-642-55345-5_7

procedure, we can plan a non-inferiority study. A non-inferiority study of two diagnostic procedures is designed to indicate that the sensitivity or specificity of the new diagnostic procedure is no more than 100Δ percent inferior compared with the sensitivity or specificity of the standard procedure, respectively, where $\Delta(0 < \Delta \leq 1)$ is a pre-specified acceptable difference between the two proportions. In general, sensitivity is defined as the probability that a result of a diagnostic procedure is positive when the subject has the disease, and specificity is defined as the probability that a result of a diagnostic procedure is negative when the subject does not have the disease. These two measures are very important to evaluate the performance of the diagnostic procedure. However, these measures are calculated on the basis of different populations of subjects. Therefore, we consider the statistical inference for the difference in sensitivities in this chapter. However, the same methods can be applied to examine the difference in the specificities using a different study population.

If two diagnostic procedures are performed on each subject, the difference in proportions for matched-pair data has a correlation between the two diagnostic procedures. Nam [10] and Tango [17] derived the same non-inferiority test for the difference in proportions for matched-pair categorical data based on the efficient score in which the pairs were independent. Tango [17] also derived the confidence interval based on the efficient score. However, these methods are only applicable to the case where the results of the two diagnostic procedures are evaluated by a single rater. Multiple independent raters often evaluate the diagnoses obtained from these diagnostic procedures (see, e.g., [6]). If multiple raters are involved in the evaluation, the differences in proportions for matched-pair data also have correlations between different raters. Although we can apply the aforementioned methods by considering consensus evaluations or majority votes to handle multiple results from the multiple raters as if there were a single rater, these methods are not recommended for the primary evaluation [1, 2, 12]. The consensus evaluations may produce a bias caused by non-independent evaluations. For example, senior or persuasive raters may affect the evaluations of junior or passive raters. Moreover, the majority votes cannot take into account the variability in results of the multiple raters. Therefore, all results from the multiple independent raters should be used in the analysis.

In this chapter, we introduce a non-inferiority test, confidence interval and sample size formula proposed by Saeki and Tango [14], for inference of the difference in correlated proportions between two diagnostic procedures on the basis of the results from the multiple independent raters where the matched pairs are independent. Furthermore, we consider a possible procedure based on majority votes and we conduct Monte Carlo simulation studies to examine the validity of the proposed methods in comparison with the procedure based on majority votes. Finally, we illustrate the methods with data from studies of diagnostic procedures for the diagnosis of oesophageal carcinoma infiltrating the tracheobronchial tree [13] and for the diagnosis of aneurysm in patients with acute subarachnoid hemorrhage [4].

7.2 Design

7.2.1 Data Structure and Model

Consider a clinical experimental design where a new diagnostic procedure (or treatment) and a standard diagnostic procedure (or treatment) that are independently performed on the same subject (or matched pairs of subjects) and independently evaluated by K raters are compared. Each rater's judgment is assumed to take on one of two values: 1 represents that the subject is diagnosed as 'positive', and 0 indicates that the subject is diagnosed as 'negative'. Suppose we have n subjects. If we consider only subjects with a pre-specified disease, we use a positive probability as a measure, that is, sensitivity. On the other hand, if we consider subjects without the disease, we use a negative probability as a measure, that is, specificity. In the following, we consider a situation on the basis of sensitivity.

For ease of explanation, let us consider the case of $K = 2$ first. The resulting types of matched observations and probabilities are naturally classified as a 4×4 contingency table shown in Table 7.1, where $+(1)$ or $-(0)$ denotes a positive or negative judgment on a procedure, respectively. For example, y_{1101} denotes the observed number of matched type $\{+$ on the new procedure by rater 1, $+$ on the new procedure by rater 2, $-$ on the standard procedure by rater 1, $+$ on the standard procedure by rater 2$\}$ and r_{1101} indicates its probability.

Let $\pi_N^{(k)}$ $(\pi_S^{(k)})$ denote the probability that rater k judges as positive on the new (standard) diagnostic procedure of a randomly selected subject. Then, it will be naturally calculated as

$$\pi_N^{(1)} = r_{11\cdot\cdot} + r_{10\cdot\cdot}, \quad \pi_N^{(2)} = r_{11\cdot\cdot} + r_{01\cdot\cdot} \tag{7.1}$$

Table 7.1 A 4×4 contingency table for matched-pair categorical data in the case of two raters

		Standard procedure				
	Judgment of (Rater 1, Rater 2)	(+,+)	(+,−)	(−,+)	(−,−)	Total
New procedure	(+,+)	r_{1111}	r_{1110}	r_{1101}	r_{1100}	$r_{11..}$
		(y_{1111})	(y_{1110})	(y_{1101})	(y_{1100})	$(y_{11..})$
	(+, −)	r_{1011}	r_{1010}	r_{1001}	r_{1000}	$r_{10..}$
		(y_{1011})	(y_{1010})	(y_{1001})	(y_{1000})	$(y_{10..})$
	(−, +)	r_{0111}	r_{0110}	r_{0101}	r_{0100}	$r_{01..}$
		(y_{0111})	(y_{0110})	(y_{0101})	(y_{0100})	$(y_{01..})$
	(−, −)	r_{0011}	r_{0010}	r_{0001}	r_{0000}	$r_{00..}$
		(y_{0011})	(y_{0010})	(y_{0001})	(y_{0000})	$(y_{00..})$
	Total	$r_{..11}$	$r_{..10}$	$r_{..01}$	$r_{..00}$	1
		$(y_{..11})$	$(y_{..10})$	$(y_{..01})$	$(y_{..00})$	(n)

and $\pi_S^{(1)}$ and $\pi_S^{(2)}$ are defined in a similar manner. Let π_N and π_S denote the probability of a positive judgment on the new and standard diagnostic procedures, respectively. Then, these probabilities can, in general, be defined as follows:

$$\pi_N = \omega^{(1)}\pi_N^{(1)} + \omega^{(2)}\pi_N^{(2)} , \tag{7.2}$$

$$\pi_S = \omega^{(1)}\pi_S^{(1)} + \omega^{(2)}\pi_S^{(2)} , \tag{7.3}$$

where $\omega^{(k)}$ $(\omega^{(1)} + \omega^{(2)} = 1)$ denotes the weight for rater k, showing the difference in the raters' evaluation skill. However, raters are usually selected among the raters with *at least equivalent skill*, and it is assumed in this paper that

$$\omega^{(k)} = 1/K \quad (k = 1, \ldots, K) . \tag{7.4}$$

Therefore, these probabilities can be defined as follows:

$$\pi_N = \frac{\pi_N^{(1)} + \pi_N^{(2)}}{2} = r_{11\cdot\cdot} + \frac{r_{10\cdot\cdot} + r_{01\cdot\cdot}}{2} , \tag{7.5}$$

$$\pi_S = \frac{\pi_S^{(1)} + \pi_S^{(2)}}{2} = r_{\cdot\cdot 11} + \frac{r_{\cdot\cdot 10} + r_{\cdot\cdot 01}}{2} . \tag{7.6}$$

On the basis of the form of the expressions of (7.5) and (7.6), the 4×4 contingency table is found to be reduced to the 3×3 contingency table shown in Table 7.2, where $p_{\ell m}$ $(x_{\ell m})$ denotes the probability (observed number of observations) that ℓ raters judge as positive on the new procedure and m raters judge as positive on the standard procedure. Then, we have

$$\begin{aligned}\pi_N &= p_{2\cdot} + \frac{1}{2}p_{1\cdot} \\ &= p_{20} + (p_{21} + \frac{1}{2}p_{10}) + (p_{22} + \frac{1}{2}p_{11}) + \frac{1}{2}p_{12} ,\end{aligned} \tag{7.7}$$

Table 7.2 A 3×3 contingency table for matched-pair categorical data in the case of two raters

		Standard procedure			
	Judgment of (Rater 1, Rater 2)	(+, +)	(+,−) or (−,+)	(−,−)	Total
New procedure	(+, +)	p_{22}	p_{21}	p_{20}	$p_{2.}$
		(x_{22})	(x_{21})	(x_{20})	$(x_{2.})$
	(+, −) or (−, +)	p_{12}	p_{11}	p_{10}	$p_{1.}$
		(x_{12})	(x_{11})	(x_{10})	$(x_{1.})$
	(−, −)	p_{02}	p_{01}	p_{00}	$p_{0.}$
		(x_{02})	(x_{01})	(x_{00})	$(x_{0.})$
	Total	$p_{.2}$	$p_{.1}$	$p_{.0}$	1
		$(x_{.2})$	$(x_{.1})$	$(x_{.0})$	(n)

$$\begin{aligned}\pi_S &= p_{\cdot 2} + \frac{1}{2}p_{\cdot 1} \\ &= p_{02} + (p_{12} + \frac{1}{2}p_{01}) + (p_{22} + \frac{1}{2}p_{11}) + \frac{1}{2}p_{21} \ .\end{aligned} \tag{7.8}$$

Let λ denote the difference in positive probabilities; that is,

$$\begin{aligned}\lambda &= \pi_N - \pi_S \\ &= p_{20} + \frac{1}{2}(p_{21} + p_{10}) - p_{02} - \frac{1}{2}(p_{12} + p_{01}) \ ,\end{aligned} \tag{7.9}$$

and its sample estimate will be

$$\tilde{\lambda} = \frac{1}{n}\left\{x_{20} + \frac{1}{2}(x_{21} + x_{10}) - x_{02} - \frac{1}{2}(x_{12} + x_{01})\right\} \ , \tag{7.10}$$

which clearly shows that the inference on λ can be made by the observed vector $\boldsymbol{x} = (x_{20}, x_{21} + x_{10}, x_{02}, x_{12} + x_{01}, x_{22} + x_{11} + x_{00})$ following a multinomial distribution with parameters n and $\boldsymbol{p} = (p_{20}, p_{21} + p_{10}, p_{02}, p_{12} + p_{01}, p_{22} + p_{11} + p_{00})$.

It should be noted that x_{20} is the frequency such that the number of raters judging as positive on the new procedure is larger than the number of raters judging as positive on the standard procedure by 2 and that $(x_{21} + x_{10})$ is the frequency such that the number of raters judging as positive on the new procedure is larger than the number of raters judging as positive on the standard procedure by 1. Similarly, x_{02} is the frequency such that the number of raters judging as positive on the standard procedure is larger than the number of raters judging as positive on the new procedure by 2 and $(x_{12} + x_{01})$ is the frequency such that the number of raters judging as positive on the standard procedure is larger than the number of raters judging as positive on the new procedure by 1. These observations lead to a generalization to K raters. The resulting types of matched observations and probabilities are classified as a $(K + 1) \times (K + 1)$ contingency table similar to Table 7.2. However, the method is reduced to the following. Let n_{Nk} denote the frequency such that the number of raters who judge as positive on the new procedure is larger than the number of raters who judge as positive on the standard procedure by k and let q_{Nk} indicate such probability. Namely, we have

$$n_{Nk} = \sum_{\ell - m = k} x_{\ell m} \ ,$$

$$q_{Nk} = \sum_{\ell - m = k} p_{\ell m} \ ,$$

where ℓ is the number of raters who judge as positive on the new procedure, and m is the number of raters who judge as positive on the standard procedure. Similarly, let n_{Sk} denote the frequency such that the number of raters who judge as positive on the standard procedure is larger than the number of raters who judge as positive on the new procedure by k and let q_{Sk} indicate such probability. Then, we have

$$n_{Sk} = \sum_{\ell - m = -k} x_{\ell m} ,$$

$$q_{Sk} = \sum_{\ell - m = -k} p_{\ell m} ,$$

and $q_{N0} = q_{S0}$ and $n_{N0} = n_{S0}$. Namely, for K raters, the inference on λ can be made by the vector of random variables $\boldsymbol{n} = (n_{N0}, n_{N1}, \ldots, n_{NK}, n_{S1}, \ldots, n_{SK})$ following a multinomial distribution with parameters n and $\boldsymbol{q} = (q_{N0}, q_{N1}, \ldots, q_{NK}, q_{S1}, \ldots, q_{SK})$. Then, we have

$$\begin{aligned}
\pi_N &= \sum_{k=1}^{K} \omega^{(k)} \pi_N^{(k)} = \frac{1}{K} \sum_{k=1}^{K} k \sum_{m=0}^{K} p_{km} = \frac{1}{K} \sum_{k=1}^{K} k p_{k\cdot} \\
&= \frac{1}{K} \sum_{k=1}^{K} k q_{Nk} + \frac{1}{K} \sum_{k=1}^{K} k p_{kk} + \frac{1}{K} \sum_{\substack{\ell, m \in K \\ \ell < m}} \ell p_{\ell m} + \frac{1}{K} \sum_{\substack{\ell, m \in K \\ m < \ell}} m p_{\ell m} , \\
\pi_S &= \sum_{k=1}^{K} \omega^{(k)} \pi_S^{(k)} = \frac{1}{K} \sum_{k=1}^{K} k \sum_{\ell=0}^{K} p_{\ell k} = \frac{1}{K} \sum_{k=1}^{K} k p_{\cdot k} \\
&= \frac{1}{K} \sum_{k=1}^{K} k q_{Sk} + \frac{1}{K} \sum_{k=1}^{K} k p_{kk} + \frac{1}{K} \sum_{\substack{\ell, m \in K \\ \ell < m}} \ell p_{\ell m} + \frac{1}{K} \sum_{\substack{\ell, m \in K \\ m < \ell}} m p_{\ell m} . \qquad (7.11)
\end{aligned}$$

Therefore, the difference in positive probabilities (7.9) is generalized to

$$\begin{aligned}
\lambda = \pi_N - \pi_S &= \left(\frac{1}{K} \sum_{k=1}^{K} k p_{k\cdot} \right) - \left(\frac{1}{K} \sum_{k=1}^{K} k p_{\cdot k} \right) \\
&= \frac{1}{K} \sum_{k=1}^{K} k (q_{Nk} - q_{Sk}) . \qquad (7.12)
\end{aligned}$$

Then, the estimate $\tilde{\lambda}$ given in (7.10) is generalized to

$$\tilde{\lambda} = \frac{1}{nK} \sum_{k=1}^{K} k (n_{Nk} - n_{Sk}) . \qquad (7.13)$$

7.2.2 Problems in Consensus Evaluations or Majority Votes

Although we can handle multiple results from the multiple raters as if there were a single rater by considering consensus evaluations or majority votes, these handlings are not recommended for the primary evaluation [1, 2, 12]. The consensus evaluations may produce a bias caused by non-independent evaluation, even if the consensus evaluations are performed after individual evaluations by the multiple raters are completed. For example, senior or persuasive raters may affect the evaluations of junior or passive raters. Moreover, the majority votes cannot take into account the variability in results of the multiple raters. For ease of explanation, let us consider the case of $K = 3$. The resulting types of matched observations are classified as a 4×4 contingency table in Table 7.3. In this case, $\tilde{\lambda}_{K=3}$ can be addressed from (7.13) as

$$\tilde{\lambda}_{K=3} = \frac{1}{n}\left\{(n_{N3} - n_{S3}) + \frac{2}{3}(n_{N2} - n_{S2}) + \frac{1}{3}(n_{N1} - n_{S1})\right\} ,$$

where $(n_{N3} - n_{S3}) = (x_{30} - x_{03})$, $(n_{N2} - n_{S2}) = \{(x_{31} + x_{20}) - (x_{13} + x_{02})\}$ and $(n_{N1} - n_{S1}) = \{(x_{32} + x_{21} + x_{10}) - (x_{23} + x_{12} + x_{01})\}$. If we adopt the majority votes, the 4×4 contingency table shown in Table 7.3 is transformed to the 2×2 contingency table shown in Table 7.4, and the estimate of the difference between π_N and π_S on the basis of the results from the majority votes will be

$$\tilde{\lambda}_{MV} = \frac{(b-c)}{n} = \frac{1}{n}\{(n_{N3} - n_{S3}) + (n_{N2} - n_{S2}) + (x_{21} - x_{12})\} .$$

We should focus on two problems in $\tilde{\lambda}_{MV}$.

Table 7.3 A 4×4 contingency table for matched-pair categorical data in the case of three raters

		Standard procedure			
	Judgment of (Rater 1, Rater 2, Rater 3)	(+, +, +)	(+, +, −) or (+, −, +) or (−, +, +)	(+, −, −) or (−, +, −) or (−, −, +)	(−, −, −)
New procedure	(+, +, +)	x_{33}	x_{32}	x_{31}	x_{30}
	(+, +, −) or (+, −, +) or (−, +, +)	x_{23}	x_{22}	x_{21}	x_{20}
	(+, −, −) or (−, +, −) or (−, −, +)	x_{13}	x_{12}	x_{11}	x_{10}
	(−, −, −)	x_{03}	x_{02}	x_{01}	x_{00}

Table 7.4 A 2×2 contingency table transformed from Table 7.3 by majority votes

		Standard procedure	
	Judgment	(+)	(−)
New procedure	(+)	a $(= x_{33} + x_{32} + x_{23} + x_{22})$	b $(= x_{30} + x_{31} + x_{20} + x_{21})$ $(= n_{N3} + n_{N2} + x_{21})$
	(−)	c $(= x_{03} + x_{13} + x_{02} + x_{12})$ $(= n_{S3} + n_{S2} + x_{12})$	d $(= x_{11} + x_{10} + x_{01} + x_{00})$

1. $\tilde{\lambda}_{MV}$ involves $(n_{N2} - n_{S2})$ and $(x_{21} - x_{12})$ without the weights of the contribution for π_N and π_S from $\pi_N^{(1)}, \pi_N^{(2)}, \pi_N^{(3)}$ and $\pi_S^{(1)}, \pi_S^{(2)}, \pi_S^{(3)}$.
2. x_{32}, x_{10} and x_{23}, x_{01} do not take part in $\tilde{\lambda}_{MV}$, because these values are involved in the cells 'a' and 'd' in Table 7.4.

Therefore, it is important that all results from the multiple independent raters are used in the analysis appropriately.

7.3 Methods for Statistical Inference

In this section, we shall introduce methods for statistical inference of the difference λ, that is, a non-inferiority test, confidence interval and formula for determination of sample size.

7.3.1 Non-inferiority Test

The non-inferiority hypothesis will be formulated as

$$H_0 : \pi_N = \pi_S - \Delta, \ H_1 : \pi_N > \pi_S - \Delta \ ,$$

where Δ $(0 < \Delta \leq 1)$ is a pre-specified acceptable difference in two probabilities. Let

$$\delta = \lambda + \Delta = \pi_N - (\pi_S - \Delta) = \frac{1}{K}\sum_{k=1}^{K} kq_{Nk} - \left(\frac{1}{K}\sum_{k=1}^{K} kq_{Sk} - \Delta\right) . \tag{7.14}$$

Then, under the null hypothesis, the log-likelihood function without constant terms is expressed as

$$
\begin{aligned}
L = L(\boldsymbol{\theta}) &= n_{N0}\log(q_{N0}) + n_{NK}\log(q_{NK}) + \sum_{k=1}^{K-1} n_{Nk}\log(q_{Nk}) + \sum_{k=1}^{K} n_{Sk}\log(q_{Sk}) \\
&= n_{N0}\log(1-\delta+\Delta-A-B-C) + n_{NK}\log(\delta-\Delta+A) \\
&\quad + \sum_{k=1}^{K-1} n_{Nk}\log(q_{Nk}) + \sum_{k=1}^{K} n_{Sk}\log(q_{Sk}) , \qquad (7.15)
\end{aligned}
$$

where $\boldsymbol{\theta} = (\delta, q_{N1}, \ldots, q_{N(K-1)}, q_{S1}, \ldots, q_{SK})^T$ is the parameter vector of dimension $2K$ and

$$
A = \frac{1}{K}\left(\sum_{k=1}^{K} k q_{Sk} - \sum_{k=1}^{K-1} k q_{Nk}\right), \quad B = \sum_{k=1}^{K-1} q_{Nk} , \quad C = \sum_{k=1}^{K} q_{Sk} .
$$

Then, the score test for testing the null hypothesis $H_0 : \delta = 0$ against $H_1 : \delta > 0$ is expressed as

$$
Z_S = \left[\frac{\partial L}{\partial \delta}\bigg|_{\delta=0,\, q_{Nk}=\hat{q}_{Nk},\, q_{Sk}=\hat{q}_{Sk}} \right] \sqrt{\left(\hat{I}^{-1}\right)_{11}\Big|_{\delta=0,\, q_{Nk}=\hat{q}_{Nk},\, q_{Sk}=\hat{q}_{Sk}}} \sim_{H_0} N(0,1) , \qquad (7.16)
$$

where $(\hat{q}_{N1}, \ldots, \hat{q}_{N(K-1)}, \hat{q}_{S1}, \ldots, \hat{q}_{SK})$ is the vector of the maximum likelihood estimators under the null hypothesis, which is the unique solution for the following equations:

$$
\frac{\partial L}{\partial q_{Nk}}\bigg|_{\delta=0} = 0, \quad (k = 1, \ldots, K-1) , \qquad (7.17)
$$

$$
\frac{\partial L}{\partial q_{Sk}}\bigg|_{\delta=0} = 0, \quad (k = 1, \ldots, K) . \qquad (7.18)
$$

These equations can be obtained iteratively using the quasi-Newton method with constraints. The R function 'constrOptim' is useful for the quasi-Newton method with constraints. Further, $(\hat{I}^{-1})_{11}$ indicates the $(1,1)$th element of the $(2K \times 2K)$ inverse Fisher information matrix evaluated at the maximum likelihood estimators. On the other hand, we can consider a test based on the sample estimate T for the difference δ

$$
T = \tilde{\lambda} + \Delta = \frac{1}{nK}\sum_{k=1}^{K} k(n_{Nk} - n_{Sk}) + \Delta . \qquad (7.19)
$$

The variance of T evaluated at the null hypothesis $\delta = 0$ is

$$
\mathrm{Var}_{H_0}(T) = \frac{1}{n}\left[\frac{1}{K^2}\sum_{k=1}^{K} k^2(q_{Nk} + q_{Sk}) - \Delta^2\right] .
$$

Therefore, the normal deviate for testing $H_0 : \delta = 0$ against $H_1 : \delta > 0$ is expressed as

$$Z_{ND} = \frac{\frac{1}{nK}\sum_{k=1}^{K} k(n_{Nk} - n_{Sk}) + \Delta}{\sqrt{\frac{1}{n}\left[\frac{1}{K^2}\sum_{k=1}^{K} k^2(\hat{q}_{Nk} + \hat{q}_{Sk}) - \Delta^2\right]}} \sim_{H_0} N(0,1) . \tag{7.20}$$

It can be shown that when $K = 1$, the normal deviate test statistic, Z_{ND}, is equivalent to the score test statistic Z_S [10, 17]. When $K = 2$ or 3, we confirmed that Z_S and Z_{ND} were approximately equal using the example data (see Sect. 7.5). However, we have not been able to show the equivalence between Z_S and Z_{ND} analytically. On the other hand, by using the observed proportions $\tilde{q}_{Nk} = n_{Nk}/n$, $\tilde{q}_{Sk} = n_{Sk}/n$ instead of the maximum likelihood estimators, we can construct a Wald-type test statistic for testing $H_0 : \delta = 0$ against $H_1 : \delta > 0$:

$$Z_W = \frac{\frac{1}{nK}\sum_{k=1}^{K} k(n_{Nk} - n_{Sk}) + \Delta}{\sqrt{\frac{1}{n}\left[\frac{1}{nK^2}\sum_{k=1}^{K} k^2(n_{Nk} + n_{Sk}) - \Delta^2\right]}} \sim_{H_0} N(0,1) . \tag{7.21}$$

When $\Delta = 0$, the Wald-type test Z_W is identical to Schouten's [15] generalized McNemar test although Schouten's test statistic is presented in a different form. When $K = 1$, the Wald-type test Z_W is identical to the unconditional test for non-inferiority of Lu and Bean [7]. When $\Delta = 0$ and $K = 1$, both the normal deviate test Z_{ND} and the Wald-type test Z_W are identical to the McNemar test [9].

7.3.2 *Confidence Interval*

Testing non-inferiority with an acceptable difference Δ at a one-sided significance level $\alpha/2$ is equivalent to judging whether the lower limit of the $1 - \alpha$ level confidence interval is greater than $-\Delta$. The score-type approximate confidence limits for the difference in two proportions, λ, are the two solutions to the equation

$$\frac{\frac{1}{nK}\sum_{k=1}^{K} k(n_{Nk} - n_{Sk}) - \lambda}{\sqrt{\frac{1}{n}\left[\frac{1}{K^2}\sum_{k=1}^{K} k^2(\hat{q}_{Nk} + \hat{q}_{Sk}) - \lambda^2\right]}} = \pm Z_{\alpha/2} , \tag{7.22}$$

where the plus and minus signs indicate the lower limit λ_{low} and the upper limit λ_{up}, respectively, and $Z_{\alpha/2}$ is the upper $\alpha/2$ percentile of the standard normal distribution. These two limits can be found using an iterative numerical method such as the secant method (see, e.g., [17]). On the other hand, we can easily derive the Wald-type confidence interval:

$$\mathrm{CI}_W : \frac{1}{nK}\left(\sum_{k=1}^{K} k(n_{Nk} - n_{Sk}) \pm Z_{\alpha/2}\sqrt{\sum_{k=1}^{K} k^2(n_{Nk} + n_{Sk})}\right) . \tag{7.23}$$

Equation (7.23) utilizes the variance evaluated under the null hypothesis and is identical to Schouten's [15] Wald-type confidence interval.

7.3.3 Sample Size

To calculate the sample size required for testing the null hypothesis $H_0 : \delta = 0$ against the alternative hypothesis $H_1 : \delta > 0$, we only have to consider the following properties of the statistic T:

$$E_{H_0}(T) = 0 ,$$

$$E_{H_1}(T) = \lambda + \Delta ,$$

$$S = \lim_{n\to\infty} n\mathrm{Var}_{H_1}(T) = \left[\frac{1}{K^2}\sum_{k=1}^{K} k^2(q_{Nk} + q_{Sk}) - \lambda^2\right] .$$

On the other hand, we have

$$R = \lim_{n\to\infty} n\mathrm{Var}_{H_0}(T) = \left[\frac{1}{K^2}\sum_{k=1}^{K} k^2(\bar{q}_{Nk} + \bar{q}_{Sk}) - \Delta^2\right] ,$$

where $(\bar{q}_{Nk}, \bar{q}_{Sk})$, $k = 0, \ldots, K$, are the asymptotic values of the maximum likelihood estimators $(\hat{q}_{Nk}, \hat{q}_{Sk})$, $k = 0, \ldots, K$. These asymptotic values are solutions to (7.17) and (7.18). From the aforementioned equations, the approximate sample size n required for $100(1-\beta)$ power of a one-sided normal deviate test at $\alpha/2$ level is given by

$$n = \left(\frac{Z_{\alpha/2}\sqrt{R} + Z_{\beta}\sqrt{S}}{\lambda + \Delta}\right)^2 . \tag{7.24}$$

When $K = 1$, the derived formula for determining the sample size agrees with that proposed by Nam [10]. The sample sizes required for 80 % power of a one-sided non-inferiority test at $\alpha/2 = 2.5\,\%$ for $K = 2, 3$, $\Delta = 0.1, 0.05$, and various values of $(q_{N3}, q_{N2}, q_{N1}, q_{S3}, q_{S2}, q_{S1})$ with $\pi_N - \pi_S = \lambda = 0$ are shown in Table 7.5.

Table 7.5 Sample sizes calculated by formula (7.24) for nominal power = 80 % of a non-inferiority test at $\alpha/2 = 2.5\,\%$ for $K = 2, 3$, $\Delta = 0.1, 0.05$, $\pi_N - \pi_S = \lambda = 0$, $q_{N3} = q_{S3}$, $q_{N2} = q_{S2}, q_{N1} = q_{S1}$

K	Δ	$q_{N3} = q_{S3}$	$q_{N2} = q_{S2}$	$q_{N1} = q_{S1}$	Sample size	
2	0.1	–	0.05	0.05	117	(81.7)
		–	0.05	0.1	132	(81.9)
		–	0.1	0.05	187	(80.7)
		–	0.1	0.1	204	(80.7)
	0.05	–	0.05	0.05	417	(80.6)
		–	0.05	0.1	487	(81.0)
		–	0.1	0.05	718	(80.8)
		–	0.1	0.1	793	(80.2)
3	0.1	0.05	0.02	0.05	120	(81.5)
		0.05	0.02	0.1	126	(81.5)
		0.05	0.05	0.1	142	(80.8)
		0.1	0.02	0.05	190	(80.4)
		0.1	0.02	0.1	197	(80.3)
		0.1	0.05	0.1	215	(79.8)
	0.05	0.05	0.02	0.05	428	(80.2)
		0.05	0.02	0.1	459	(80.0)
		0.05	0.05	0.1	536	(80.2)
		0.1	0.02	0.05	730	(81.1)
		0.1	0.02	0.1	763	(80.7)
		0.1	0.05	0.1	844	(80.5)

The parenthetical values are empirical power (%) based on 10,000 replicates

7.4 Simulation

We have indicated here the results of simulation studies for the methods at a one-sided 2.5 % level for the case of $K = 3$ and sample size $n = 25$, 50 or 100 with 10,000 replicates. Simulation data were generated on the basis of a multinomial distribution by considering typical situations for parameter values $(q_{N3}, q_{N2}, q_{N1}, q_{S3}, q_{S2}, q_{S1})$ and non-inferiority margin $\Delta = 0.1$. In assessing the performance of the methods based on the majority votes, we transformed the simulation data based on the following definitions: $q_N = (q_{N3} + q_{N2} + \frac{1}{3} \times q_{N1})$, $q_S = (q_{S3} + q_{S2} + \frac{1}{3} \times q_{S1})$.

7.4.1 Non-inferiority Test

We performed Monte Carlo simulation studies to assess the empirical size and power of the normal deviate test statistic Z_{ND}, the Wald-type test statistic Z_W and the test

Table 7.6 Empirical sizes of the normal deviate test Z_{ND}, the Wald-type test Z_W and the test based on majority votes Z_{MV} at $\alpha/2 = 2.5\,\%$ for $K = 3$, $\pi_N - \pi_S = \lambda = -0.1$, $\Delta = 0.1$ based on 10,000 replicates

							Size (%)		
n	q_{N3}	q_{N2}	q_{N1}	q_{S3}	q_{S2}	q_{S1}	Z_{ND}	Z_W	Z_{MV}
100	0.01	0.02	0.05	0.11	0.02	0.05	2.2	4.6	1.7
	0.01	0.02	0.1	0.11	0.02	0.1	2.3	4.3	1.3
	0.01	0.05	0.1	0.11	0.05	0.1	2.2	3.7	1.6
50	0.01	0.02	0.05	0.11	0.02	0.05	2.0	5.9	1.5
	0.01	0.02	0.1	0.11	0.02	0.1	2.2	5.5	1.3
	0.01	0.05	0.1	0.11	0.05	0.1	2.2	4.6	1.4
25	0.01	0.02	0.05	0.11	0.02	0.05	1.6	8.0	1.1
	0.01	0.02	0.1	0.11	0.02	0.1	1.9	7.3	0.9
	0.01	0.05	0.1	0.11	0.05	0.1	2.4	5.9	1.2

Table 7.7 Empirical powers of the normal deviate test Z_{ND}, the Wald-type test Z_W and the test based on majority votes Z_{MV} at $\alpha/2 = 2.5\,\%$ for $K = 3$, $\pi_N - \pi_S = \lambda = 0$, $\Delta = 0.1$ based on 10,000 replicates

							Power (%)		
n	q_{N3}	q_{N2}	q_{N1}	q_{S3}	q_{S2}	q_{S1}	Z_{ND}	Z_W	Z_{MV}
100	0.01	0.02	0.05	0.01	0.02	0.05	97.2	99.3	85.8
	0.01	0.02	0.1	0.01	0.02	0.1	95.7	98.4	78.6
	0.01	0.05	0.1	0.01	0.05	0.1	89.5	93.2	62.2
50	0.01	0.02	0.05	0.01	0.02	0.05	70.6	89.6	45.8
	0.01	0.02	0.1	0.01	0.02	0.1	68.4	85.2	38.1
	0.01	0.05	0.1	0.01	0.05	0.1	60.1	72.7	29.7
25	0.01	0.02	0.05	0.01	0.02	0.05	22.7	69.6	15.2
	0.01	0.02	0.1	0.01	0.02	0.1	22.9	65.5	10.0
	0.01	0.05	0.1	0.01	0.05	0.1	25.0	50.6	10.8

statistic based on the majority votes Z_{MV}. Z_{MV} was calculated using the method of Nam [10] and Tango [17]. Table 7.6 presents the empirical sizes. For the set of parameter values $(q_{N3}, q_{N2}, q_{N1}, q_{S3}, q_{S2}, q_{S1})$ considered here, the empirical sizes for the normal deviate test Z_{ND} are generally closer to the nominal $\alpha/2$-level of 2.5 % than those for the Wald-type test Z_W or the test based on the majority votes Z_{MV}. The empirical sizes of Z_W tend to be quite inflated. The empirical sizes of Z_{MV}, on the other hand, tend to be quite reduced. Table 7.7 presents the empirical powers for the alternative hypothesis $H_1 : \pi_N = \pi_S$ for the case of $\Delta = 0.1$. The differences in powers between Z_{ND} and Z_W are generally small. When the sample size is small, however, the empirical powers of Z_W are far greater than those of Z_{ND}. On the other hand, the empirical powers of Z_{MV} are far smaller than those of Z_{ND} under all situations.

Table 7.8 Coverage probabilities of the score-type 95 % confidence interval, the Wald-type 95 % confidence interval and the 95 % confidence interval based on the majority votes for $K = 3$ based on 10,000 replicates generated under the null hypothesis $\pi_N - \pi_S = \lambda = -0.1$

							Coverage prob. (%)		
n	q_{N3}	q_{N2}	q_{N1}	q_{S3}	q_{S2}	q_{S1}	score-type	CI_W	CI_{MV}
100	0.01	0.02	0.05	0.11	0.02	0.05	95.0	94.8	96.4
	0.01	0.02	0.1	0.11	0.02	0.1	94.9	94.9	97.3
	0.01	0.05	0.1	0.11	0.05	0.1	94.7	95.2	96.7
50	0.01	0.02	0.05	0.11	0.02	0.05	94.7	94.2	96.7
	0.01	0.02	0.1	0.11	0.02	0.1	94.7	94.4	97.7
	0.01	0.05	0.1	0.11	0.05	0.1	95.0	95.1	97.1
25	0.01	0.02	0.05	0.11	0.02	0.05	95.3	93.7	97.7
	0.01	0.02	0.1	0.11	0.02	0.1	95.4	93.9	98.4
	0.01	0.05	0.1	0.11	0.05	0.1	95.9	94.6	97.9

7.4.2 *Confidence Interval*

We performed Monte Carlo simulation studies to evaluate the coverage probability of the score-type confidence interval, the Wald-type confidence interval CI_W and the confidence interval based on the majority votes CI_{MV}. CI_{MV} was calculated using the method of Tango [17]. Table 7.8 shows the empirical coverage probabilities of the score-type 95 % confidence interval, the Wald-type 95 % confidence interval and the 95 % confidence interval based on the majority votes under the hypothesis $\pi_N - \pi_S = \lambda = -0.1$. It shows that the score-type confidence interval and the Wald-type confidence interval both generally perform very well. However, when $n = 25$, the score-type confidence interval outperforms the Wald-type confidence interval. On the other hand, the confidence interval based on the majority votes shows a conservative property.

7.5 Example

7.5.1 *Study of Diagnostic Procedures for the Diagnosis of Oesophageal Carcinoma Infiltrating the Tracheobronchial Tree*

Here, we shall consider the data presented by Rapp-Bernhardt et al. [13]. They compared the sensitivities between axial computed tomography (CT) slices and minimal intensity projection (MIP) in 21 patients with oesophageal carcinoma infiltrating the tracheobronchial tree. The bronchoscopic findings were determined as the gold standard. Three radiologists, working independently of each other and without knowledge of the findings on the gold standard, assessed separately the

Table 7.9 A 4 × 4 contingency table ($K = 3$) of the assessments of MIP and axial CT slices by three radiologists (True positive (TP: +) and false negative (FN: −) by three radiologists (1, 2, 3): I (+, +, +), II (+, +, − or +, −, + or −, +, +), III (+, −, − or −, +, − or −, −, +), IV (−, −, −)) (Rapp-Bernhardt et al. [13])

		Axial CT slices				
	TP and FN by three radiologists	I	II	III	IV	Total
MIP	I	14	2	1	0	17
	II	0	0	0	0	0
	III	0	0	2	0	2
	IV	0	0	2	0	2
	Total	14	2	5	0	21

CT, computed tomography; FN, false negative;
MIP, minimal intensity projection; TP, true positive

axial CT slices and MIP. In these diagnostic procedures, stenoses were localized, and the degree of stenosis was assessed as in real bronchoscopy. The resulting type of matched observations was classified as a 4 × 4 contingency table for MIP versus axial CT slices and is shown in Table 7.9 (similar to Table 7.3), where '+' indicates a true positive and '−' indicates a false negative based on binary assessment where 0–50 % of total occlusion was considered as negative and 50–100 % of total occlusion was considered as positive. MIP is one of the reconstruction techniques of making three-dimensional images. MIP images make it easier to appreciate the condition of the whole tracheobronchial tree than axial CT slices. Therefore, we are interested in the non-inferiority of MIP to axial CT slices where the non-inferiority margin is set as $\Delta = 0.1$. From Table 7.9, we have $\tilde{p}_{3.} = 17/21$, $\tilde{p}_{2.} = 0/21$, $\tilde{p}_{1.} = 2/21$, $\tilde{p}_{.3} = 14/21$, $\tilde{p}_{.2} = 2/21$ and $\tilde{p}_{.1} = 5/21$. Then, the sensitivities of MIP and axial CT slices are estimated as $\tilde{\pi}_{MIP} = (17 + 2/3 \times 0 + 1/3 \times 2)/21 = 0.841$ and $\tilde{\pi}_{CT} = (14 + 2/3 \times 2 + 1/3 \times 5)/21 = 0.810$, respectively. Moreover, we have $\tilde{q}_{N3} = 0/21$, $\tilde{q}_{N2} = (1 + 0)/21$, $\tilde{q}_{N1} = (2 + 0 + 0)/21$, $\tilde{q}_{S3} = 0/21$, $\tilde{q}_{S2} = (0+0)/21$ and $\tilde{q}_{S1} = (0+0+2)/21$. Then, the difference in the sensitivities between MIP and axial CT slices based on the three raters is $\tilde{\lambda}_{K=3} = 0.032$, and the normal deviate test has $Z_{ND} = 1.753 \approx Z_S$ (one-sided p-value $= 0.040$). The score-type 95 % confidence interval is −0.141 to 0.181 where the lower limit is not greater than $-\Delta = -0.1$. These results suggest that the non-inferiority of MIP to axial CT slices cannot be claimed at the one-sided 2.5 % significance level. The Wald-type test statistic, on the other hand, suggests non-inferiority because $Z_W = 3.358$ with one-sided p-value <0.001 and because the Wald-type 95 % confidence interval under the null hypothesis is −0.056 to 0.120. However, the simulation study suggests that the Wald-type test result here is not reliable because of its inflated empirical sizes for a quite small sample size such as $n = 21$. The result of the normal-deviate test, on the other hand, may or may not be reliable because its empirical sizes for $\Delta = 0.1$ and $n = 25$ are shown to be around $1.6 \sim 2.4$.

7.5.2 Study of Diagnostic Procedures for the Diagnosis of Aneurysm in Patients with Acute Subarachnoid Hemorrhage

Jäger et al. [4] performed a blinded multi-rater study comparing magnetic resonance angiography (MRA) and digital subtraction angiography (DSA) in 34 prospectively enrolled patients who presented with acute subarachnoid hemorrhage (SAH). Two raters independently evaluated the MRA and DSA images. The presence of an aneurysm was evaluated on a 4-point ordinal scale (1, absent; 2, probably absent; 3, probably present; 4, definitely present). Additionally, all aneurysms for which the two raters had given different evaluations on the 4-point scale were subsequently reviewed by consensus evaluations. Because the authors intended to study the inter-rater and inter-procedure agreement, neither method was a priori taken as the gold standard. However, they showed the data of evaluation of the MRA and DSA images by the two raters with details of the clinical follow-up of all patients. Therefore, we considered comparing the difference in sensitivities between MRA and DSA on the basis of the data of 27 patients with aneurysms among the patients with SAH. Data were analyzed on a patient-basis, taking into account only the aneurysm with the highest ranking on the 4-point scale in each patient. We assigned the rating of true positive ('+') for scores of 3 and 4 or false negative ('−') for scores of 1 and 2. The resulting types of matched observations based on the two independent raters and the consensus evaluations were classified as a 3×3 and 2×2 contingency tables, respectively (Tables 7.10 and 7.11). DSA is a procedure in which radiographic images of blood vessels filled with a contrast agent are digitized and then subtracted from images obtained before administration of the contrast agent. This method increases the contrast between the vessels and the background. However, as a catheter (a long, thin, flexible tube) is inserted into an artery, DSA is considered to be invasive. MRA is a procedure to image blood vessels based on MRI. Unlike DSA that involves placing a catheter into the body, MRA is considered noninvasive. Therefore, we are interested in the non-inferiority of MRA to DSA where the non-inferiority margin is set as $\Delta = 0.1$. From Table 7.10 based on the multiple raters, we have $\tilde{p}_{2.} = 20/27$, $\tilde{p}_{1.} = 5/27$, $\tilde{p}_{.2} = 22/27$ and $\tilde{p}_{.1} = 2/27$. Then, the sensitivities of MRA and DSA are estimated as $\tilde{\pi}_{MRA} = (20 + 1/2 \times 5)/27 = 0.833$ and $\tilde{\pi}_{DSA} = (22 + 1/2 \times 2)/27 = 0.852$, respectively. Moreover, we have $\tilde{q}_{N2} = 1/27$, $\tilde{q}_{N1} = (0 + 2)/27$, $\tilde{q}_{S2} = 0/27$ and $\tilde{q}_{S1} = (3 + 2)/27$. Then, the difference in the sensitivities between MRA and DSA based on the two raters is $\tilde{\lambda}_{K=2} = -0.019$, and the normal deviate test has $Z_{ND} = 1.393 \approx Z_S$ (one-sided p-value $= 0.082$). The score-type 95 % confidence interval is −0.141 to 0.144 where the lower limit is not greater than $-\Delta = -0.1$. Furthermore, the Wald-type test has $Z_w = 1.397$ (one-sided p-value $= 0.081$) and the Wald-type 95 % confidence interval under the null hypothesis is −0.139 to 0.102. From Table 7.11 based on the consensus evaluations, on the other hand, the sensitivities of MRA and DSA are estimated as $\tilde{\pi}_{MRA_{CE}} = 0.926$ and $\tilde{\pi}_{DSA_{CE}} = 0.889$, respectively. Then, the difference in the sensitivities between MRA and DSA based on the

Table 7.10 A 3 × 3 contingency table ($K = 2$) of the assessments of MRA and DSA by two neuroradiologists (True positive (TP: +) and false negative (FN: −) by two neuroradiologists (1, 2): I (+, +), II (+, − or −, +), III (−, −)) (Jäger et al. [4])

		DSA			
	TP and FN by two radiologists	I	II	III	Total
MRA	I	19	0	1	20
	II	3	0	2	5
	III	0	2	0	2
	Total	22	2	3	27

DSA, digital subtraction angiography; FN, false negative;
MRA, magnetic resonance angiography; TP, true positive

Table 7.11 A 2 × 2 contingency table of the assessments of MRA and DSA by consensus evaluations (True positive (TP: +) and false negative (FN: −)) (Jäger et al. [4])

		DSA		
	TP and FN by consensus evaluations	+	−	Total
MRA	+	22	3	25
	−	2	0	2
	Total	24	3	27

DSA, digital subtraction angiography; FN, false negative;
MRA, magnetic resonance angiography; TP, true positive

consensus evaluations is $\tilde{\lambda}_{CE} = 0.037$, and the score test derived from Nam [10] and Tango [17] has $Z_S = 1.510$ (one-sided p-value $= 0.066$). Moreover, the score-based 95 % confidence interval derived from Tango [17] is −0.150 to 0.227. These results suggest that the non-inferiority of MRA to DSA cannot be claimed at the one-sided significance level. However, although the difference in the sensitivities based on the two raters $\tilde{\lambda}_{K=2}$ is a negative value, the difference in the sensitivities based on the consensus evaluations $\tilde{\lambda}_{CE}$ is a positive value. We consider that bias from the consensus evaluations caused this phenomenon.

7.6 Conclusion

A non-inferiority trial of diagnostic procedures is generally evaluated on the basis of the results from multiple independent raters who are independent of the study centers. However, consensus evaluations or majority votes to handle multiple results from the multiple raters are not recommended in terms of bias or loss of information [1, 2, 12]. Therefore, it is important that all of the results from the multiple raters are utilized appropriately in the statistical analysis. The methods addressed in this chapter are available for inference of the difference in correlated proportions between the two diagnostic procedures based on the multiple raters. In this chapter, we introduced methods on the basis of sensitivity. However, the methods can be applied to inference of the difference in specificity. Furthermore, if we need to consider the simultaneous non-inferiority of a new diagnostic procedure to the standard diagnostic procedure in sensitivity and specificity, we can extend the methods using an approach proposed by Lu et al. [8]. Lu et al. extended the score test proposed by Nam [10] and Tango [17] for a single proportion to a simultaneous test for both sensitivity and specificity based on the principle of intersection-union test.

We carried out Monte Carlo simulation studies to evaluate the performance of these methods. The normal deviate test for non-inferiority was shown to have an empirical size closer to a nominal significance level of one-sided 2.5 % than the Wald-type test or the test based on the majority votes. Moreover, the score-type confidence interval had better performance than the Wald-type confidence interval under the null-hypothesis in terms of coverage probability, when the sample size was small. On the other hand, the confidence interval based on the majority votes shows a conservative property.

When we plan a clinical trial to compare the efficacies between two diagnostic procedures, it is very important to take into account the study design. The methods addressed in this chapter are only useful for a study design in which two diagnostic procedures are applied to each subject and all raters evaluate all subjects, that is, paired-patient, paired-rater design. Zhou et al. [18] provided information on study designs for diagnostic procedures in detail. Moreover, it is noted that these methods may not be appropriate for clustered matched-pair data. Schwenke and Busse [16] proposed a Wald-type test for clustered matched-pair data based on multiple raters. However, the test of Schwenke and Busse is a so-called *test for superiority* and cannot be used as a test for non-inferiority. If the results of the two diagnostic procedures are evaluated by a single rater, we can apply several non-inferiority tests for clustered matched-pair data [3, 5, 11]. Therefore, we expect that a non-inferiority test for clustered matched-pair data on the basis of the results from multiple raters will be developed. If there are missing data among the results from the multiple raters in some subject, we would have to apply some kind of imputation method, which would require future research. Furthermore, if the presence of a qualitative interaction between the two diagnostic procedures and the multiple raters is demonstrated, we would not be able to apply these methods for those data. However, this problem could probably be solved by a non-statistical study, for example, by training all of the raters on the criteria of judgment about diagnostic procedures before the start of evaluation.

7.7 Program

The R programs for the methods of this chapter can be downloaded at http://www.medstat.jp/downloadsaeki.html.

References

1. Guidance for industry. Developing medical imaging drugs and biological products. Part 3: design, analysis, and interpretation of clinical studies (2004). URL http://www.fda.gov/downloads/Drugs/GuidanceComplianceRegulatoryInformation/Guidances/ucm071604.pdf. Cited 21 May 2012
2. Appendix 1 to the guideline on clinical evaluation of diagnostic agents (CPMP/EWP/1119/98 REV. 1) on imaging agents (Doc. Ref. EMEA/CHMP/EWP/321180/2008) (2009). URL http://www.ema.europa.eu/docs/en_GB/document_library/Scientific_guideline/2009/09/WC500003581.pdf. Cited 21 May 2012
3. Durkalski, V., Palesch, Y., Lipsitz, S., Rust, P.: Analysis of clustered matched-pair data for a non-inferiority study design. *Statistics in Medicine* **22**, 279–290 (2003). DOI 10.1002/sim.1385
4. Jäger, H., Mansmann, U., Hausmann, O., Partzsch, U., Moseley, I., Taylor, W.: MRA versus digital subtraction angiography in acute subarachnoid haemorrhage: a blinded multireader study of prospectively recruited patients. *Neuroradiology* **42**, 313–326 (2000)
5. Jin, H., Lu, Y.: Comparison of correlated proportions based on paired binary data from clustered samples. *Journal of Statistical Planning and Inference* **139**, 4206–4212 (2009). DOI 10.1016/j.jspi.2009.06.005
6. Lehr, R., Kashanian, F.: Three persistent issues in analysis of clinical trials involving diagnostic contrast agents. *Drug Information Journal* **43**, 525–532 (2009). DOI 10.1177/009286150904300501
7. Lu, Y., Bean, J.: On the sample size for one-sided equivalence of sensitivities based upon McNemar's test. *Statistics in Medicine* **14**, 1831–1839 (1995). DOI 10.1002/sim.4780141611
8. Lu, Y., Jin, H., Genant, H.: On the non-inferiority of a diagnostic test based on paired observations. *Statistics in Medicine* **22**, 3029–3044 (2003). DOI 10.1002/sim.1569
9. McNemar, Q.: Note on the sampling error of the difference between correlated proportions or percentages. *Psychometrika* **12**, 153–157 (1947). DOI 10.1007/BF02295996
10. Nam, J.: Establishing equivalence of two treatments and sample size requirements in matched-pairs design. *Biometrics* **53**, 1422–1430 (1997)
11. Nam, J., Kwon, D.: Non-inferiority tests for clustered matched-pair data. *Statistics in Medicine* **28**, 1668–1679 (2009). DOI 10.1002/sim.3580
12. Obuchowski, N., Lieber, M.: Statistics and methodology. *Skeletal Radiology* **37**, 393–396 (2008). DOI 10.1007/s00256-008-0448-1
13. Rapp-Bernhardt, U., Welte, T., Budinger, M., Bernhardt, T.: Comparison of three-dimensional virtual endoscopy with bronchoscopy in patients with oesophageal carcinoma infiltrating the tracheobronchial tree. *The British Journal of Radiology* **71**, 1271–1278 (1998)
14. Saeki, H., Tango, T.: Non-inferiority test and confidence interval for the difference in correlated proportions in diagnostic procedures based on multiple raters. *Statistics in Medicine* **30**, 3313–3327 (2011). DOI 10.1002/sim.4364
15. Schouten, H.: Estimating kappa from binocular data and comparing marginal probabilities. *Statistics in Medicine* **12**, 2207–2217 (1993). DOI 10.1002/sim.4780122306
16. Schwenke, C., Busse, R.: Analysis of differences in proportions from clustered data with multiple measurements in diagnostic studies. *Methods of Information in Medicine* **46**, 548–552 (2007). DOI 10.1160/ME0433
17. Tango, T.: Equivalence test and confidence interval for the difference in proportions for the paired-sample design. *Statistics in Medicine* **17**, 891–908 (1998). DOI 10.1002/(SICI)1097-0258(19980430)17:8⟨891::AID-SIM780⟩3.0.CO;2-B
18. Zhou, X., Obuchowski, N., McClish, D.: *Statistical Methods in Diagnostic Medicine*, 2nd edn. Wiley & Sons, New York (2011)

References

Chapter 8
Design and Analysis of Clinical Trial Simulations

Kazuhiko Kuribayashi

Abstract Clinical trial simulation is a powerful tool for supporting decision making in designing clinical trials, and plays an important role in clinical research and drug development. In clinical trial simulation, however, the design is often not well-considered and the results are empirically assessed. In this chapter, we present points to consider when planning a clinical trial simulation, and discuss how to design a clinical trial simulation employing a fractional factorial design and how to analyze the simulation results.

8.1 Introduction

Clinical trial simulation (CTS) is a process to mimic the conduct of a clinical trial on computers by generating the outcomes for each virtual patient based on the prespecified models and/or assumptions. CTS is a powerful tool for supporting decision making in designing clinical trials, and plays an important role in clinical research and drug development. The primary objective of CTS is to investigate the validity and robustness of study designs under various design scenarios and/or assumptions.

When planning clinical trials, complicated study designs such as adaptive designs are considered to achieve the objectives efficiently. Trial operating characteristics should be assessed at the planning stage of such complicated study designs. In particular, assessments of operating characteristics and factors that may influence them would help not only to select an optimal study design, but also to provide a guidance for trial monitoring. Since statistical theory for such study designs is often complicated and their operating characteristics are assessed analytically only under relatively strong assumptions, we usually rely on Monte Carlo simulations. CTS is relatively easily conducted to evaluate the operating characteristics under various practical settings. CTS is also useful for traditional fixed designs. In actual clinical

K. Kuribayashi (✉)
Pfizer Japan Inc., Tokyo, Japan
e-mail: kazuhiko.kuribayashi@pfizer.com

K. van Montfort et al. (eds.), *Developments in Statistical Evaluation of Clinical Trials*, DOI 10.1007/978-3-642-55345-5_8

trials, it is not unusual to deviate from the study protocol, and assessments of the effects of such deviations on the outcomes would be a key to study success.

In CTS, the number of simulations is often not objectively determined and the results are empirically assessed. Moreover, the design of factor arrangements is often not well-considered. It seems to be practical to perform simulations at all possible combinations of levels across all factors, which is a full factorial design. CTS generates virtual patient responses under a number of scenarios, which are combinations of levels of various factors. The number of combinations increases greatly with the increase in the number of factors and their levels. We often encounter difficulties to conduct simulations for all possible combinations of the levels with sufficient numbers of replications within a reasonable time. In such cases, if simulations are conducted with insufficient replications, then it is important to evaluate the Monte Carlo error. On the other hand, we can reduce the number of combinations of levels of factors by employing a fractional factorial design, which is a factorial design in which only an adequately chosen subset of the combinations required for the full factorial design is selected to be run (e.g., [6]).

In this chapter, we present points to consider when planning CTS and discuss how to design CTS and how to analyze the results. In Sect. 8.2, protocol development of CTS and how to determine the number of simulations based on the Monte Carlo error are described. In Sect. 8.3, the design of CTS using orthogonal array and the analysis of simulation results are presented. An example of an adaptive group sequential design is illustrated in Sect. 8.4. Finally, some remarks are provided in Sect. 8.5.

8.2 Planning of Clinical Trial Simulations

8.2.1 Protocol Preparation

As poorly designed and poorly conducted clinical trials produce questionable results, poorly designed and poorly conducted CTS also make inappropriate choices of study designs and statistical methods. Hence, CTS should be planned with similar rigor as clinical trials, in particular, if the purpose of CTS is to provide information on decision making in designing clinical trials. Planning the CTS, "protocol", which describes what the objectives of the simulation are, how the simulation is to be performed and how the results are assessed, should be prepared as clinical trials [2, 5, 12]. The protocol also includes the rationale for all the specifications of the CTS plan. An example of the contents of the protocol is as follows.

Objectives of the Simulation Study Clearly defined objectives of the simulation study should be stated in the protocol. This includes how to assess questions of interest by simulation and how to leverage the simulation results to decision making.

Scenarios and Factors to Investigate with Rationale Scenarios of the clinical outcome to be investigated by simulation should be described along with some rationale. The scenarios include favorable, unfavorable and highly possible ones. Factors and their levels to be examined should be also described.

Simulation Study Design CTS usually generates virtual patient responses under combinations of levels of various factors. This is considered as a factorial experiment. The design of factor arrangements should be well-considered. The factor arrangements in the simulation, such as full factorial design, fractional factorial design or split-plot design (e.g., [6]), should be explained.

Data Generation Method A thorough description of data generation methods should be provided. This includes the rationale for selections of assumed distributions, required parameters for statistical models and correlation structure of the covariates.

The random number generation method should be described. The quality of simulation depends very much on the quality of the pseudorandom numbers. Unreliable algorithms should not be employed.

The data generated should simulate situations that enable to generalize the simulation results, and should be checked by using some statistics, such as summary statistics for distributions of the covariates and Kaplan-Meier estimates for time-to-event data.

It might be useful to simulate data by bootstrapping or permutation from real clinical trial data for creating resemblance to reality.

It is also useful to apply the inclusion and exclusion criteria of the clinical trial to generated data.

Assessments The operating characteristics quantifying the performance of the study design, such as power, expected sample size and so on, to be evaluated in CTS, should be defined.

Determination of the Number of Simulation Replications The rationale for the number of simulation replications should be stated. The number of simulations can be determined based on the Monte Calro error. Details are described in the next section.

Statistical Evaluation The analysis methods for the simulation results should be stated. How to handle ill-conditioned cases, such as failure to estimate parameters of interest due to non-convergence and/or infrequent events, should be described.

8.2.2 Determination of the Number of Simulation Replications

The estimated accuracy of operating characteristics, which is the amount of the Monte Carlo error, depends on the number of simulation replications R. Once the target amount of the Monte Carlo error is chosen, the number of simulation replications is determined using the inversely proportional relationship between the Monte Carlo error and the square root of the number of replications [8].

Let θ be an operating characteristic to be evaluated by simulation, and $\hat{\theta}^{(R)}$ the estimate based on the R simulations. For instance, when the operating characteristic to be evaluated by simulation is the power or the probability of type I error, letting $I[\cdot]$ be an indicator function which equals 1 when the argument is true, 0 otherwise, $z^{(r)}$ the test statistics at the rth simulation and c the critical value, the estimate of the power or the probability of type I error is provided by

$$\hat{\theta}^{(R)}_{\text{power}} = \frac{1}{R}\sum_{r=1}^{R} I[z^{(r)} > c] \,.$$

The estimate of the expected sample size based on the R simulation replications is provided by

$$\hat{\theta}^{(R)}_{N} = \frac{1}{R}\sum_{r=1}^{R} N^{(r)} \,,$$

where $N^{(r)}$ denotes the sample size at the rth simulation. The variability of the estimated operating characteristics is quantified by the Monte Carlo error

$$\text{MCE}(\hat{\theta}^{(R)}) = \sqrt{V(\hat{\theta}^{(R)})} \,,$$

where $V(\cdot)$ denotes the variance [8]. To estimate the Monte Carlo error, the R simulation replications need to be replicated a sufficient number of times. This would be impractical since an additional investment of time is required. If $\hat{\theta}^{(R)}$ is asymptotically normal, then the estimated Monte Carlo error is obtained as

$$\widehat{\text{MCE}}_{\text{asym}}(\hat{\theta}^{(R)}) = \frac{\hat{\sigma}_{\theta}}{\sqrt{R}} = \frac{1}{\sqrt{R}}\sqrt{\frac{1}{R-1}\sum_{r=1}^{R}\left(S^{(r)} - \frac{1}{R}\sum_{r=1}^{R} S^{(r)}\right)^2} \,, \tag{8.1}$$

where $S^{(r)}$ denotes an outcome related to the operating characteristic at the rth simulation, such as $S^{(r)} = I[z^{(r)} > c]$ for the power or the probability of type I error and $S^{(r)} = N^{(r)}$ for the expected sample size. If $\hat{\theta}^{(R)}$ is not asymptotically normal, the bootstrap method can be employed. B sets of bootstrap samples with size R, $\boldsymbol{S}^*_1, \boldsymbol{S}^*_2, \ldots, \boldsymbol{S}^*_B$, are drawn with replacement from $\boldsymbol{S} = \{S^{(1)}, S^{(2)}, \ldots, S^{(R)}\}$ generated by R simulations, and $\hat{\theta}^{(R)}(\boldsymbol{S}^*_1)$, $\hat{\theta}^{(R)}(\boldsymbol{S}^*_2)$,..., $\hat{\theta}^{(R)}(\boldsymbol{S}^*_B)$ are calculated for each bootstrap sample. A bootstrap estimate of the Monte Carlo error is provided by

$$\widehat{\text{MCE}}_{\text{boot}}(\hat{\theta}^{(R)}) = \sqrt{\frac{1}{B-1}\sum_{b=1}^{B}\left(\hat{\theta}^{(R)}(\boldsymbol{S}^*_b) - \frac{1}{B}\sum_{b=1}^{B}\hat{\theta}^{(R)}(\boldsymbol{S}^*_b)\right)^2} \,.$$

The number of simulation replications R can be determined by the target amount of the Monte Carlo error and the variation between simulations σ_θ in (8.1). When an operating characteristic to be evaluated is the binomial proportion, such as the power or the probability of type I error, the variation between simulations is obtained as

$$\sigma_\theta = \sqrt{Q(1-Q)}\,,$$

where Q denotes the assumed value of the proportion. Letting MCE$'$ be a target amount of the Monte Carlo error, the required number of simulations is

$$R' = \left(\frac{\sigma_\theta}{\text{MCE}'}\right)^2 . \tag{8.2}$$

For example, when estimating the probability of type I error with the Monte Carlo error 0.001 in a one-sided test with significance level 0.025, 24,375 simulations are required. In the case of Monte Carlo error 0.005, 975 simulations are required. It is not unusual to have much uncertainty in the assumed value of the power. In such case, the calculation using $Q = 0.5$, which gives the largest variation between simulations, is on the safe side. When $Q = 0.5$, 10,000 simulations are required to achieve a 0.005 for the Monte Carlo error. This means that the Monte Carlo error of the binomial probability estimated by 10,000 simulations is at most 0.005.

When the variation between simulations is unknown, such as the expected sample size, it can be estimated by simulation. First, R simulations are tentatively conducted and $\{S^{(1)}, S^{(2)}, \ldots, S^{(R)}\}$ are obtained. Next, $R_1, R_2, \ldots, R_p$ samples are drawn with replacement. That is, $\boldsymbol{S}_1^* = \{S^{(1)}, \ldots, S^{(R_1)}\}$, $\boldsymbol{S}_2^* = \{S^{(1)}, \ldots, S^{(R_2)}\}$, ..., $\boldsymbol{S}_p^* = \{S^{(1)}, \ldots, S^{(R_p)}\}$ are generated. The Monte Carlo error is estimated in each set, and the variation between simulations σ_θ is estimated as the slope by applying the least-squares method to the paired data, $\left(\frac{1}{\sqrt{R_1}}, \widehat{\text{MCE}}_{\text{boot}}(\hat{\theta}^{(R_1)})\right)$, $\left(\frac{1}{\sqrt{R_2}}, \widehat{\text{MCE}}_{\text{boot}}(\hat{\theta}^{(R_2)})\right), \ldots, \left(\frac{1}{\sqrt{R_p}}, \widehat{\text{MCE}}_{\text{boot}}(\hat{\theta}^{(R_p)})\right)$.

8.2.3 Determination of the Number of Bootstrap Samples

When employing the bootstrap method to estimate the Monte Carlo error, the accuracy depends on the number of bootstrap samples B. The number of bootstrap samples is determined so that the probability that the relative error of the bootstrap estimates of the Monte Carlo error falls within a certain range is ensured [1]. That is, we choose B such that

$$1 - \omega = \Pr\left(1 - \gamma < \frac{\widehat{\text{MCE}}_{\text{boot}}(\hat{\theta}^{(R)})^2}{\hat{V}(\hat{\theta}^{(R)})} < 1 + \gamma\right),$$

where ω and γ denote a small probability and a small positive value, respectively. Suppose that the distribution of $\hat{\theta}^{(R)}$ is approximately normal and $\chi^2_{B-1} = (B-1)\widehat{\mathrm{MCE}}_{\mathrm{boot}}(\hat{\theta}^{(R)})^2 \,/\, \hat{V}(\hat{\theta}^{(R)})$ has approximately a chi-squared distribution with $(B-1)$ degrees of freedom. Since B is large enough to ignore the difference between B and $B-1$, the approximation

$$
\begin{aligned}
\Pr&\left(1-\gamma < \frac{\widehat{\mathrm{MCE}}_{\mathrm{boot}}(\hat{\theta}^{(R)})^2}{\hat{V}(\hat{\theta}^{(R)})} < 1+\gamma\right) \\
&\approx \Pr\left(B(1-\gamma) < \chi^2_B < B(1+\gamma)\right) \\
&\approx \Pr\left(B(1-\gamma) < B + \sqrt{2B}\frac{\hat{\theta}^{(R)}}{\sqrt{\hat{V}(\hat{\theta}^{(R)})}} < B(1+\gamma)\right) \\
&= 1 - 2\Phi\left(-\sqrt{\frac{B}{2}}\gamma\right)
\end{aligned}
$$

can be obtained. The number of bootstrap samples to achieve a relative error less than γ with probability $1-\omega$ is approximately

$$
B \approx \frac{2\left(\Phi^{-1}\left(\frac{\omega}{2}\right)\right)^2}{\gamma^2}.
$$

For example, 769 bootstrap samples are required to achieve a relative error ranged from 0.9 to 1.1 with probability 0.95.

8.3 Design and Analysis of CTS by Orthogonal Arrays

In CTS, operating characteristics quantifying the performance of the study design are evaluated under a number of scenarios, which are combinations of the levels across various factors. This is considered as a factorial experiment. In practice, CTS is often conducted at all possible combinations of levels across all factors, which is a full factorial experiment. In that case, the number of combinations increases greatly with an increase in the number of factors and their levels. For example, with ten factors each taking two levels, a full factorial experiment would have $2^{10} = 1{,}024$ combinations in total. This means R simulations at each combination have to be replicated 1,024 times. It can easily be imagined that such a full factorial experiment with a sufficient number of simulations for each combination requires a great deal of time. In that case, it might be difficult to perform CTS with sufficient replications. On the other hand, we can try to reduce of the number of combinations of the levels of the factors.

Table 8.1 Orthogonal array for 2 levels, $L_8(2^7)$

	Columns						
Run	1	2	3	4	5	6	7
1	1	1	1	1	1	1	1
2	1	1	1	2	2	2	2
3	1	2	2	1	1	2	2
4	1	2	2	2	2	1	1
5	2	1	2	1	2	1	2
6	2	1	2	2	1	2	1
7	2	2	1	1	2	2	1
8	2	2	1	2	1	1	2
	a	b	a b	c	a c	b c	a b c

A full factorial experiment evaluates the main effect of each factor as well as the effects of interactions between factors. For ten factors, each taking two levels, the full factorial experiment requires 1,024 simulation runs and allows to evaluate 1,013 interactions including ${}_{10}C_2 = 45$ two-factor interactions, ${}_{10}C_3 = 120$ three-factor interactions, . . ., ${}_{10}C_{10} = 1$ ten-factor interactions. However, usually it is very difficult to interpret higher-order interactions, such as more than three factors. Such higher-order interactions could be negligible. If so, there is no need to employ a full factorial experiment. Rather a fractional factorial experiment, which is a factorial experiment in which only an adequately chosen subset of the combinations required for the full factorial experiment is selected to be run, may be useful and the factors are easily assigned by Taguchi's orthogonal array (e.g., [6]).

Table 8.1 shows an example of an orthogonal array for 2 levels. This table is represented by $L_8(2^7)$, where "L" stands for Latin squares because orthogonal array is an expansion of Latin squares, "8" indicates the number of rows, "2" means the number of levels and "7" is the number of columns. When selecting any two columns from this table, they include four types of combinations, (1,1), (1,2), (2,1) and (2,2), with the same frequency. We allocate a factor to one of the columns and assign 1 for one level and 2 for the other level, and then conduct simulations for eight combinations of the levels of the factors.

When the number of factors is three, this is equivalent to the full factorial experiment. But if some interactions are negligible, then we can allocate more than three factors. Consider a simulation study with four factors, say A, B, C and D, each taking two levels, and no interactions between the factors. The full factorial experiment requires 16 simulation runs. In contrast, a fractional factorial design using the orthogonal array presented in Table 8.2 requires 8 simulation runs. The four factors, A, B, C and D are allocated to 4 columns out of 7 and 8 combinations of the levels of the factors are determined. We can examine the main effects of the factors based on results of the 8 simulation runs.

Table 8.2 Assignment of factors in an orthogonal array

	Columns							
Run	1	2	3	4	5	6	7	Combinations
1	1	1	1	1	1	1	1	$A_1B_1C_1D_1$
2	1	1	1	2	2	2	2	$A_1B_1C_1D_2$
3	1	2	2	1	1	2	2	$A_1B_2C_2D_2$
4	1	2	2	2	2	1	1	$A_1B_2C_2D_1$
5	2	1	2	1	2	1	2	$A_2B_1C_2D_2$
6	2	1	2	2	1	2	1	$A_2B_1C_2D_1$
7	2	2	1	1	2	2	1	$A_2B_2C_1D_1$
8	2	2	1	2	1	1	2	$A_2B_2C_1D_2$
	A	B	C				D	

In the example above, we used an array with 2 levels in each factor and 7 columns for simplicity. If each factor takes the same number of levels, then corresponding orthogonal arrays are available. For factors with three levels, $L_{27}(3^{13})$, which has 27 rows and 13 columns, is available. Orthogonal arrays can handle factors taking different number of levels. For example, when allocating a factor taking 4 levels to $L_8(2^7)$, we choose any two columns and allocate the 4 levels to each of 4 types of combinations, (1,1), (1,2), (2,1) and (2,2).

Simulation results based on the orthogonal array can be analyzed as a factorial experiment since all the factors are orthogonal. In the case of Table 8.2, the total sum of squares S_T is the summation of the sum of squares of the factors, A, B, C, D and the error:

$$S_T = S_A + S_B + S_C + S_D + S_e \ ,$$

where S_e denotes the sum of squares of the error, and the effects of the factors are evaluated by analysis of variance.

8.4 An Illustrative Example: Adaptive Group Sequential Trial

We describe a process of CTS using an example, that applies an adaptive group sequential trial.

Consider a one-sided test with significance level $\alpha(= 0.025)$ of the null hypothesis $H_0 : \mu_x = \mu_y$ against the alternative hypothesis $H_1 : \mu_x > \mu_y$ in a confirmatory trial with two treatments. Now suppose the response of the test treatment $x \sim N(\mu_x, \sigma^2)$, that of the control $y \sim N(\mu_y, \sigma^2)$, and $\delta = (\mu_x - \mu_y)/\sigma$.

This trial employs a group sequential design with the sample size $2n_0$ allowing an interim analysis with $2tn_0$ $(0 < t < 1)$ subjects. When the test statistic doesn't cross the boundary at the interim analysis, the sample size is re-estimated based on the conditional power and increased up to $2rn_0$ $(r > 1)$. Let $2n$ be the re-estimated sample size, $\bar{x}_1$ and $\bar{y}_1$ denote the sample means at the interim analysis in each treatment group, respectively, and $\hat{\delta}_1 = (\bar{x}_1 - \bar{y}_1)/\sigma$. The test statistic at the interim analysis is given by

$$z_1 = \frac{\hat{\delta}_1}{\sqrt{\frac{2}{tn_0}}} .$$

At the final analysis, the weighted Wald statistic

$$z = z_1\sqrt{t} + z_2\sqrt{1-t}$$

is used as the test statistic [3], where

$$z_2 = \frac{\hat{\delta}_2}{\sqrt{\frac{2}{n-tn_0}}} = \frac{\bar{x}_2 - \bar{y}_2}{\sigma\sqrt{\frac{2}{n-tn_0}}}$$

denotes the Wald statistic based on $2(n - tn_0)$ subjects entered after the interim analysis. Cui et al. [3] showed that the weighted Wald statistic z has the same distribution as with the original sample size $2n_0$ under the null hypothesis. So we can use the original boundary without inflation of the probability of type I error even when increasing the sample size.

The conditional power given z_1 at the interim analysis is provided by

$$CP_{\hat{\delta}_1} = \Pr(z > c \mid z_1) = 1 - \Phi\left(c\sqrt{\frac{1}{1-t}} - z_1\sqrt{\frac{t}{1-t}} - \frac{\hat{\delta}_1}{\sqrt{\frac{2}{n-tn_0}}}\right) ,$$

where c denotes the boundary at the final analysis and $\Phi(\cdot)$ denotes the cumulative distribution function of the standard normal distribution. The sample size to achieve the conditional power CP is obtained as

$$n = \frac{2\left(c\sqrt{\frac{1}{1-t}} - z_1\sqrt{\frac{t}{1-t}} - u_{1-CP}\right)^2}{\hat{\delta}^2} + tn_0 ,$$

where $u_{1-CP} = \Phi^{-1}(1 - CP)$. The boundaries for efficacy and futility stopping are calculated based on O'Brien-Fleming type α-spending function [9].

Suppose that we would like to assess the influence of the minimum requirement for sample size increase (A), target conditional power (B), upper limit of sample size (C) and timing of interim analysis (D) on the overall power, and to estimate the optimal combination of the levels of the factors, and also evaluate the expected

Table 8.3 Factors and their levels

Factor	Levels
Minimum requirement for sample size increase (A)	$z_1 >$ lower boundary, $CP_{\hat{\delta}_1} > 0.5$
Target conditional power (B)	$CP = 0.8, CP = 0.9$
Upper limit of sample size (C)	$r = 2, r = 3$
Timing of interim analysis (D)	$t = 0.3, t = 0.5$

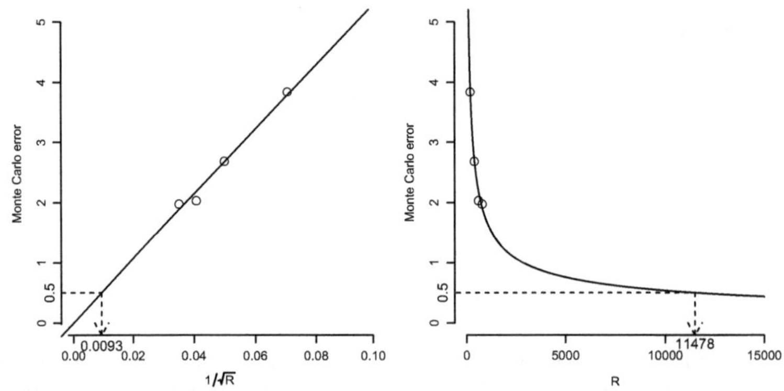

Fig. 8.1 Plots of the four pair of the estimated Monte Carlo error and the size of bootstrap samples and the line fitted by the least-squares method

sample size at that combination. Table 8.3 shows the levels of interest of the factors. In addition, we have interests in 2 two-factor interactions, $A \times B$ and $B \times C$, while the others are negligible.

The number of simulations is determined based on the Monte Carlo error in estimating the power and the expected sample size. For the power, 10,000 simulations are required to estimate it with 0.005 of Monte Carlo error when the variation between simulations $\sigma_\theta = 0.5$, which is largest. The variation between simulations for the expected sample size is estimated by simulation. Thousand simulations are conducted using the following levels of the factors shown in Table 8.3: $A : z_1 >$ Lower boundary, $B : CP = 0.9$, $C : r = 2$, $D : t = 0.5$. From the simulation results $\{S^{(1)}, \ldots, S^{(1,000)}\}$, four sets of bootstrap samples with the size $\{R_1, R_2, R_3, R_4\} = \{200, 400, 600, 800\}$ are drawn with replacement, and the Monte Carlo error for each bootstrap sample is calculated. The variation between simulations is estimated as $\hat{\sigma}_\theta = 53.57$ by the least-squares method applied to the four pairs of the estimated Monte Carlo error and the size of the bootstrap samples. Figure 8.1 shows the plots of the four pair values and the fitted line. The number of simulations required to estimate the expected sample size with 0.5 of the Monte Carlo error is calculated by assigning $\hat{\sigma}_\theta = 53.57$ and $\text{MCE}' = 0.5$ to (8.2). This provides $R' = 11{,}478$. Taking into consideration the above calculations, we determine to conduct 10,000 simulations.

Table 8.4 Assignment of factors and simulation results

	Columns								Simulation results[a]	
Run	1	2	3	4	5	6	7	Combinations	Power	ESS
1	1	1	1	1	1	1	1	$A_1B_1C_1D_1$	0.9508	167.97
2	1	1	1	2	2	2	2	$A_1B_1C_2D_2$	0.9422	218.28
3	1	2	2	1	1	2	2	$A_1B_2C_1D_2$	0.9156	149.38
4	1	2	2	2	2	1	1	$A_1B_2C_2D_1$	0.9840	251.75
5	2	1	2	1	2	1	2	$A_2B_1C_1D_2$	0.9197	136.25
6	2	1	2	2	1	2	1	$A_2B_1C_2D_1$	0.9877	216.65
7	2	2	1	1	2	2	1	$A_2B_2C_1D_1$	0.9525	147.34
8	2	2	1	2	1	1	2	$A_2B_2C_2D_2$	0.9425	200.95
	A	B	$A \times B$	C		$B \times C$	D			

[a] Based on 10,000 replications

Table 8.5 Analysis of variance for the simulation result

Factors	Df	Sum Sq	Mean Sq	F value	Pr(>F)	Prop SS
A	1	0.00001200	0.00001200	29.6420	0.115641	0.002501
B	1	0.00000421	0.00000421	10.3827	0.191572	0.000876
C	1	0.00173460	0.00173460	4282.9753	0.009727	0.361407
D	1	0.00300313	0.00300313	7415.1235	0.007393	0.625704
$A \times B$	1	0.00004512	0.00004512	111.4198	0.060132	0.009402
$B \times C$	1	0.00000012	0.00000012	0.3086	0.677171	0.000026
Residuals	1	0.00000040	0.00000040			0.000084
Total	7	0.00479960				

Table 8.6 The point estimates and 95 % confidence intervals of means at each combination of the upper limit of the sample size (C) and the timing of the interim analysis (D)

		Estimate	95 % C.I.	
C_1	$r = 2$	0.9347	0.9335	0.9358
C_2	$r = 3$	0.9641	0.9630	0.9652
D_1	$t = 0.3$	0.9688	0.9676	0.9699
D_2	$t = 0.5$	0.9300	0.9289	0.9311

The allocation of the factors and the simulation results are shown in Table 8.4 and the analysis of variance (ANOVA) table is shown in Table 8.5. This indicates that the upper limit of the sample size (C) and timing of the interim analysis (D) have some effect on the power.

The point estimates and 95 % confidence intervals of means at each combination of the upper limit of the sample size (C) and the timing of the interim analysis (D) are calculated based on the fitted ANOVA model, and shown in Table 8.6. This table suggests that the combination of C_2 ($r = 3$) and D_1 ($t = 0.3$) is optimal. The point estimates and 95 % confidence intervals of means at the optimal combination based on the fitted ANOVA model are 0.9835 for the power with 95 % confidence interval (0.9821, 0.9848) and 231.76 for the expected sample size with 95 % confidence interval (220.23, 243.30).

This example was implemented by R [11].

8.5 Concluding Remarks

Clinical trial simulations are a statistical experiment, and should be appropriately performed with careful planning. Even if advanced methodologies/technologies are employed, incomplete inputs produce incomplete outputs or, as it is often said, "garbage in, garbage out." CTS should be planned with similar rigor as clinical trials, and conducted with the following two points in mind:

1. To achieve the given purpose of the simulation study, what is the best way to obtain appropriate information with the smallest number of simulations in total?
2. To draw the accurate conclusion, how should the simulation results including the Monte Carlo error be analyzed?

In reporting clinical trials, standard errors and 95 % confidence intervals are routinely presented with point estimates. In reporting CTS, only point estimates are presented in practice. As a guidance for reporting simulation studies for statistical methods, it is pointed out that all reporting should make it easy for the reader to assess the quality of the experimental work and the accuracy of the results [7]. In the same way, reporting CTS should routinely include the Monte Carlo error and 95 % confidence intervals. The 95 % confidence interval is given by

$$\left(\hat{\theta}^{(R)} - 1.96\, \widehat{\mathrm{MCE}}_{\mathrm{asym}}(\hat{\theta}^{(R)}),\quad \hat{\theta}^{(R)} + 1.96\, \widehat{\mathrm{MCE}}_{\mathrm{asym}}(\hat{\theta}^{(R)})\right),$$

where $\hat{\theta}^{(R)}$ is asymptotically normal. If it is not normal, but the distribution is symmetric about $\hat{\theta}^{(R)}$, the 95 % confidence interval is estimated by the 2.5 and 97.5 percentile of $\hat{\theta}^{(R)}(\boldsymbol{S}_1^*), \hat{\theta}^{(R)}(\boldsymbol{S}_2^*), \ldots, \hat{\theta}^{(R)}(\boldsymbol{S}_B^*)$,

$$\left(\hat{\theta}^{(R)}_{B[0.025]},\quad \hat{\theta}^{(R)}_{B[0.975]}\right).$$

If it is not symmetric, the interval is given by

$$\left(2\hat{\theta}^{(R)} - \hat{\theta}^{(R)}_{B[0.975]},\quad 2\hat{\theta}^{(R)} - \hat{\theta}^{(R)}_{B[0.025]}\right)$$

(e.g., [4]). In addition, the limitations of the conclusion and recommendation from the simulation study should be addressed in the reporting of CTS.

In this chapter, we discussed CTS with factors, which each takes fixed level values. Taking into account uncertainties, including randomness in the sampling of subjects, uncertainty about the baseline characteristics of the subject population and uncertainty about the treatment's clinical effects, we can consider Bayesian CTS, which simulates parameter values from probability distributions that represent the current state of knowledge about the parameters [10]. Bayesian CTS accounts for all sources of uncertainty and allows more realistic assessments of the outcomes of individual clinical trials and sequences of clinical trials for the purpose of decision making. In Bayesian CTS as well, the concept of the experimental design discussed here is important. This concept is applicable not only to CTS, but also to assessment of statistical methodologies.

References

1. Booth, J.G., Sarkar, S.: Monte Carlo approximation of bootstrap variances. *The American Statistician* **52**, 354–357 (1998)
2. Burton, A., Altman, D.G., Royston, P., Holder, R.L.: The design of simulation studies in medical statistics. *Statistics in Medicine* **25**, 4279–4292 (2006)
3. Cui, L., Hung, H.M.J., Wang, S.J.: Modification of sample size in group sequential clinical trials. *Biometrics* **55**, 853–857 (1999)
4. Davison, A.C., Hinkley, D.V.: *Bootstrap Methods and Their Applications*. Cambridge University Press, Cambridge, UK (1997)
5. Gaydos, B., Anderson, K.M., Berry, D., Burnham, N., Chuang-Stein, C., Dudinak, J., Fardipour, P., Gallo, P., Givens, S., Lewis, R., Maca, J., Pinheiro, J., Pritchett, Y., Krams, M.: Good practices for adaptive clinical trials in pharmaceutical product development. *Drug Information Journal* **43**, 539–556 (2009)
6. Giesbrecht, F.G., Gumpertz, M.L.: *Planning, Construction, and Statistical Analysis of Comparative Experiments*. John Wiley & Sons: Hoboken, NJ (2004)
7. Hoaglin, D.C., Andrews, D.F.: The reporting of computation-based results in statistics. *The American Statistician* **29**, 122–126 (1975)
8. Koehler, E., Brown, E., Haneuse, S.J.P.A.: On the assessment of Monte Carlo error in simulation-based statistical analysis. *The American Statistician* **63**, 155–162 (2009)
9. Lan, K.K.G., DeMets, D.L.: Discrete sequential boundaries for clinical trials. *Biometrika* **70**, 659–663 (1983)
10. O'Hagan, A., Stevens, J.W., Campbell, M.J.: Assurance in clinical trial design. *Pharmaceutical Statistics* **4**, 187–201 (2005)
11. R Core Team: *R: A Language and Environment for Statistical Computing*. R Foundation for Statistical Computing, Vienna, Austria (2012). URL http://www.R-project.org/
12. Smith, M.K., Marshall, A.: Importance of protocols for simulation studies in clinical drug development. *Statistical Methods in Medical Research* **20**, 613–622 (2011)

Chapter 9
Causal Effect Estimation and Dose Adjustment in Exposure-Response Relationship Analysis

Jixian Wang

Abstract Determining causal exposure effects is often a challenging task even with randomized clinical trials. Confounding factors may cause bias in: (1) the pharmacokinetic exposure-response relationship and (2) dose-response relationship when dose-adjustment depends on potential responses. Dose adjustment often happens in clinical trials either designed for therapeutic dose monitoring, or spontaneously due to, for example, adverse events. It makes causal effect inference difficult since it often relates to potential response. On the other hand, dose adjustment in some trials such as the randomized concentration controlled (RCC) trials are designed to reduce confounding bias in exposure-response relationship. We review different types of dose-adjustment mechanisms and their impact on causal effect estimation with a number of dose-exposure and exposure response models. Following the concept of sequential randomization and approaches for missing data analysis, we examine a number of approaches for causal effect estimation including the classical joint modeling based on joint likelihood functions and instrumental variable and control function methods. We explore simplified approaches for joint modeling with sequential randomization conditional on potentially confounded subject effects and alternatives to the joint modeling approaches. Performance of these approaches in typical practical scenarios was assessed with a simulation study.

9.1 Introduction

In a randomized clinical trial the differences of mean responses between groups are unbiased estimates for the causal treatment effects since potential confounding factors between treatment and response are eliminated by randomization. However, it is more complex to determine exposure-response relationships since drug exposure may not be fully controlled even in a randomized clinical trial, and is often affected by confounding factors. The Food and Drug Administration (FDA) have issued a technical document for exposure-response analysis [3] with an emphasis on

J. Wang (✉)
Novartis Pharma AG, Basel, Switzerland
e-mail: jixwang@celgene.com

K. van Montfort et al. (eds.), *Developments in Statistical Evaluation of Clinical Trials*, DOI 10.1007/978-3-642-55345-5_9

the importance of dealing with confounding bias. Recent statistical developments in causal effect estimation make available several approaches for determining causal drug effects in exposure-response relationships. See, for example, Robins and Hernán [16]. This chapter reviews relevant current developments and discusses potential applications of classical approaches and recent technical advances for determining causal drug effects in exposure-response relationships.

We consider drug exposure as a general measure of drug strength. It is referred to as either drug dose or pharmacokinetic (PK) exposure as in [3]. With this general definition of exposure, we will consider causal effect estimation in both dose-response and PK exposure-response relationships in a uniform way, and will use "exposure" to refer to PK exposure when there is no ambiguity. In some simple situations such as randomized dose controlled trials in which patients are randomized into different dose levels, it is straightforward to determine the causal effect of dose level changes. However, it is no longer the case when the dose changes are made during the trial by dose adjustment, since the adjustment is often related to potential exposure or response to the drug. An example is that a dose reduction is more likely for patients at higher risk of adverse events (AE) than those at lower risk.

The exposure-response relationship is more difficult to determine than the dose-response relationship since in general drug exposure cannot be randomized to a specific value and it is often affected by factors relating to the response. Partial control may be achieved by random concentration controlled (RCC) trials [17] in which patients are randomized into two or three exposure ranges and the PK exposure is measured repeatedly and the dose adjusted, if necessary, until the individual exposure level is within the range the patient is randomized to. Kraiczi et al. [9] reviewed applications of RCC trials focusing on practical considerations. Design and analysis issues in RCC trials can be found in [8]. This approach only controls the exposure level within a range, hence is still affected by confounding factors. Therefore, routine analysis procedures may still lead to confounding bias in the estimated drug effect. Karlsson et al. [8] pointed out that most analyses for RCC trials did not adjust for confounding bias, but they used a joint modeling approach, which showed advantages over the approaches without any adjustment. To control the exposure in a target range for highly variable drugs, a similar approach is the therapeutic dose monitoring. With a given target exposure range, a therapeutic dose monitoring uses a dose adjustment algorithm to adjust individual doses to force individual exposure levels to lie in the range.

Rapid development in determining causal treatment effects allowing treatment changes has been seen in the last 10–20 years. The development that closely relates to dose-adjustment is known as dynamic treatment regimens (DTR) [12], referring to response related treatment changes in a general sense. The development has been mainly focused on simple treatment switching between a few alternatives, since dealing with continuous changes is not only more complex in theory, but also more difficult in practice. A key assumption for determining causal effects is that treatment changes can be considered as the consequence of a sequential randomization at any time point given observed history, or more intuitively no unobserved confounding factors. Since some unknown subject level characteristics are often the main source of confounding between dose adjustment, exposure and response, it may be necessary to find less restrictive assumptions than sequential randomization to estimate causal effects. We consider a number of approaches including the classical joint modeling under slightly different assumptions to sequential randomization. Particularly we borrow approaches for missing data analysis based on joint likelihood functions of dose, exposure and response under a few classes of dose adjustment mechanism, one being similar to the missing data mechanism classified as missing at random (MAR) [10] and we call it dose adjustment at random (DAR), when the dose changes can be considered as sequentially randomized.

Another important tool for causal effect estimation is the instrumental variable (IV), which was originated in econometrics but has found increasing applications in clinical trials [1]. The approach takes the advantage of randomized clinical trials using randomization as the IV to deal with, for example, non-compliance. Randomization in RCC trials can also be used as an IV. We will show a two-stage implementation for the IV approach and also discuss differences between IV based estimates for RCC trials and for randomized clinical trial.

This chapter is organized as follows. The next section introduces concepts of causal effects and confounding bias with dynamic treatment adjustments. Then models for dose-exposure, exposure-response and dose-response relationships are introduced in Sect. 9.3. The following section examines dose adjustment mechanisms, the definition of sequential randomization and introduces conditional sequential randomization, the class when the sequential randomization condition is satisfied by conditioning on subject effects in the above models, and the use of directional acyclic graphs (DAG) [15] to determine dependence between dose, exposure, response and other factors. Joint modeling approaches and their alternatives are introduced in Sects. 9.5 and 9.6 followed by using IV based approaches for RCC trials. In Sect. 9.9 the performance of a number of approaches are examined by simulation under some typical practical scenarios.

9.2 Causal Effects, Confounding Bias and Dynamic Treatment Regimen

In this section we briefly review the concept of confounding bias, response related treatment changes, and approaches for causal effect determination and confounding adjustment. To focus on the topic of dose adjustment, we will use dose changes to represent treatment changes and consider response related dose changes as a special case of dynamic treatment regimens. Pre-determined dose changes are not under our consideration.

Let y be the response of interest, the causal effect of a dose change is referred to as the change in y it causes. For example, the effect of dose change from 0 to 1 on response y can be defined as $E_{(1,0)} = E(y|1) - E(y|0)$, where $E(y|d)$ is the conditional mean of y under dose d. Suppose that patient i was exposed to dose $d_i = 1$ and had response $y_i(1)$ and patient $j, j \neq i$ had dose $d_j = 0$ and response $y_j(0)$. Then $E(y_i(1) - y_j(0))$ may not be an unbiased estimate for $E_{(1,0)}$, since the baseline means $E(y_i(0))$ and $E(y_j(0))$ may be different unless the dose is randomized. Therefore, the naive estimate $\hat{E}_{(1,0)} = \bar{y}.(1) - \bar{y}.(0)$, where $\bar{y}.(d)$ is the mean of all patients with dose d, is often biased. The focus of causal effect determination is to eliminate or reduce this bias. One approach is to use response comparisons between different doses within a unit, e.g., a patient. This approach is not possible when the dose does not change in the same unit. However, we can use this idea to introduce the counterfactual framework to dose adjustment. Suppose that patient i was given dose $d_i = 1$ and had response $y_i(1)$. To determine the causal exposure effect on i, we need to compare $y_i(1)$ with $y_i(0)$. They are known as counterfactuals since often only one is realized. Although one cannot compare them directly, a model (known as a structure model) can be used to describe them and their relationship with dose and other factors. We can write a structural model for $y_i(d_i)$ as:

$$y_i(d_i) = \mu + d_i\beta + u_i + e, \tag{9.1}$$

where μ is the overall mean, β is a parameter and u_i is an unknown subject effect and e is a random term with $E(e) = 0$. The causal exposure effect is determined by β since $E(y_i|1) - E(y_i|0) = \beta$. Hence determining the causal effect can be simplified to the estimation of β.

Confounding bias can be adjusted in a number of ways if confounding factors are observed. For example, a direct adjustment is to add the confounding factors in the dose-response or exposure-response model, given that a correct model can be determined. Stratification can be used if there is one or a few confounding factors which can be stratified. The inverse probability weighting (IPW) approach makes weighted data between different exposure levels comparable (Robins and Hernán [16]). The key assumption needed for all these approaches is no unobserved confounding factor. Dynamic treatment regimen adds more complexity to causal effect estimation, since it is a dynamic process during which data are sequentially observed and confounding factors may also be introduced sequentially. An extension to the key assumption of no unobserved confounding factor for dynamic treatment regimen is sequential randomization, that is, the treatment change at each time point only depends on observed data in its history, hence there is no unobserved confounding factor at each time point.

Due to high complexity of general dynamic treatment regimen framework, researches in this area have been mostly focused on situations when treatment changes are among a few pre-determined alternatives. Some procedures such as the IPW can only be applied to these situations. Theoretical complexity for dealing with continuous treatment changes such as dose changes can be seen in [4]. Most commonly used models for dynamic treatment regimen are marginal mean models [13]. We use a slightly different approach based on specific models and assumptions for dose adjustment and revisit the joint likelihood function and approaches for missing data analysis. However, we will also look at alternatives to joint modeling based on the joint likelihood function. To avoid complexity of dealing with continuous time (see, e.g., [11]), we will consider discrete times such as, for example, study visits, since they are sufficient in most clinical scenarios.

9.3 Dose-Exposure and Exposure-Response Models

Consider a trial with multiple visits $j = 1, \ldots, J$ at which exposure measurements such as the drug concentration are taken. Let c_{ij} be the drug concentration and d_{ij} be the dose level of subject $i, i = 1, \ldots, n$ at visit j. The following power model is widely used in clinical pharmacology to describe typical dose-exposure relationships:

$$\log(c_{ij}) = \alpha \log(d_{ij}) + \mathbf{X}_{ij}^T \boldsymbol{\alpha}_x + v_i + e_{ij}, \tag{9.2}$$

where $\mathbf{X}_{ij}$ contains the intercept and a set of covariates such as age and body surface area but may also include pre-determined time varying covariates, $e_{ij} \sim N(0, \sigma_e^2)$ represents within subject variations including, for example, assay error, and $v_i \sim N(0, \sigma_v^2)$ is a zero-mean subject effect representing inter-subject variations in the exposure. When $\alpha = 1$ the exposure is proportional to the dose and the dose-exposure relationship is known as dose-proportional. Although this model may be considered as empirical, some models derived from PK mechanisms may reduce to this form, particularly when v_i represent factors such as age that affect drug clearance and volume of distribution in a single compartment model. In fact, the log-linear relationship with dose in model (9.2) is not necessary for our purpose, the following general dose-exposure model for K doses is sufficient:

$$\log(c_{ij}) = \boldsymbol{\alpha}^T \mathbf{d}_{ij} + \mathbf{X}_{ij}^T \boldsymbol{\alpha}_x + v_i + e_{ij}, \tag{9.3}$$

where $\mathbf{d}_{ij} = (d_{ij1}, \ldots, d_{ijK})$ is a K-vector with $d_{ijk} = 1$ if subject i is given the kth dose at visit j and $d_{ijk} = 0$ otherwise. $\boldsymbol{\alpha}$ is also a K-vector containing the corresponding log-mean exposures of these doses. The key feature of this model is that v_i is also additive, as in model (9.2).

In contrast to the dose-exposure model, exposure-response models may take many forms depending on factors such as the type of responses. We will mainly consider the response as a continuous variable, e.g., a biomarker with a linear relationship with the exposure, but most approaches in this chapter also applies to other models with a linear structure such as generalized linear models for categorical response variables. Let y_{ij} be the response of patient i at visit j. Assume that

$$y_{ij} = \beta c_{ij} + \mathbf{X}_{ij}^T \boldsymbol{\beta}_x + u_i + \varepsilon_{ij}, \tag{9.4}$$

where u_i is the patient's baseline characteristics and ε_{ij} is a measurement error. Here we assume that $\mathbf{X}_{ij}$ is the maximum set of all covariates in this model and in model (9.3). For elements in $\mathbf{X}_{ij}$ not applicable to one model, we set the corresponding coefficients to zero. Transformations may apply to one or both sides of the model.

Dose-exposure models have wide applications as often PK exposure may not be measured. A linear dose-response model may have a similar structure to (9.4):

$$y_{ij} = \theta d_{ij} + \mathbf{X}_{ij}^T \boldsymbol{\theta}_x + u_i^* + \varepsilon_{ij}, \tag{9.5}$$

where the parameters and factors have similar interpretation as those in (9.4). Sometimes a dose-exposure model can be derived as a combination of dose-exposure and exposure-response models. Suppose that

$$y_{ij} = \beta \log(c_{ij}) + \mathbf{X}_{ij}^T \boldsymbol{\beta}_x + u_i + \varepsilon_{ij} \tag{9.6}$$

and the power model (9.2) does not contain covariates. Then taking $\log(c_{ij})$ in (9.2) into (9.6) leads to, after dropping the covariates, a model

$$y_{ij} = \alpha\beta \log(d_{ij}) + \mathbf{X}_{ij}^T \boldsymbol{\beta}_x + \beta(v_i + e_{ij}) + u_i + \varepsilon_{ij}, \tag{9.7}$$

which has the same form as model (9.5) with $\theta = \alpha\beta$ and $u_i^* = \beta v_i + u_i$. The (log)linear relationship in model (9.2) makes it easy to derive the corresponding dose-response model from the dose-exposure and exposure-response models in many common settings. For example, if the exposure-response model is a generalized linear mixed model (GLMM) with $\log(c_{ij})$ and the dose-response model is (9.2), the dose-response model is also a GLMM.

9.4 Dose Adjustment Mechanisms

There are many types of dose adjustment, some are not planned and may be difficult to describe exactly. Nevertheless it is important to understand and to model, if necessary, the mechanism for causal inference. Two common types of dose adjustment are adjustments based on drug exposure and adjustments based on drug response. The former is often planed and serves purposes such as therapeutic dose monitoring, while the latter is often spontaneous.

9.4.1 Exposure Dependent Dose Adjustment

Often the purpose of dose adjustment is to keep individual exposure levels within a target exposure range (L, U), for example, in a trial with therapeutic dose monitoring. For this purpose a simple adjustment rule is to escalate the dose one level higher when $c_{ij} < L$ and to de-escalate one level lower when $c_{ij} > U$. This adjustment is a trial-and-error approach and needs no dose-exposure model. Formally this mechanism can be written as

$$d_{ij+1} = \begin{cases} d_{ij} + \Delta & c_{ij} < L, \\ d_{ij} & L \leq c_{ij} \leq U, \\ d_{ij} - \Delta & c_{ij} > U, \end{cases} \tag{9.8}$$

where Δ is the dose adjustment size, which may also be variable. In practice, often the dose adjustment stops when the exposure becomes stable within the target range. This adjustment is widely used for therapeutic dose monitoring because of its simplicity.

Knowing the dose-exposure model is sometimes beneficial as dose adjustment can be made more efficiently and safer by using this model. For example, with the power model the required dose to achieve exposure level c_0 can be found as [2]

$$\hat{d} = (\log(c_0) - \mathbf{X}_i^T \hat{\boldsymbol{\alpha}}_x - \hat{v}_i)/\hat{\alpha}, \tag{9.9}$$

where $\hat{\boldsymbol{\alpha}}_x$, $\hat{v}_i$ and $\hat{\alpha}$ are estimated based on dosing and exposure data observed until visit $j-1$, and $\mathbf{X}_{ij} = \mathbf{X}_i$, $j = 1, \ldots, J$. Another model-based approach using an empirical model was proposed by O'Quigley et al. [14]. They all examined asymptotic properties of the dose adjustments and showed consistency of sequential estimation of α under some technical conditions. Since often there are only a few available doses, e.g., when the drug formulation is tablet or capsule, the simple algorithm (9.8) is the most commonly used. We will concentrate on the simple adjustments based (9.8).

In practice, the adjustment described by (9.8) may not be followed exactly, since some factors that are often unknown or not recorded may also have an impact on dose adjustment. To model this mechanism, one has to consider them as random factors. A straightforward approach is to use two binary (e.g., logistic) models. The probability of dose increase at step j may be expressed as

$$P(d_{ij+1} = d_{ij} + \Delta) = 1/\big(1 + \exp(-\eta(L - c_{ij}) - \eta_0 + s_i)\big), \tag{9.10}$$

where L is the lower bound of the target exposure and η and η_0 are parameters. In addition to the part $\eta(L - c_{ij})$ depending on the exposure, s_i is a factor representing the impact of subject level characteristics on dose escalation. For example, a subject may have a lower probability of a dose increase if he has a higher risk of AE than someone with the same c_{ij} but having a lower risk of AE. One can set up the model for dose reduction in the same way. Alternative models treating dose levels as an ordered categorical variable may also be used if appropriate. Since there are unknown parameters in the model, if the dose adjustment mechanism is to be considered in causal effect estimation, one needs to fit the dose adjustment model as well.

9.4.2 Dose Adjustment for Causal Effect Determination: RCC Trials

Under some situations dose adjustment may be used for the purpose of causal effect determination, e.g., in RCC trials. The dose adjustment in RCC trials is essentially exposure-dependent. In an RCC trial patients are randomized into $k = 1, \ldots, K$ groups with exposure ranges (L_k, U_k), and often with $L_k = U_{k-1}$. Then the dose for a patient is adjusted so that the exposure falls into the range corresponding to the group that the patient is randomized to. For RCC the simple adjustment (9.8) can be written as

$$d_{ij+1} = \begin{cases} d_{ij} + \Delta & c_{ij} < L_k, \\ d_{ij} & U_k \geq c_{ij} \geq L_k, \\ d_{ij} - \Delta & c_{ij} > U_k, \end{cases} \tag{9.11}$$

where subject i is randomized to the kth group. Δ is the step size for dose adjustment, which may not necessarily be constant.

9.4.3 Response Dependent Dose-Adjustment

A common dose response-dependent adjustment in clinical practice is dose reduction due to AEs. Here we introduce the following model assuming the probability of AE occurrence depending on the exposure level. Let $y_{ij} = 1$ if an AE occurs between visits j and $j-1$ and $y_{ij} = 0$ otherwise. A dose reduction is triggered when $y_{ij} = 1$, i.e., $d_{ij+1} = d_{ij} - \Delta$ if $y_{ij} = 1$. We assume that the risk of the AE relates to the exposure via a logistic model with

$$P(y_{ij} = 1 | c_{ij-1}) = 1/(1 + \exp(-\beta(c_{ij-1}) - \beta_0 + u_i)), \tag{9.12}$$

where $u_i \sim N(0, \sigma_u^2)$ is a subject level effect. With this model we link drug exposure to dose adjustment, hence the exposure-response model forms a part of the dose adjustment mechanism.

A similar dose adjustment to (9.8), but depending on efficacy measurements, may also be used. For example, when y_{ij} is the blood pressure level, then a dose increase of an anti-hypertension drug may be granted when y_{ij} is higher than a certain level. Again we find the exposure-response model in the dose adjustment mechanism.

9.4.4 Dose Adjustment and Sequential Randomization

We consider an important class of dose adjustment with which the analysis of dose-dose and exposure-response relationship can be simplified. In particular we are interested in a class of dose adjustments that satisfy the condition of sequential randomization in the dynamic treatment regimen framework to characterize dose adjustments that only depend on observed previous dosing, exposure and response history, and known factors. Specifically,

$$d_{ij+1} \sim g(d; \bar{F}_{ij}), \tag{9.13}$$

where $g(d; \bar{F}_{ij})$ is an arbitrary density function, conditional on $\bar{F}_{ij}$ containing the history of dosing and exposure information till visit j, and known constant or time varying covariates in $\mathbf{X}_{ij}$. For convenience, we call a dose adjustment mechanism a dose adjustment at random (DAR), named after missing at random (MAR) in missing data analysis, when it satisfies the condition of sequential randomization. The key feature of DAR is that d_{ij} can be considered as sequentially randomized given previous history. Note that this assumption may not hold if the dose adjustment is also based on unobserved exposure, as we will find later.

Although condition (9.13) is very general, we only consider it in the context of the specific models given in the previous section. With c_{ij} observed, adjustments (9.8) and (9.11) are a DAR, while (9.10) is a DAR only when s_i is independent of u_i and v_i. The key characteristic of DAR is that it allows fitting the dose-response model separately from the dose adjustment model, given that a proper adjustment for factors in $\bar{F}_{ij}$ is made in the model fitting. The assumption of DAR appears similar to MAR in missing data mechanisms. Sometimes dose adjustment may depend on models sequentially estimated. These adjustments may also be a DAR.

Since sometimes s_i depends on u_i or v_i, we may introduce a class that is less restrictive than DAR: dose adjustments which satisfy the sequential randomization condition conditional on s_i

$$d_{ij+1} \sim g(d; \bar{F}_{ij}, s_i) \tag{9.14}$$

and we call this class conditional DAR. Note that the sequential randomization condition is not satisfied since s_i may not be observed and cannot be counted as a part of the history. This class excludes change of d_{ij} due to y_{ij} directly. The exclusion is not trivial since it is not uncommon that a dose adjustment is not even a conditional DAR, for example, when the doctor makes a decision based on factors not reflected in the history nor the constant characteristics. For example, when the patient's characteristics change very quickly, there may be a time varying factor connecting y_{ij+1} and d_{ij+1}.

We can also define a similar condition for the exposure c_{ij} as

$$c_{ij+1} \sim h(c; \bar{G}_{ij}), \tag{9.15}$$

where $h(.)$ is an arbitrary density function for c_{ij} and $\bar{G}_{ij}$ includes observed history as $\bar{F}_{ij}$ does but also d_{ij+1}. Similarly the conditional sequential randomization for c_{ij} conditioning on v_i is

$$c_{ij+1} \sim h(c; \bar{G}_{ij}, v_i). \tag{9.16}$$

These conditions will play a key role in determining the modeling strategy.

The sequential randomization condition can be utilized in the estimation of causal effects. First confounding factors in $\bar{F}_{ij}$ can be controlled directly by including appropriate terms in $\mathbf{X}_{ij}$, given the model is correctly specified. Other approaches such as stratification by these factors can also be used. Note that including u_i as unknown parameters is equivalent to stratification by subject and results in estimation by within subject comparisons.

9.4.5 A View in Directional Acyclic Graphs

The development of directional acyclic graphs (DAG) [15] provides a powerful and intuitive tool for investigating potential confounding and finding approaches to deal with it. A DAG is a graph consisting of nodes, each represents a variable, linked by directional edges (indicated by arrows), but the directional edges do not form directional cycles. For each arrow the starting node is a parent of the ending node, and the ending node is the descendant of the starting nodes. A DAG represents how the distribution of one variable depends on the others. Under the assumption of Markov factorization, the joint distribution of all variables in a DAG can be factorized as a product of distributions of descendants conditional on their parents, while for those without any parent, their distributions are unconditional. Figure 9.1 shows the DAG with conditional sequential randomization for both d_{ij} and c_{ij}, and L_i as the source of correlation between potential confounding factors s_i, u_i and v_i. The arrows to d_{ij+1} represent typical dose adjustments but are by no means exhaustive. From the DAG, it can be found that for the distribution of d_{ij+1}, $\bar{F}_{ij}$ should include c_{ij} and y_{ij}, while for the distribution of c_{ij}, $\bar{G}_{ij}$ should include d_{ij}. The assumption of stationary L_i is often an approximation. For example, when the patient's characteristics change very quickly, there may be a time varying L_{ij} and paths from its node to y_{ij+1} and d_{ij+1}.

A DAG can also be used to determine the confounding on the causal effect of one variable on another one. Intuitively confounding occurs when a factor affects both the exposure (d_{ij} or c_{ij}) and the response y_{ij} as indicated by paths (regardless of direction) linking the exposure and the response via u_i, s_i and v_i. It also shows that conditional on u_i the path is blocked hence a causal effect can be identified by conditioning. A useful result in DAG theory is the back-door criterion for a set of nodes blocking confounding paths [15]. Specifically for Fig. 9.1, a node set blocks confounding paths between c_{ij} and y_{ij} if: (1) no node in the set is a descendant of

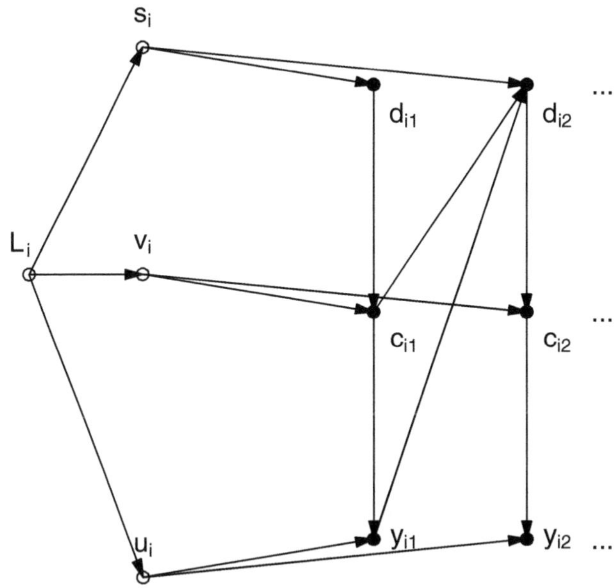

Fig. 9.1 Directional acyclic graph with longitudinal observations and exposure/response dependent dose adjustment and potential subject level confounding factors

c_{ij} and (2) the set blocks every path between c_{ij} and y_{ij} that has an arrow into c_{ij}. It can be found that the block set is v_i and s_i. A practical implication of the back-door criterion is that when all confounding paths between, e.g., c_{ij} and y_{ij}, are blocked by a set of nodes, causal effects of c_{ij} on y_{ij} can be estimated by conditioning on elements in the blocking set.

9.5 Joint Modeling and Likelihood Function

In this section we consider estimation of the dose-response and exposure-response models for different scenarios of dose adjustment and patterns of confounding with the assumption of conditional sequential randomization for d_{ij} and c_{ij}. We use y_{ij} as a general notation for the measurement of a safety or efficacy response, and use $f(y_{ij}, c_{ij}, d_{ij})$ as the joint distribution of y_{ij}, c_{ij} and d_{ij} with $\mathbf{X}_{ij}$ omitted. The joint distribution for y_{ij}, c_{ij} and d_{ij}, conditioning on subject level factors, can be factorized into

$$f(y_{ij}, c_{ij}, d_{ij}|s_i, u_i, v_i) = l(y_{ij}|c_{ij}, u_i)h(c_{ij}|\bar{G}_{ij-1}, v_i)g(d_{ij}|\bar{F}_{ij-1}, s_i), \tag{9.17}$$

where $l(y|c,u)$, $h(c|\bar{G}, v)$ and $g(d|\bar{F}, s)$ are conditional distributions for y given c, for c given $\bar{G}$ and v and for d given $\bar{F}$ and s, respectively, and we drop the parameters in the models for simplicity. u_i, v_i and s_i are subject level random variables as defined in the exposure-response, dose exposure and dose adjustment models, respectively, and $\bar{F}_{ij-1}$ and $\bar{G}_{ij-1}$ are defined as in the previous section. The factorization is in fact the Markov factorization for the DAG in Fig. 9.1, which underlines some assumptions we make based on our models. For example, one is that given c_{ij} and s_i, y_{ij} is independent of d_{ij}. The marginal likelihood with parameters $\boldsymbol{\Omega}$ for y_{ij}, c_{ij} and d_{ij} is then

$$L(\boldsymbol{\Omega}) = \prod_{i=1}^{n} \int_{(s_i,u_i,v_i)} \prod_{j=1}^{J} f(y_{ij}, c_{ij}, d_{ij}|s_i, u_i, v_i)\, dH(s_i, u_i, v_i), \tag{9.18}$$

where $\int_{\mathbf{c}}$ is a shorthand for a multiple integration with respect to all the components in $\mathbf{c}$ and $H(\mathbf{c})$ is their joint distribution. The potential correlations between s_i, u_i and v_i play an important role in fitting these models. To estimate the parameters in the models using the maximum likelihood estimate (MLE), one needs to evaluate (9.18), which is generally difficult except under some simple situations, e.g., when all models are linear. The approach also needs correct specification of the joint distributions of s_i, u_i and v_i. In the following we consider situations when (9.18) can be simplified and alternatives to the full joint modeling approach can be found.

The MLE approach based on the joint marginal likelihood can be simplified by examining the DAG in Fig. 9.1. When the exposure-response model is the only consideration, it is possible to separate the dose-exposure part even when s_i is correlated with u_i and v_i. By conditioning on d_{ij+1} (as a part of $\bar{G}_{ij}$) we block the path from s_i to c_{ij+1}, and consequently can fit a joint model for y_{ij} and c_{ij} only. However, in doing so we eliminate the contribution of the exposure change due to the change in d_{ij} even if it is not confounded and get a safe but less efficient parameter estimate in the joint model.

Obviously, when s_i, u_i and v_i are independent, hence d_{ij} and c_{ij} are unconditional sequential randomization, the likelihood can be factored into three parts for y_{ij}, c_{ij} and d_{ij} separately. Therefore, the three models can be fitted separately. Note that this is a very strong assumption even when d_{ij}s are randomized. For example, an old patient may have impaired liver or renal function leading to a higher exposure (hence a higher u_i value), while he also has a higher risk of having an AE (hence higher s_i value). The positive correlation will give a false positive correlation between y_{ij} and c_{ij}. Therefore, dose randomization does not control the bias in the exposure-response relationship, unless an appropriate approach is used. Some approaches will be discussed in Sect. 9.7.

When s_i is independent of u_i and v_i, at least one can factor out $f(d_{ij}|\bar{F}_{ij-1}, s_i)$ from the marginal likelihood (9.18) since d_{ij} is a DAR. Formula (9.18) becomes

$$L(\boldsymbol{\Omega}) = \prod_{i=1}^{n} L_d \int_{(u_i,v_i)} \prod_{j=1}^{J} l(y_{ij}|c_{ij}, u_i) h(c_{ij}|\bar{G}_{ij-1}, v_i)\, dH(u_i, v_i), \tag{9.19}$$

where L_d is the part relating to d_{ij}s only. Therefore, joint modeling of the exposure-response and dose-exposure models only is a valid approach when v_i and u_i are correlated. For a large range of model combinations, it can be implemented in standard softwares such as SAS proc MIXED and proc GLIMMIX without extra programming. More general models can be fitted jointly using proc NLMIXED with some extra programming. A sample program is given in the Appendix.

When the c_{ij}s are not observed and the dose-response relationship is the only concern, the likelihood function can also be simplified. As discussed in Sect. 9.3, for some types of exposure-response models, one may be able to combine it with the dose-exposure model (9.2) to get a single model such as (9.7). Therefore, the marginal likelihood (9.18) can be written as

$$L(\boldsymbol{\Omega}) = \prod_{i=1}^{n} \int_{(u_i^*, s_i)} \prod_{j=1}^{J} l^*(y_{ij}|d_{ij}, u_i^*) g(d_{ij}|\bar{G}_{ij-1}, s_i) dH(u_i^*, s_i), \tag{9.20}$$

where $u_i^* = u_i + \beta v_i$, $l^*(.)$ is the likelihood function for y_{ij} as given in model (9.7). When s_i is independent of u_i^* the dose adjustment is a DAR and the dose exposure model can be fitted separately. This situation includes response-related dose adjustments and, as a trivial example, randomized dose trials. However, exposure-related dose adjustments do not satisfy the DAR condition, since c_{ij}s are not observed. Consequently $g(d_{ij}|\bar{G}_{ij-1}, s_i)$, cannot be factored out if $\bar{G}_{ij-1}$ contains c_{ij-1}. When d_{ij} is a sequential randomization conditionally, i.e. s_i is not independent of u_i^*, a joint modeling approach can be used based on (9.20).

9.6 Alternatives to Joint Modeling

Although joint modeling provides a general approach to dealing with subject level confounding factors, it has a number of drawbacks, for example, the need for specifying the joint distribution and complexity of model fitting techniques. An easy alternative when using longitudinal exposure-response models is conditioning on individual subject effects. This approach has an intuitive interpretation. It leads to using within subject comparisons for estimating β, which is free from individual level confounding factors. A simple way of conditioning on, say, u_i when J is sufficiently large is to treat u_i as an unknown parameter and estimate it. Note that the condition of sufficiently large J is needed in general. Although under some situations, such as when case-control studies or trials are designed in such a way that u_i and exposure are orthogonal, this is unnecessary, dose adjustments in general do not lead to similar situations. Therefore, small sample (i.e., small J) properties of this approach are of technical and practical interest.

With this approach it is also important to count the total variation in exposure that can be used for the conditioning approach, since between-subject variation is eliminated by conditioning. The variance of $\hat{\beta}_c$, the β estimate treating u_is as unknown parameters can be found in the text on panel data analysis with fixed individual intercepts [7]:

$$\text{var}(\hat{\beta}_c) = \sigma_e^2 \Big(\sum_{i=1}^{n} \sum_{j=1}^{J} (c_{ij} - \bar{c}_{i.})^2 \Big)^{-1}, \tag{9.21}$$

where $\bar{c}_{i.} = \sum_{j=1}^{J} c_{ij}/J$. Therefore, it is straightforward to estimate the precision of $\hat{\beta}_c$ given the dose-adjustment mechanism and variability of the exposure, if available, at the design stage.

Sometimes the exposure in a trial may be measured repeatedly, e.g., for therapeutic dose monitoring, but the response may only be measured once. Although joint modeling can be used, a simple alternative known as the control function (CF) approach [20] also exists. The idea is that since the confounding in u_i is due to its correlation with v_i, if one can estimate v_i then one may use a direct adjustment by including the estimates in the exposure-response model. A key assumption for this approach is that the conditional mean of u_i given v_i can be written as $E(u_i|v_i) = av_i$, where a is a constant. It is satisfied when u_i and v_i are jointly normally distributed, or $u_i = av_i + w_i$ in which v_i is a shared latent variable between the dose-exposure and exposure-response models. The efficiency of this approach depends on the prediction of v_i, while repeated measurements provides data for the prediction. The approach consists of two stages of simple model fitting:

- Fit the mixed effect dose-exposure model to repeated exposure data and obtain prediction $\hat{v}_i$ for each subject, using common approaches for mixed models such as the best linear unbiased prediction (BLUP).
- Fit the exposure-response model adding $\hat{v}_i$ from the first step as a covariate.

This approach has been widely used in combination with IV in social science. For an application in dose-exposure modeling, see [19]. As a general approach it can also be used for estimation of the dose-response relationship, given that a good prediction for s_i can be obtained by fitting the dose-adjustment model, and based on the prediction, a proper approach to fitting the dose-exposure model (9.7), e.g., a direct adjustment including the predicted s_i in the model, is implemented.

9.7 Instrumental Variable Approach for RCC Trials

The IV approach is also an alternative to joint modeling but it is rather special and different from those in the last section as it does not need the assumption of no unobserved confounding factors. We assume that the exposure is measured repeatedly but the response is measured only until the target exposure has been achieved, e.g. at the jth visit. Let y_i and c_i be the response and exposure of patient i at visit j and we suppress the subscript j. We assume the following exposure-response model:

$$y_i = \boldsymbol{\beta}_x^T \mathbf{X}_i + \beta c_i + u_i + \varepsilon_i, \tag{9.22}$$

where $\mathbf{X}_i$ is a set of covariates including the intercept, $\boldsymbol{\beta}_x$ are the corresponding parameters. Here $u_i + \varepsilon_i$ is treated as a single error term, but potential confounding comes from u_i only. Although patients are randomized into these ranges and a part of variation in c_i is indeed randomized, there is often still significant variation potentially confounded within each range. The reason is mainly due to feasibility; most common RCC designs use only two ranges, hence the sizes of the ranges are very large and leave variation within them uncontrolled. Therefore, a least squares estimate for β is still biased.

The IV method is a powerful tool to eliminate the confounding bias. An IV is a variable relating to the exposure but not directly relating to the potential outcome. In RCC trials randomization is a natural IV, since randomization is independent of the response, and dose adjustment to achieve a certain range of exposure makes it strongly related to the exposure. For linear and a few special nonlinear models, the following two-stage IV (2SIV) approach is very easy to implement [18].

- Fit the randomization-exposure model

$$\log(c_i) = \boldsymbol{\alpha}^T \mathbf{R}_i + v_i + e_i, \tag{9.23}$$

  where $\mathbf{R}_i = (R_{i1}, \ldots, R_{iK})$ with $R_{ik} = 1$ if subject i is randomized to group k and $R_{ik} = 0$ otherwise, to the exposure data and obtain the predicted mean exposure for each group.
- Fit the exposure-response model $y_i = \boldsymbol{\beta}_x^T \mathbf{X}_i + \beta c_i + u_i + \varepsilon_i$ with c_i replaced by the predicted mean exposure of the group subject i is randomized to. The IV estimate $\hat{\beta}_{IV}$ is the coefficient for the predicted exposure.

$\hat{\beta}_{IV}$ does not have confounding bias; but its small sample bias depends mainly on how closely the dose is correlated to the exposure.

$\hat{\beta}_{IV}$ is also the solution to the following estimating equation (EE)

$$S(\boldsymbol{\beta}) = \sum_{i=1}^{N} \mathbf{Z}_i (y_i - \mathbf{X}_i^T \boldsymbol{\beta}_x - c_i \beta), \tag{9.24}$$

where $\boldsymbol{\beta} = (\boldsymbol{\beta}_x, \beta)$ and $\mathbf{Z}_i = (\mathbf{X}_i, \mathbf{R}_i)$. For simplicity, one may consider non-confounders $\mathbf{X}_i$ as their own IVs so we refer to $\mathbf{Z}_i$ as a set of IVs. Although the two-stage approach is more convenient than solving the EE (9.24), the EE plays a key role in checking if $\hat{\beta}_{IV}$ is unbiased, since the key condition for unbiasedness of the solution to the EE is $E(S(\boldsymbol{\beta})) = 0$, which is satisfied here since $\sum_{i=1}^{N} E(\mathbf{Z}_i(u_i + \varepsilon_i)) = 0$, when R_i is randomized. However, the IV approach can not eliminate the bias due to confounding factors in treatment heterogeneity u_{bi} in the following model.

$$y_i = \boldsymbol{\beta}_x^T \mathbf{X}_i + (\beta + u_{bi})c_i + u_i + \varepsilon_i, \tag{9.25}$$

although in general it helps to reduce the bias. For more details and the impacts of confounding in u_{bi}, see [18].

9.8 Testing Confounding Factors

Although in general it is not possible to test for the existence of confounding factors, with the model assumptions the correlations between u_i, v_i and s_i can be tested and their impact on confounding can be judged based on the DAG, the conditions of sequential randomization and approaches for estimation of β and θ. Since often s_i is independent of u_i and v_i, we consider this situation and test the correlation between u_i and v_i. The idea is to predict v_i based on the dose-exposure model, then test if it has a significant impact in the exposure-response model as a surrogate for u_i. Based on the dose-exposure and dose adjustment models, the following two-stage approach can be used.

- First fit the dose-exposure model and obtain prediction $\hat{v}_i$.
- Fit the exposure-response model including $\hat{v}_i$ as a fixed effect in model

$$y_{ij} = \beta c_{ij} + \mathbf{X}_{ij}^T \boldsymbol{\beta}_x + u_i + \xi \hat{v}_i + \varepsilon_{ij}. \tag{9.26}$$

- Test $H_0 : \xi = 0$ vs $H_a : \xi \neq 0$.

For fitting the dose-exposure model to predict v_i, we have assumed independence between s_i and v_i. This test does not test the confounding bias directly. One obvious example is that when the estimation is based on within subject comparisons, hence subject level confounding does not cause bias in the estimate. Nevertheless the relationship between u_i and v_i as latent variables may still be useful information.

There are other ways to test independence between u_i and v_i. For example, one can fit the joint dose-exposure and exposure-response models, then use the likelihood ratio test to test the correlation between u_i and v_i. However, this approach involves fitting a complex model, and will not be discussed here.

There is a natural link between the two-stage test procedure and the control function approach for estimating β. The test procedure is similar to the approach of using the control function method for estimating β with dose as an IV. The test is equivalent to the test for the difference between the control function adjusted estimate (with $\hat{v}_i$ in the model) and that without adjustment, known as the Hausman test [5]. The two-stage approach in some specific situations is equivalent to a test by comparing the two estimates [6].

9.9 Simulation Study

The performance of different approaches for estimating the causal effect β in the simple exposure-response model (9.4) was investigated by a simulation study. The simulation setting was based on typical practical scenarios. Suppose that for drug A the exposure is dose-proportional and the dose-exposure model can be written as

$$\log(c_{ij}) = \log(d_{ij}) + v_i + e_{ij}. \tag{9.27}$$

In addition we assume that c_{ij} has 20 and 50 % intra- and inter-subject coefficient of variation (CV), which is equivalent to $\sigma_v = 0.47$ and $\sigma_e = 0.2$. The response is the change from baseline of a continuous measurement (e.g., a biomarker) and the exposure-response model is

$$\log(y_{ij}) = \log(c_{ij}) + u_i + \varepsilon_{ij}. \tag{9.28}$$

We further assume that 50 % of u_i is from v_i and another 50 % is independent, i.e. $u_i = v_i + w_i$ and w_i is independent, and $\text{var}(\varepsilon_{ij}) = 0.3$. Note that this model does not allow zero concentration and we assume that there is no placebo arm. A more flexible model may include the baseline, but we opt to use model (9.28) for simplicity. Suppose that the target exposure level is 5 ng/mL, hence the starting dose is set 5 mg. Due to high inter-subject CV, a therapeutic dose monitoring with repeated exposure measurements at a number of visits is considered to allow a dose increase to 10 mg if $c_{ij} < 5$ ng/mL at any visit j. An RCC trial was considered as an alternative for a more reliable causal effect estimation. In the RCC trial patients were randomized into two groups, one with exposure lower than 5 ng/mL while the other higher than 5 ng/mL. Since a dose reduction would be needed when $c_{ij} > 5$ ng/mL, a 2.5 mg dose was introduced, but the starting dose for all patients was 5 mg. A subject randomized to the lower range group would have dose reduced by one level if at the current dose $c_{ij} > 5$ ng/mL, and who randomized into the higher range group would have dose increased by one level if at the current dose $c_{ij} < 5$ ng/mL, subject to availability of the increased/decreased dose level.

For the response measurement two scenarios were considered, one with repeated response measurements y_{ij} and the other only measured at one visit when the drug effect has become stable. Therefore, in the latter scenario, there is only one response measurement available. We first consider the performance of the approaches and models for estimating β with the therapeutic dose monitoring design. For each scenario and a number of J values and sample sizes, 500 simulation data sets were generated and β in the exposure-response model (9.4) was estimated with a number of approaches. The approaches we used are (1) joint modeling with the dose-exposure model, (2) fitting a mixed model without adjustment, (3) fitting a fixed subject effect model (treating subject effects as unknown parameters), (4) unadjusted exposure-response model at visit J only, (5) control function estimate at visit J only (Sect. 9.6). The mean, variance, minimum and maximum of the

Table 9.1 Summary of parameter estimates using different methods and models. The true values of β equals to 1. The five models are (1) joint modeling, (2) mixed model, (3) fixed subject effect model, (4) unadjusted exposure-response model at visit J only, and (5) control function estimate at visit J only

n	Model/approach	Mean	Var	Min	Max	Mean	Var	Min	Max
		$J = 2$				$J = 3$			
50	1	0.9983	0.0098	0.8021	1.1858	1.0080	0.0063	0.8404	1.2786
	2	1.1609	0.0138	0.9014	1.4881	1.1175	0.0066	0.9423	1.3923
	3	1.0008	0.0100	0.7911	1.1761	1.0081	0.0064	0.8366	1.2934
	4	1.7040	0.0834	0.8626	2.4319	1.7748	0.0420	1.3137	2.4891
	5	1.0005	0.0704	0.4665	1.9738	0.9958	0.0761	0.2487	1.6911
100	1	0.9917	0.0042	0.8222	1.1869	0.9967	0.0032	0.8313	1.1342
	2	1.1531	0.0061	0.9820	1.4141	1.1049	0.0034	0.9087	1.2355
	3	0.9916	0.0043	0.8314	1.1850	0.9967	0.0031	0.8314	1.1276
	4	1.7171	0.0282	1.2994	2.0981	1.7536	0.0206	1.3831	2.1865
	5	1.0002	0.0338	0.5742	1.4631	1.0049	0.0374	0.4701	1.5912
200	1	0.9960	0.0028	0.8596	1.1424	0.9936	0.0016	0.8946	1.1004
	2	1.1610	0.0040	1.0207	1.3137	1.1041	0.0018	0.9784	1.2028
	3	0.9961	0.0029	0.8652	1.1515	0.9940	0.0016	0.8933	1.1047
	4	1.7186	0.0191	1.3995	2.0604	1.7437	0.0103	1.4799	1.9753
	5	1.0065	0.0178	0.7164	1.3231	0.9838	0.0185	0.7242	1.3318
		$J = 5$				$J = 7$			
50	1	1.0011	0.0043	0.8437	1.1857	0.9925	0.0024	0.8771	1.1049
	2	1.0572	0.0044	0.8886	1.2624	1.0309	0.0023	0.9241	1.1379
	3	1.0007	0.0042	0.8428	1.1749	0.9920	0.0023	0.8841	1.1071
	4	1.7477	0.0359	1.2309	2.1110	1.7268	0.0418	1.1078	2.2733
	5	1.0157	0.0559	0.4479	1.5261	1.0217	0.0588	0.4196	1.5487
100	1	0.9951	0.0022	0.8849	1.1164	0.9930	0.0012	0.9013	1.0818
	2	1.0521	0.0023	0.9406	1.1765	1.0331	0.0012	0.9407	1.1209
	3	0.9949	0.0022	0.8864	1.1167	0.9929	0.0012	0.9010	1.0821
	4	1.7316	0.0194	1.3217	2.0334	1.7144	0.0198	1.3438	2.0059
	5	1.0166	0.0239	0.6577	1.3729	0.9882	0.0333	0.2860	1.4424
200	1	1.0010	0.0011	0.9130	1.1057	0.9952	0.0007	0.9291	1.0560
	2	1.0584	0.0011	0.9666	1.1694	1.0352	0.0007	0.9721	1.0995
	3	1.0010	0.0011	0.9128	1.1081	0.9951	0.0007	0.9283	1.0551
	4	1.7323	0.0091	1.5099	1.9279	1.7190	0.0101	1.4267	1.9275
	5	1.0078	0.0152	0.5942	1.3455	1.0012	0.0164	0.6738	1.4102

estimates are present in Table 9.1. The joint modeling approach resulted in estimates with almost no bias and the lowest variance among all the estimates for all scenarios. The mixed model estimate had considerable bias when $J = 2$ but the bias reduced with increasing J. The fixed subject effect model estimates also had almost no bias with variance slightly higher than the joint modeling ones when J is small. The estimates based on response data at the end visit $j = J$ showed that the unadjusted

Table 9.2 Summary of parameter estimates using different methods and models with RCC design. The true values of β equals to 1. The six models or approaches are (1) joint modeling, (2) mixed model, (3) fixed subject effect model, (4) unadjusted exposure-response model at visit J only, (5) IV estimate at visit J only and (6) control function estimate at visit J only

n	Model/approach	Mean	Var	Min	Max	Mean	Var	Min	Max
		$J = 2$				$J = 3$			
50	1	0.9965	0.0090	0.7179	1.2812	0.9992	0.0049	0.7929	1.1782
	2	1.1582	0.0121	0.8154	1.6076	1.0629	0.0054	0.8299	1.2514
	3	0.9975	0.0094	0.7017	1.2808	0.9996	0.0052	0.7957	1.1855
	4	1.3762	0.0435	0.7971	1.9150	1.1495	0.0341	0.5279	1.8336
	5	0.9589	0.0979	−0.0013	1.9576	0.9914	0.0516	0.3541	1.6394
	6	0.8715	0.0760	−0.1622	1.6152	0.8949	0.0383	0.3273	1.5345
100	1	1.0017	0.0056	0.7261	1.2507	0.9983	0.0024	0.8614	1.1725
	2	1.1626	0.0072	0.9191	1.4596	1.0605	0.0026	0.8976	1.2239
	3	1.0010	0.0058	0.7354	1.2483	0.9973	0.0026	0.8554	1.1824
	4	1.3903	0.0208	1.0180	1.8623	1.1548	0.0164	0.7364	1.5553
	5	0.9961	0.0486	0.3257	1.6628	0.9946	0.0228	0.4590	1.4026
	6	0.8698	0.0373	0.3470	1.4778	0.8860	0.0177	0.4517	1.2806
		$J = 5$				$J = 7$			
50	1	0.9994	0.0028	0.7994	1.1359	0.9987	0.0017	0.8839	1.1177
	2	1.0256	0.0029	0.8388	1.1834	1.0133	0.0019	0.8890	1.1326
	3	0.9981	0.0029	0.7985	1.1311	0.9978	0.0019	0.8751	1.1261
	4	1.0835	0.0290	0.6188	1.6406	1.0437	0.0231	0.5292	1.4270
	5	1.0151	0.0364	0.4580	1.6954	1.0005	0.0261	0.3967	1.4267
	6	0.9223	0.0298	0.4806	1.5948	0.9211	0.0233	0.4726	1.3618
100	1	1.0001	0.0013	0.8933	1.1033	0.9980	0.0009	0.9042	1.0871
	2	1.0257	0.0014	0.9158	1.1382	1.0129	0.0010	0.9359	1.1091
	3	0.9996	0.0014	0.8901	1.1094	0.9977	0.0010	0.9152	1.0878
	4	1.0764	0.0148	0.7228	1.3897	1.0455	0.0129	0.7058	1.3709
	5	1.0043	0.0166	0.6618	1.3566	0.9935	0.0163	0.6253	1.3453
	6	0.9111	0.0142	0.5154	1.2139	0.9165	0.0138	0.5846	1.2272

one was severely biased, while the control function approach eliminated the bias successfully even with very small J and sample sizes.

Next we examine the performance of the approaches and models for estimating β for the same scenarios when the RCC design is applied. In addition to those used with the therapeutic dose monitoring design, an additional one is the two-stage IV estimates. To using the control function approach for the analysis based on response at the end visit only one can estimate v_i using repeated exposure data or the exposure data at the end visit only. The latter is in fact equivalent to the two-stage IV estimate for the linear exposure-response model [20]. The results are presented in Table 9.2. The results are generally similar to those when the therapeutic dose monitoring design was used. For example, when $J = 2$ the joint modeling estimate had the lowest variance, but only 5 % lower than the estimate with the fixed subject effect model. Special features of the results were that the bias of unadjusted estimate was significantly reduced due to the RCC. The IV estimate was generally unbiased except with slight bias when $J = 2$ and $n = 50$. This is likely to be due to the fact that the exposure was not well controlled when $J = 2$ hence randomized ranges as an IV was weak. With a weak IV the IV estimate may behave badly particularly when the sample size is small. The bias in the control function approach using repeated exposure data was, however, unexpected.

In general we found that the joint modeling approach provided a very good estimate for β. When repeated response measurements were available, the estimate with fixed subject effect model was also good, while the mixed model approach might subject to some bias. When the response was only measured at the end visit, the control function estimate provided a good estimate for the therapeutic dose monitoring design but bias occurred when the RCC design was used. In this case, the two-stage IV estimate provided an unbiased estimate as long as J and the samples size were not all small.

9.10 Discussion

In this chapter we have considered the role and impacts of dose adjustments, particularly individual exposure-dependent and response-dependent dose adjustments in the analysis of dose-exposure and exposure-response relationships. Exposure-response modeling is a key part of modeling and simulation during drug development with increasing applications in industry as well as in academic research. Recent research in causal effect determination in econometrics and medical statistics has led useful techniques, and some are closely related to causal effect estimation in studies with dose adjustments. This chapter has shown a number of possible combinations of approaches from both areas and potential applications in drug development.

A number of interesting topics can not be covered by this chapter due to the space limit. Exposure may be considered as a mediator of dose, and the exposure-response relationship as representing indirect effects of dose changes. We have assumed that there is not a direct effect of the dose on the response in any of the models. This is a reasonable assumption in common cases since a drug is normally absorbed in the central system (blood or plasma) then transported to target organs where drug effects take place. Therefore, PK samples taken from the central system is a good surrogate for the change in exposure at target organs due to dose change in general. However, it is possible a drug may bypass the central system. A well known example is the first pass scenario in clinical pharmacology. Another interesting topic is the heterogeneity in exposure-response relationship among patients with different observed characteristics. When the characteristics are observed and included in the models, the g-computation is a powerful tool to calculate causal effects in the population, e.g., the average diastolic blood pressure reduction by 1 unit concentration increase in a mixed population of different characteristic values. In practice the g-computation may be implemented by simulation.

We have concentrated on individual response-dependent and exposure-dependent dose adjustments. The approaches can be adapted for some other scenarios. For example, in a typical phase I dose escalation trial with a $3 + 3$ design, a number of doses are tested with cohorts of three subjects. At the beginning, a dose is given to one cohort. If no AE occurs, the next higher dose will be tested on another cohort; if there is one AE, the same dose will be tested, and if there is more than one AE the next lower dose will be tested. In this case the cohort can be considered as an individual and the exposure-AE model (9.12) can be used to model the number of subjects with AEs. Then the model can be used to describe the dose adjustment mechanism.

A variant of model (9.4) has c_{ij} replaced by the right hand side of model (9.2) without e_{ij}, a model often used in PK/PD modeling. If the variant is the true model, model (9.4) is a classical measurement error model and fitting the model with observed c_{ij} results in inconsistent parameter estimates even without confounding. This situation is beyond the scope of this chapter.

One key feature of dynamic treatment regimen is the optimal treatment selection and adjustment, which we have not considered in this chapter. The omission is partially due to the technical complexity of the method and partially the feasibility of implementing complex dosing formulae in practice. Nevertheless, some quasi-optimal approaches might be feasible. In some special situations optimal dose adjustments using complex algorithms can also find applications. Some general principles discussed here, e.g. when the dose-adjustment process can be ignored, also apply. Some approaches, such as the control function approach, that needs fitting the dose-adjustment model, may depend on specific situations.

Appendix

This section provides a SAS program to fit the following joint model:

$$\begin{aligned} \log(c_{ij}) &= \theta \log(d_{ij}) + v_i + e_{ij}, \\ y_{ij} &= Emax/(1 + EC_{50}/c_{ij}) + u_i + \varepsilon_{ij}, \end{aligned} \tag{9.29}$$

where the first one is the power model (9.2) and the second one is known as Emax model. u_i and v_i are correlated, hence u_i is a confounding factor. For simplicity no other random effect is included. This model cannot be fitted with SAS proc MIXED or GLIMMIX due to nonlinearity in the Emax model.

In dataset “joint” below, one variable `rij` contains both the exposure and response variables as two records identified by an indicator `ind`. When `ind` = `"pk"`, `rij` $= \log(c_{ij})$ and `logdose` $= \log(d_{ij})$ in the power model. Otherwise (when `ind` = `"resp"`), `rij` $= y_{ij}$ in the Emax model, and `logcij` $= \log(c_{ij})$ in the power model. In the program, variable `i` is the subject identifier, `siguv` is the covariance between u_i and v_i, and sigp and sigr are var(e_{ij}) and var(ε_{ij}), respectively.

```
proc nlmixed data=simu qpoints=6;
  parms sigu=1 sigv=1 sigr=1 sigp=1 sige=1 theta=1 emax=1
     ec50=1;
  bounds sigu sigv sigr sigp sige >0;
   if ind="pk" then do;
      pred=theta*logdose +vi; g=sigp;
   end;
   else if ind="resp" then do;
      pred=emax/(1+ec50/exp(logcij))+ui; g=sigr;
   end;
  model rij~normal(pred,g);
  random ui vi~normal([0,0],[sigu,siguv,sigv]) subject=i;
run;
```

The program is illustration only. Adjustments on starting parameter values and options in the procedure is often necessary to fit real data. The program can be adapted to fit other types of response by specifying the likelihood function.

References

1. Angrist, J.D., Imbens, G., Rubin, D.B.: Identification of causal effects using instrumental variables. *Journal of the American Statistical Association* **94**, 444–455 (1996)
2. Diaz, F.J., Rivera, T.E., Josiassen, R.C., de Leon J.: Individualizing drug dosage by using a random intercept linear model. *Statist. Med.* **26**, 2052–2073 (2007)
3. Food and Drug Administration: *Exposure-Response Relationships – Study Design, Data Analysis, and Regulatory Applications* (2003)
4. Gill, R., Robins, J.: Causal inference in complex longitudinal studies: the continuous case. *Annals of Statistics* **29**, 1785–1811 (2001)
5. Hausman, J.: Specification tests in econometrics. *Econometrica* **46**, 1251–1271 (1978)
6. Hausman, J., Taylor, W.: Panel data and unobservable individual effects. *Econometrica* **49**, 1377–1399 (1981)
7. Hsiao, C.: *Analysis of Panel Data*. Cambridge University Press. Cambridge (1989)
8. Karlsson, K.E., Grahnen, A., Karlsson, M.O., Jonsson, E.N.: Randomized exposure-controlled trials; impact of randomization and analysis strategies. *Br. J. Clin. Pharmacol.* **64**, 266–77 (2007)
9. Kraiczi, H., Jang, T., Ludden, T., Peck, C.C.: Randomized concentration-controlled trials: motivations, use, and limitations. *Clin. Pharmacol. Ther.* **74**, 203–214 (2003)
10. Little, R.J.A., Rubin, D.B.: *Statistical analysis with missing data*, 2nd edn. Wiley. Hoboken, NJ (2002)
11. Lok, J.: Statistical modeling of causal effects in continuous time. *Annals of Statistics* **36**, 1464–1507 (2008)
12. Murphy, S.: Optimal dynamic treatment regimes. *Journal of the Royal Statistical Society: Series B* **65**, 331–584, (2003)
13. Murphy, S., van der Laan, M.J., Robins, J.M.: Marginal mean models for dynamic regimes. *Journal of the American Statistical Association* **96**, 1410–1423 (2001)
14. O'Quigley, J., et al.: Dynamic calibration of pharmacokinetic parameters in dose-finding studies. *Biostatistics* **11**, 537–545 (2010)
15. Pearl, J.: Causal diagrams for empirical research. *Biometrika* **82**, 669–710 (1995)

16. Robins, J.M., Hernán, M.A.: Estimation of the causal effects of time-varying exposures. In: G. Fitzmaurice, M. Davidian, G. Verbeke, G. Molenberghs (eds.) *Longitudinal Data Analysis*. Chapman & Hall/CRC. Boca Raton (2009)
17. Sanathanan, L.P., Peck, C.C.: The randomized concentration-controlled trial: An evaluation of its sample size efficiency. *Controlled Clinical Trials* **12**, 780–94 (1991)
18. Wang, J.: Determining causal effect in exposure-response relationship with randomized concentration controlled trials. *Journal of Biopharmaceutical Statistics* **24**, 874–92 (2014)
19. Wang, J.: Dose as instrumental variable in exposure-safety analysis using count models. *Journal of Biopharmaceutical Statistics* **22**, 565–581 (2012)
20. Wooldridge, J.: Control function and related methods. In: *What's New in Econometrics? Lecture Notes 6*. National Bureau of Economic Research (2007). URL http://www.nber.org/WNE/lect-6-controlfuncs.pdf

Chapter 10
Different Methods to Analyse the Results of a Randomized Controlled Trial with More Than One Follow-Up Measurement

Jos W.R. Twisk

Abstract In this chapter, an overview is given of different methods to analyse data from a randomised controlled trial (RCT) with more than one follow-up measurement. For a continuous outcome variable, a classical GLM for repeated measurements can be used to analyse the difference in development over time between the intervention and control group. However, because GLM for repeated measurements has some major disadvantages (e.g., only suitable for complete cases), it is advised to use more advanced statistical techniques such as mixed model analysis or Generalised Estimating Equations (GEE). The biggest problem with the analysis of data from an RCT with more than one follow-up measurement is the possible need for an adjustment for baseline differences. To take these differences into account a longitudinal analysis of covariance, an autoregressive analysis or a 'combination' approach can be used. The choice for a particular method depends on the characteristics of the data. For dichotomous outcome variables, an adjustment for baseline differences between the groups is mostly not necessary. Regarding the more advanced statistical techniques it was shown that the effect measures (i.e. odds ratios) differ between a logistic mixed model analysis and a logistic GEE analysis. This difference between these two methods was not observed in the analysis of a continuous outcome variable. Based on several arguments (e.g., mathematical complexity, unstable results, etc.), it was suggested that a logistic GEE analysis has to be preferred above a logistic mixed model analysis.

10.1 Introduction

Randomized controlled trials (RCT's) are considered to be the gold standard for evaluating the effect of a certain intervention [10]. In a randomized controlled trial, the population under study is randomly divided into an intervention group and a

J.W.R. Twisk (✉)
Department of Epidemiology and Biostatistics, VU Medical Centre, Amsterdam, The Netherlands

e-mail: jwr.twisk@vumc.nl

K. van Montfort et al. (eds.), *Developments in Statistical Evaluation of Clinical Trials*, DOI 10.1007/978-3-642-55345-5_10

non intervention or control group (e.g., a placebo group or a group with 'usual' care). Regarding the analysis of RCT data a distinction must be made between studies with only one follow-up measurement and studies with more than one follow-up measurement. When there is only one follow-up measurement relatively simple statistical techniques can be used to evaluate the effect of the intervention, while when more than one follow-up measurement is considered, in general, more advanced statistical techniques are necessary.

In the past decade, an RCT with only one follow-up measurement has become very rare. At least one short-term follow-up measurement and one long-term follow-up measurement 'must' be performed. However, more than two follow-up measurements are usually performed in order to investigate the 'development over time' of the outcome variable, and to compare the 'developments over time' among the intervention and control group. Sometimes these more complicated experimental designs are analysed with simple cross-sectional methods, mostly by analysing the outcome at each follow-up measurement separately, or sometimes even by ignoring the information gathered from the in-between measurements, i.e. only using the last measurement as outcome variable to evaluate the effect of the intervention. Besides this, summary statistics are often used. The general idea behind a summary statistic is to capture the longitudinal development of an outcome variable over time into one value; the summary statistic. With a relative simple cross-sectional analysis these summary statistics can be compared between the intervention and control group in order to analyse the effect of the intervention. One of the most frequently used summary statistics is the area under the curve (AUC) [14]. However, nowadays mostly more advanced statistical methods such as mixed model analysis or generalised estimating equations (GEE analysis) [8, 19] are used to analyze RCT data with more than one follow-up measurement. In this chapter, the different methods will be discussed by using an example dataset in which the effect of a new therapy (i.e. intervention) for low back pain is evaluated. The example dataset is a manipulated dataset from an RCT in which patients who seek care in a private physical therapy clinic with low back pain as primary complaint were included. Besides a baseline measurement, three follow-up measurements were performed at 6, 12 and 18 months respectively. In the present example, two outcome variables will be considered: one continuous outcome variable and one dichotomous outcome variable. The continuous outcome variable is a score on a questionnaire aiming to measure complaints, while the dichotomous outcome variable reflects whether the patient is recovered or not; this is based on subjective self-report.

10.2 Continuous Outcome Variables

Table 10.1 shows descriptive information for both the intervention and control group at baseline and at the three follow-up measurements.

Table 10.1 Descriptive information regarding the example with a continuous outcome variable

	Intervention		Control	
Complaints	Mean (sd)	N	Mean (sd)	N
Baseline	3.25 (0.40)	74	3.47 (0.43)	82
Time-point 1	3.03 (0.45)	68	3.25 (0.48)	71
Time-point 2	2.89 (0.51)	64	3.18 (0.57)	73
Time-point 3	2.83 (0.47)	67	3.12 (0.55)	73

Table 10.2 Results of a GLM for repeated measurements performed on the example dataset with a continuous outcome variable

Overall time effect	Overall intervention effect	Intervention*time interaction
$p < 0.001$	$p < 0.001$	$p = 0.74$

10.2.1 Generalised Linear Model (GLM) for Repeated Measurements

Although GLM for repeated measurements can not be seen as a new (more advanced) statistical technique to analyse longitudinal data, it can be used to analyse a continuous outcome variable measured in an RCT with more than one follow-up measurement. The basic idea behind GLM for repeated measurements (which is also known as (multivariate) analysis of variance ((M)ANOVA) for repeated measurements) is the same as for the well known paired t-test. The statistical test is carried out for the $T - 1$ absolute differences between subsequent measurements. In fact, GLM for repeated measurements is a multivariate analysis of these $T - 1$ absolute differences between subsequent time-points. Multivariate refers to fact that $T - 1$ differences are used simultaneously as outcome variable. Besides the 'multivariate' approach, the same research question can also be answered with a 'univariate' approach. This 'univariate' procedure is comparable to the procedures carried out in simple analysis of variance (ANOVA) and is based on the 'sum of squares', i.e. squared differences between observed values and average values. In most software packages, the results of both the 'multivariate' and 'univariate' approach are provided at the same time. From a GLM for repeated measurements with one dichotomous determinant (i.e. intervention versus control), basically three 'effects' can be derived [14]. An overall time-effect (i.e. is there a change over time, independent of the different groups), an overall group effect (i.e. is there a difference between the groups on average over time) and, most important, a group*time interaction effect (i.e. is there a difference between the groups in development over time). Table 10.2 shows the results of a GLM for repeated measurements performed on the example dataset, while Fig. 10.1 shows the so called 'estimated marginal means' resulting from the GLM for repeated measurements.

From Table 10.2 it can be seen that there is an overall time effect, an overall intervention effect but no intervention*time interaction effect. From Fig. 10.1 (and also from Table 10.1), however, it can be seen that the baseline values of

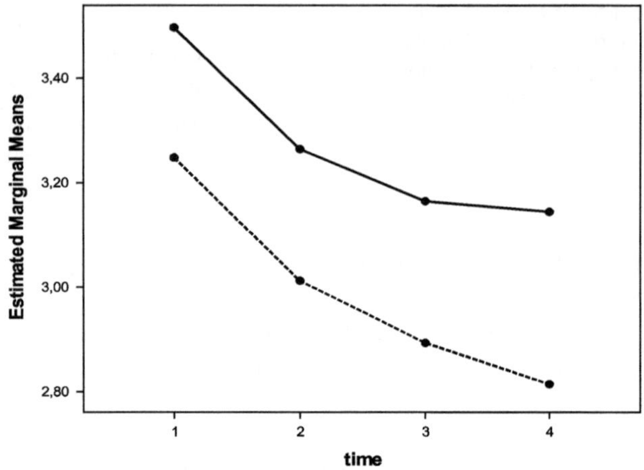

Fig. 10.1 Estimated marginal means derived from a GLM for repeated measurements performed on the example dataset with a continuous outcome variable (____ control, ... intervention)

both groups are different. This is a problem that often occurs in RCT data which should be taken into account in the analysis evaluating the effect of the intervention. Different baseline values for the therapy and the control group causes ‘regression to the mean’. If the outcome variable at a certain time-point $t = 1$ is a sample of random numbers, and the outcome variable at the next time-point $t = 2$ is also a sample of random numbers, then the subjects in the upper part of the distribution at $t = 1$ are less likely to be in the upper part of the distribution at $t = 2$, compared to the other subjects. In the same way, the subjects in the lower part of the distribution at $t = 1$ are less likely than the other subjects to be in the lower part of the distribution at $t = 2$. The consequence of this is that, just by chance, the change between $t = 1$ and $t = 2$ is correlated with the initial value. For the group with higher baseline values, a decrease in the outcome variable is much easier to achieve than for the group with the lower baseline value. It is clear that this problem arises in the analysis of the example dataset. Therefore, the consequence is that when the intervention group and control group differ at baseline, a comparison of the changes between the groups can lead to either an overestimation or an underestimation of the intervention effect [15].

There is, however, a nice way to adjust for the phenomenon of regression to the mean. This approach is known as ‘analysis of covariance’. With this technique the value of the outcome variable Y at the second measurement is used as outcome variable in a linear regression analysis, with the observation of the outcome variable Y at the first measurement as one of the covariates:

$$Y_{it2} = \beta_0 + \beta_1 Y_{it1} + \beta_2 X_i + \cdots + \varepsilon_i, \tag{10.1}$$

where Y_{it2} = observations for subject i at time-point $t = 2$; β_1 = regression coefficient for Y_{it1}; Y_{it1} = observations for subject i at time-point $t = 1$; β_2 = regression coefficient for X_i; X_i = intervention variable and ε_i = error for subject i.

Table 10.3 Results of a GLM for repeated measurements adjusted for the baseline differences performed on the example dataset with a continuous outcome variable

Overall time effect	Overall intervention effect	Intervention*time interaction
$p < 0.001$	$p = 0.04$	$p = 0.14$

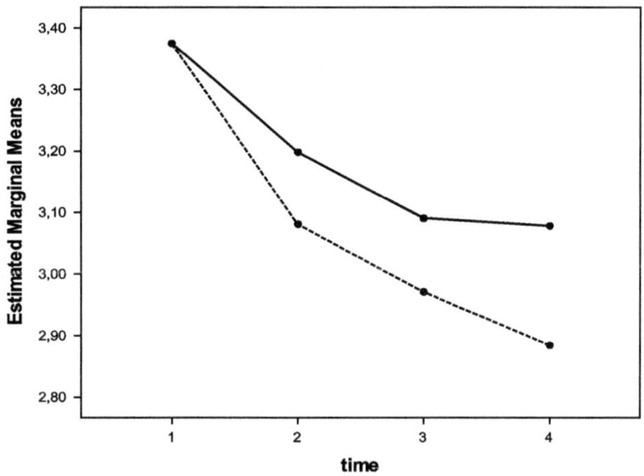

Fig. 10.2 Estimated marginal means derived from a GLM for repeated measurements adjusted for the baseline differences performed on the example dataset with a continuous outcome variable (____ control, ... intervention)

In the analysis of covariance, the change is defined relative to the value of Y at $t = 1$. This relativity is expressed in the regression coefficient β_1 and, therefore, it is assumed that this method adjusts for the phenomenon of regression to the mean. In fact the effect of the intervention is evaluated assuming the same baseline value for both groups. The same idea can be used in a GLM for repeated measurements; i.e. the analysis can be adjusted for the baseline value. This approach is also known as (M)ANCOVA for repeated measurements. Table 10.3 and Fig. 10.2 show the results of a GLM for repeated measurements adjusting for the baseline value performed on the example dataset.

Although GLM for repeated measurements is often used, it has a few major drawbacks. First of all, it can only be applied to complete cases; all subjects with one or more missing observation are not part of the analyses. Secondly, GLM for repeated measurements is mainly based on statistical significance testing, while there is more interest in effect estimation. Because of this, nowadays, new more advanced statistical techniques, such as mixed model analysis and GEE analysis are mostly used.

10.2.2 More Advanced Analysis

The questions answered by a GLM for repeated measurements could also be answered by more advanced methods, such as mixed model analysis and GEE analysis [14]. The advantage of the more advanced methods is that all available data is included in the analysis, while with GLM for repeated measurements only those subjects with complete data are included. Another important advantage of the more advanced analyses is that they are basically regression techniques, from which the effect estimates (i.e. the magnitude of the effect of the intervention) and the corresponding confidence intervals can be derived.

The general idea behind all statistical techniques to analyse longitudinal data is that because of the dependency of observations within a subject an adjustment must be made for 'subject'. The problem, however, is that the variable 'subject' is a categorical variable that must be represented by dummy variables. Suppose there are 200 subjects in a particular study. This means that 199 dummy variables are needed to adjust for subject. Because this is practically impossible, the adjustment for 'subject' has to be performed in a more efficient way and the different longitudinal techniques differ from each other in the way they perform that adjustment [14].

Mixed model analysis is also known as multilevel analysis [4, 13], hierarchical linear modeling or random effects modeling [6]. As has been mentioned before, the general idea behind all longitudinal statistical techniques is to adjust for 'subject' in an efficient way. Adjusting for 'subject' actually means that for all subjects in the longitudinal study, different intercepts are estimated. The basic principle behind the use of mixed model analysis in longitudinal studies is that not all separate intercepts are estimated, but that (only one) variance of those intercepts is estimated, i.e. a random intercept. It is also possible that not only the intercept is different for each subject, but that also the development over time is different for each subject, in other words, there is an interaction between 'subject' and time. In this situation the variance of the regression coefficients for time can be estimated, i.e. a random slope for time. In fact, these kind of individual interactions can be added to the regression model for all covariates. In a regular RCT, however, assuming a random slope for the intervention effect is not possible, because the intervention variable is time-independent [13]. When a certain subject is assigned to either the intervention or control group, that subject stays in that group along the intervention period. An exception is the cross-over trial, in which the subject is his own control and the intervention variable is time-dependent. In this situation the intervention effect can be different for each subject and therefore a random slope for the intervention variable can be assumed. For mixed model analysis, one has to choose which coefficients have to be assumed random. This choice can be based on the result of a likelihood ratio test.

Within GEE, the adjustment for the dependency of observations is done in a slightly different way, i.e. by assuming (a priori) a certain 'working' correlation structure for the repeated measurements of the outcome variable [8, 19]. Depending on the software package used to estimate the regression coefficients, different correlation structures are available. They basically vary from an 'exchangeable' (or 'compound symmetry') correlation structure, i.e. the correlations between subsequent measurements are assumed to be the same, irrespective of the length of the measurement interval, to an 'unstructured' correlation structure. In this structure no particular structure is assumed, which means that all possible correlations between repeated measurements have to be estimated.

In the literature it is assumed that GEE analysis is robust against a wrong choice for a correlation structure, i.e. it does not matter which correlation structure is chosen, the results of the longitudinal analysis will be more or less the same [9, 12]. However, when the results of analysis with different working correlation structures are compared to each other, the magnitude of the regression coefficients are different [14]. It is therefore important to realize which correlation structure should be chosen for the analysis. Although the unstructured working correlation structure is always the best, the simplicity of the correlation structure also has to be taken into account. The number of parameters (in this case correlation coefficients) which needs to be estimated differs for the various working correlation structures. The best option is therefore to choose the simplest structure which fits the data well. The first step in choosing a certain correlation structure can be to investigate the observed within-person correlation coefficients for the outcome variable. It should be kept in mind that when analyzing covariates, the correlation structure can change (i.e. the choice of the correlation structure should better be based conditionally on the covariates).

Within the framework of the more advanced statistical techniques, several models are available to evaluate the effect of an intervention. In an RCT with more than one follow-up measurement, the simplest model that can be used is

$$Y_{it} = \beta_0 + \beta_1 X_i + \cdots + \varepsilon_{it}, \tag{10.2}$$

where Y_{it} = observations for subject i at time t; β_0 = intercept; β_1 = regression coefficient for X_i; X_i = intervention variable and ε_{it} = error for subject i at time t.

With this model the outcome variable at the different follow-up measurements is compared between the therapy and control group simultaneously. This is comparable to the comparison of the post-test value between two groups in a pre-post test design. It should be noted that with this model, the influence of possible differences at baseline between the two groups is ignored. In the example dataset, however, it was seen that there is a (big) difference in baseline values between the intervention and control group and that this difference can cause regression to the mean. The intervention effect estimated with the standard model shown in (10.2), is therefore not correct. To adjust for differences at baseline, a longitudinal analysis of covariance can be used:

$$Y_{it} = \beta_0 + \beta_1 X_i + \beta_2 Y_{it0} + \cdots + \varepsilon_{it}, \tag{10.3}$$

where $Y_i t$ = observations for subject i at time t; β_0 = intercept; β_1 = regression coefficient for X_i; X_i = intervention variable; β_2 = regression coefficient for observation at t_0; Y_{it0} = observation for subject i at time t_0 and ε_{it} = error for subject i at time t.

The general idea behind a longitudinal analysis of covariance is that the outcome variable at each of the follow-up measurements is adjusted for the baseline value. The regression coefficient of interest, i.e. the regression coefficient for

Table 10.4 Average effect of the intervention over time estimated with both mixed model analysis and GEE analysis with a standard analysis, a longitudinal analysis of covariance and an autoregressive analysis

	Effect	95 % confidence interval	p-value
Mixed models			
Standard analysis	−0.23	−0.38 to −0.09	<0.01
Longitudinal analysis of covariance	−0.14	−0.27 to −0.01	0.03
Autoregressive analysis	−0.17	−0.26 to −0.07	<0.01
GEE analysis			
Standard analysis	−0.23	−0.38 to −0.09	<0.01
Longitudinal analysis of covariance	−0.14	−0.27 to −0.02	0.03
Autoregressive analysis	−0.15	−0.24 to −0.06	<0.01

the intervention variable reflects the overall 'adjusted' difference between the intervention and control group over time.

Another possible way to analyse RCT data with more than one follow-up is to use a so-called autoregressive analysis. In an autoregressive analysis the outcome variable is not adjusted for the baseline value, but each measurement of the outcome variable is adjusted for the value of the outcome variable one time-point earlier:

$$Y_{it} = \beta_0 + \beta_1 X_i + \beta_2 Y_{it-1} + \cdots + \varepsilon_{it}, \tag{10.4}$$

where Y_{it} = observations for subject i at time t; β_0 = intercept; β_1 = regression coefficient for X_i; X_i = intervention variable; Y_{it-1} = observation for subject i at time $t - 1$; β_2 = regression coefficient for observation at $t - 1$ (autoregression coefficient) and ε_{it} = error for subject i at time t.

The idea underlying an autoregressive analysis is that the value of an outcome variable at each time-point is primarily influenced by the value of this variable one measurement earlier. To estimate the 'real' influence of the intervention variable on the outcome variable, the model should therefore adjust for the value of the outcome variable at time-point $t - 1$. In fact, with an autoregressive analysis, the 'adjusted' changes between subsequent measurements are compared between the therapy and the control group. Table 10.4 shows the results of the three analyses performed on the example dataset. For all analyses the results of both a mixed model analysis and a GEE analysis are shown.

From Table 10.4 it can first be seen that the results derived from a mixed model analysis and the results derived from a GEE analysis are more or less the same. Furthermore, it can be seen that the standard analysis gives a higher effect measure compared to the other two methods. This has to do with the fact that with the standard analysis, the differences at baseline between the intervention and control group are not taken into account. Because the intervention group has lower values all over the follow-up period, the intervention effect obtained from the standard analysis is overestimated. In the example dataset the differences between an analysis of covariance and an autoregressive analysis are small. Slightly higher effect estimates for the autoregressive analysis and slightly smaller 95 % confidence intervals.

Although the longitudinal analysis of covariance is mostly used, it is questionable whether or not this is correct. In fact, the adjustment for baseline for all the follow-up measurements can overestimate the overall therapy effect. This is especially true when the effect of the therapy is particularly found in the first part of the follow-up period [14]. In the present example this is not really the case, so therefore, the longitudinal analysis of covariance and the autoregressive analysis gave more or less the same results. It is sometimes argued that both analyses are not correct. This has to do with the fact that in a RCT only the differences at baseline are caused by chance. Differences between the groups at the follow-up measurements are probably mostly caused by the intervention and should therefore not be adjusted for. To take that into account, a so-called 'combination' approach is suggested [17]. In this 'combination' approach, the first follow-up measurement is adjusted for the baseline differences, but the next follow-up measurements are not adjusted anymore for either the baseline differences (as in the longitudinal analysis of covariance) or the value of the outcome one time-point earlier (as in the autoregressive analysis). Although this approach makes sense, it is not much used in practice.

Up to now, the more advanced analyses performed were aimed to estimate an overall intervention effect. Sometimes, however, one is more interested in the estimation of effects at the different follow-up measurements. This can be done in a simple way by performing separate analyses at the different follow-up measurements, either by comparing the change between the baseline measurements and the three follow-up measurements or by performing three separate analyses of covariance (see Tables 10.5 and 10.6).

As expected, the results derived from the analysis of change scores are totally different from the results derived from the analyses of covariance. This has to do with the fact that the analyses of change scores not adjust for the difference at baseline. The analyses of covariance take into account these differences and because the intervention group has a lower value for the outcome variable at baseline, the effect derived from analyses of covariance are much higher than the ones derived from the analyses of change scores. Performing separate analyses, however, is theoretically wrong because the separate analyses are highly dependent on each

Table 10.5 Effects of the intervention at different time-point estimated with three separate analyses of change scores

	Effect	95 % confidence interval	p-value
Time-point 1	−0.01	−0.16 to 0.15	0.90
Time-point 2	−0.02	−0.18 to 0.14	0.79
Time-point 3	−0.06	−0.21 to 0.08	0.38

Table 10.6 Effects of the intervention at different time-point estimated with three separate analyses of covariance

	Effect	95 % confidence interval	p-value
Time-point 1	−0.12	−0.26 to 0.03	0.11
Time-point 2	−0.18	−0.34 to −0.01	0.05
Time-point 3	−0.19	−0.35 to −0.03	0.03

Table 10.7 Effects of the intervention at different time-point derived from one longitudinal analysis estimated with a mixed model analysis with a longitudinal analysis of covariance and an autoregressive analysis

	Effect	95 % confidence interval	p-value
Longitudinal analysis of covariance			
Time-point 1	−0.10	−0.26 to 0.05	0.19
Time-point 2	−0.14	−0.30 to 0.01	0.07
Time-point 3	−0.18	−0.34 to −0.03	0.02
Autoregressive analysis			
Time-point 1	−0.13	−0.27 to 0.01	0.07
Time-point 2	−0.16	−0.31 to −0.02	0.03
Time-point 3	−0.22	−0.36 to −0.07	<0.01

other. To obtain the separate effects in one analysis, time and the interaction between the intervention variable and time can be added to the longitudinal analysis of covariance and the autoregressive analysis.

Table 10.7 shows the results of the analyses performed on the example dataset in order to obtain the effects of the intervention at the three follow-up measurements.

From Table 10.7 it can be seen that the differences between the results obtained from a longitudinal analysis of covariance and the ones obtained from an autoregressive analysis are comparable to the differences between the two analyses in the estimation of the overall effect over time (see Table 10.4). Table 10.7 only shows the results from a mixed model analysis. It is obvious that the results obtained from a GEE analysis are comparable.

An approach to evaluate the effect of an intervention at different time-points is provided by Fitzmaurice et al. [3]. In this approach all measurements are used as outcome (including the baseline measurement). The following model (which is basically an extension of the standard model shown in (10.2)) is then used:

$$Y_{it} = \beta_0 + \beta_1 X_i + \beta_2 \,\text{time}_1 + \beta_3 \,\text{time}_2 + \beta_4 \,\text{time}_3 + \beta_5 X_i \times \text{time}_1 + \beta_6 X_i \times \text{time}_2 + \beta_7 X_i \times \text{time}_3 + \cdots + \varepsilon_{it}, \tag{10.5}$$

where Y_{it} = observations for subject i at follow-up time t, β_1 = the regression coefficient for X_i; X_i = intervention variable and time_1, time_2, time_3 = dummy variables for time and ε_{it} = error for subject i at time t.

In this model, the β_1 coefficient reflects the differences between the two groups at baseline, $\beta_1 + \beta_5$ reflects the differences between the two groups at the first follow-up measurement, while $\beta_1 + \beta_6$ reflects the differences between the two groups at the second follow-up measurement and $\beta_1 + \beta_7$ the differences between the two groups at the third follow-up measurement. Although, this is a nice way of analysing the effect of the intervention at the different time-points, it does not adjust for the differences between the groups observed at baseline, or in other words, it does not adjust for possible regression to the mean.

Table 10.8 Effects of the intervention at different time-point derived from one longitudinal analysis estimated with a mixed model analysis based on (10.5) and (10.6)

	Effect	95 % confidence interval	p-value
Standard analysis			
Time-point 1	−0.22	−0.38 to −0.06	<0.01
Time-point 2	−0.27	−0.43 to −0.11	<0.01
Time-point 3	−0.29	−0.46 to −014	<0.001
Standard analysis without the intervention variable			
Time-point 1	−0.11	−0.24 to 0.03	0.13
Time-point 2	−0.16	−0.30 to −0.02	0.03
Time-point 3	−0.19	−0.32 to −0.05	<0.01

Table 10.8 shows the results of the two analyses performed on the example dataset estimated with a mixed model analysis

An alternative approach to tackle this problem is to use the same model but without the intervention variable:

$$Y_{it} = \beta_0 + \beta_1 \text{time}_1 + \beta_2 \text{time}_2 + \beta_3 \text{time}_3 + \beta_4 X_i \times \text{time}_1 \\ + X_i \times \text{time}_2 + \beta_6 X_i \times \text{time}_3 + \cdots + \varepsilon_{it}, \tag{10.6}$$

where Y_{it} = observations for subject i at follow-up time t, X_i = intervention variable and $\text{time}_1, \text{time}_2, \text{time}_3$ = dummy variables for time and ε_{it} = error for subject i at time t.

Because the intervention variable is not in the model, the baseline values for both groups are assumed to be equal and are reflected in the intercept of the model (i.e. β_0). In this model, the coefficients of interest are the same as in the model with the intervention variable. The only difference is that now the effects of intervention at the different time-points are adjusted for the differences at baseline (Fig. 10.8).

The analyses based on (10.6) (i.e. the model without the intervention variable) are basically the same as a longitudinal analysis of covariance. The difference in the effect estimates between the two approaches is caused by a different number of observations (due to missing values). When the two analyses would have been performed on a full dataset without any missing values, the results of the two analyses would have been exactly the same.

Although longitudinal analysis of covariance is mostly used to analyse the effect of an RCT with more than one follow-up measurement one should be careful with the interpretation of the results of such an analysis. In some situations, it is better to use an autoregressive analysis. However, when differences at baseline occur between the groups, they must be taken into account in the analysis.

Table 10.9 Number of subjects recovered at different time-points for both the intervention and control group

	Intervention		Control	
	Recovered	Not recovered	Recovered	Not recovered
Time-point 1	10	58	5	66
Time-point 2	21	43	14	59
Time-point 3	25	42	12	61

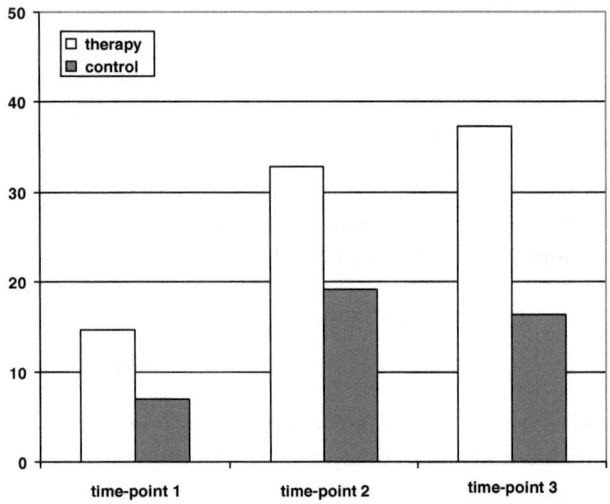

Fig. 10.3 Percentage of subjects recovered at different time-points for both the intervention and control group

10.3 Dichotomous Outcome Variables

The statistical models used for the analysis of dichotomous outcome variables derived from an RCT with more than one follow-up measurement are somewhat less complex than the models discussed for the analysis of a continuous outcome variable. This has to do with the fact that in general an adjustment for differences in baseline values is not necessary, because all subjects have the same value at baseline (e.g. all subjects are not recovered). As has been mentioned before, the example dataset used in this section is derived from the same RCT that has been used in the example with a continuous outcome variable. However, in this section the outcome is dichotomous reflecting whether the patient is recovered or not. Table 10.9 shows the number of subjects recovered versus the number of subjects not recovered in the intervention and in the control group at the three follow-up measurements, while Fig. 10.3 shows the percentages over time for both groups.

The classical way to analyse the results of such an RCT is to analyse the difference in proportion of patients experiencing recovery between the intervention and the control group at each of the three follow-up measurements, by simply applying a Chi-square test. Furthermore, at each of the follow-up measurements, the effect of the intervention can be estimated by calculating the relative risk

Table 10.10 Effects (expressed as relative risks) of the intervention at different time-points estimated with three separate analyses

	Relative risk	95 % confidence interval	p-value
Time-point 1	1.60	0.77–3.33	0.15
Time-point 2	1.45	0.93–2.24	0.07
Time-point 3	1.83	1.12–2.99	0.01

Table 10.11 Effects (expressed as odds ratios) of the intervention at different time-points estimated with three separate analyses

	Odds ratio	95 % confidence interval	p-value
Time-point 1	2.28	0.74–7.05	0.15
Time-point 2	2.06	0.94–4.50	0.07
Time-point 3	3.03	1.37–6.68	0.01

(and corresponding 95 % confidence interval). The relative risk is defined as the proportion of subjects recovered in the intervention group, divided by the proportion of subjects recovered in the control group [10]. Table 10.10 summarizes the results of the analyses.

From Table 10.10 it can be seen that the effect of the intervention at the first and second follow-up measurement is more or less the same, while at the third follow-up measurement the effect of the intervention is somewhat greater and also statistically significant.

It is, of course, also possible to estimate the effect of the intervention with a more advanced longitudinal technique. Because of the nature of the outcome variable, a logistic mixed model analysis or a logistic GEE analysis should be used instead of a linear mixed model analysis or a linear GEE analysis. It should be noted that for a dichotomous outcome variable GLM for repeated measurements is not possible. Furthermore, it should be realized that as a result of a logistic longitudinal analysis, odds ratios are calculated. Odds ratios are often interpreted as relative risks, but they are not the same. Owing to the mathematical background of the odds ratios and relative risks, the odds ratios are always an overestimation of the 'real' relative risk. This overestimation becomes stronger as the proportion of 'cases' (i.e. recovered patients) increases. To illustrate this, the odds ratios for intervention versus control were calculated at each of the follow-up measurements (see Table 10.11).

From the results in Table 10.11 it can be seen that the calculated odds ratios are bigger than the relative risks shown in Table 10.10, and that the confidence intervals are wider, but that the significance levels are the same. So, when a logistic GEE analysis is carried out, one must realize that the results (i.e. odds ratios) obtained from such an analysis cannot be interpreted as relative risks.

Table 10.12 presents the results of a logistic mixed model analysis and a logistic GEE analysis in which the average effect of the intervention over time is analysed.

The most intriguing finding regarding the comparison of the two analyses is that the odds ratio obtained from a logistic mixed model analysis is much higher

Table 10.12 Average effect of the intervention over time estimated with both mixed model analysis and GEE analysis

	Odds ratio	95 % confidence interval	p-value
Mixed model analysis	3.94	1.29–12.04	0.02
GEE analysis	2.15	1.13–4.10	0.02

compared to the odds ratio obtained from a logistic GEE analysis. This is not just a coincidence, but this has a theoretical background; i.e. the odds ratio obtained from a logistic mixed model analysis will always be bigger than the one obtained from a logistic GEE analysis.

Basically, both 'longitudinal' techniques take all measurements into account, and use a logistic regression approach with an adjustment for the dependency of the observations. This is done either by assuming a certain 'working' correlation structure (GEE analysis) or by allowing random regression coefficients (mixed model analysis). The difference between the two techniques is that GEE analysis is a so-called population average approach, while mixed model analysis is a so-called subject specific approach [14]. The different estimation procedures cause the difference in the magnitude of the odds ratios, which is always in favour of the mixed model analysis, i.e. the effects estimated with a logistic mixed model analysis are always bigger than the effects estimated with a logistic GEE analysis. Because the standard errors are also bigger for a logistic mixed model analysis (and therefore the 95 % confidence intervals are wider), the corresponding p-values are not much different and when conclusions are based on these p-values, they will be more or less the same. However, when the conclusions are based on the magnitude of the odds ratios, the conclusions will differ remarkably between the two techniques.

It should further be noted that the estimations of the regression coefficients (i.e. odds ratios) with logistic mixed model analysis can be very complicated and often lead to instable results. Furthermore, the results of these analyses can differ between software packages [7, 14, 18].

It is of course also possible to estimate the effects of the intervention at the three follow-up measurements in one analysis. This can be done in exactly the same way as has been described for continuous outcome variables, i.e. by adding dummy variables for time and the interaction between these dummy variables and the intervention variable to the model. Again, this is far less complicated as for a continuous outcome variable because in general an adjustment for differences in baseline values is not necessary.

Table 10.13 shows the results derived from a both a logistic mixed model analysis and a logistic GEE analysis.

From Table 10.13 it can be seen again that the odds ratios derived from a logistic mixed model analysis are much higher than the ones derived from a logistic GEE analysis. It can also be seen that the odds ratios derived from a logistic GEE analysis are much closer to the ones obtained from the three separate analyses than the odds ratios derived from a logistic mixed model analysis (Table 10.11). This suggests that

Table 10.13 Effects of the intervention at different time-point estimated with one analysis

	Odds ratio	95 % confidence interval	p-value
GEE-analysis			
Time-point 1	1.96	0.68–5.70	0.21
Time-point 2	1.83	0.85–3.91	0.12
Time-point 3	2.99	1.35–6.59	0.01
Mixed model analysis			
Time-point 1	3.84	0.54–27.46	0.18
Time-point 2	3.63	0.75–17.65	0.11
Time-point 3	9.98	1.87–53.09	0.01

regarding the more advanced longitudinal techniques, logistic GEE analysis has to be preferred above logistic mixed model analysis.

The data used in the present example is an example of 'recurrent event' data. To analyse 'recurrent event' data, also some other approaches are available. Based on a survival approach, Cox proportional hazards regression for recurrent events can be performed [5, 11, 16]. Although there are different estimation procedures available the general idea behind Cox proportional hazards regression for recurrent events is that the different time periods are analysed separately adjusted for the fact that the time periods within one patient are dependent. The idea of this adjustment is that the standard error of the regression coefficient of interest is increased proportional to the correlation of the observations within one subject. One of the problems using Cox proportional hazards regression for recurrent events for RCT data is that it is assumed that the events under study are short lasting, which means that after an event the particular subject is directly at risk to get another event. This assumption does not hold for most RCT's, including the example RCT used in this chapter. Although the events can be recurrent, most of the events are long lasting. So in this situation, Cox proportional hazards regression for recurrent events is not very suitable.

There are also other possibilities to model recurrent events data, such as the continuous-time Markov process model for panel data [1] or the conditional frailty model [2]. However, most of those alternative methods are mathematically complicated and not much used in practice.

10.4 Discussion

In this chapter several methods were discussed that can be used to analyse data from an RCT with more than one follow-up measurement. The data of the examples were analysed with different software packages. GLM for repeated measurements was performed with SPSS, while both mixed model analysis and GEE analysis were performed with STATA. Nowadays, it is possible to perform both linear and logistic mixed model analysis as well as linear and logistic GEE analysis with all popular (commercial) software packages such as SPSS, STATA, SAS and R. It should be realised that the results of linear mixed model analysis and linear GEE analysis are very stable; i.e. there is no difference in results obtained from the different software packages. This also holds for logistic GEE analysis. However, for logistic mixed model analysis this is not the case. The use of different software packages lead to different results as well as the use of different estimation procedures within a software package [14]. Therefore, the results obtained from a logistic mixed model analysis should be interpreted with great caution.

References

1. Berkhof, J., Knol, D., Rijmen, F., Twisk, J., Uitdehaag, B., Boers, M.: Relapse - remission and remission - relapse switches in rheumatiod arthritis patients were modelled by random effects. *Journal of Clinical Epidemiology* **62**, 1085–1094 (2009)
2. Box-Steffensmeier, J., De Boef, S.: Repeated events survival models: the conditional frailty model. *Statistics in Medicine* **25**, 3518–3533 (2006)
3. Fitzmaurice, G.M., Laird, N.M., Ware, J.H.: *Applied longitudinal data analysis*. Wiley, Hoboken, New Jersey, USA (2004)
4. Goldstein, H.: *Multilevel statistical models*, 3rd edn. Edward Arnold, London (2003)
5. Kelly, P.J., Lim, L.Y.: Survival analysis for recurrent event data: An application to childhood infectious diseases. *Statistics in Medicine* **19**, 13–33 (2003)
6. Laird, N.M., Ware, J.H.: Random effects models for longitudinal data. *Biometrics* **38**, 963–974 (1982)
7. Lesaffre, E., Spiessens, B.: On the effect of the number of quadrature points in a logistic random-effects model: an example. *Annals of Applied Statistics* **50**, 325–335 (2001)
8. Liang, K., Zeger, S.L.: Longitudinal data analysis using generalized linear models. *Biometrika* **73**, 45–51 (1986)
9. Liang, K.Y., Zeger, S.L.: Regression analysis for correlated data. *Annual Review of Public Health* **14**, 43–68 (1993)
10. Rothman, K.J., Greenland, S.: *Modern Epidemiology*. Lippincott-Raven Publishers, Philidelphia (1998)
11. Stürmer, T., Glynn, R.J., Kliebsch, U., Brenner, H.: Analytic strategies for recurrent events in epidemiologic studies: background and application to hospitalization risk in the elderly. *Journal of Clinical Epidemiology* **53**, 57–64 (2000)
12. Twisk, J.W.R.: Longitudinal data analysis. A comparison between generalized estimating equations and random coefficient analysis. *European Journal of Epidemiology* **19**, 769–776 (2004)
13. Twisk, J.W.R.: *Applied multilevel analysis*. Cambridge University Press, Cambridge, UK (2006)
14. Twisk, J.W.R.: *Applied longitudinal data analysis for epidemiology: a practical guide*, 2nd edn. Cambridge University Press, Cambridge, UK (2013)
15. Twisk, J.W.R., Proper, K.: Evaluation of the results of a randomized controlled trial: how to define changes between baseline and follow-up. *Journal of Clinical Epidemiology* **57**, 223–228 (2004)
16. Twisk, J.W.R., Smidt, N., de Vente, W.: Applied analysis of recurrent events: a practical overview. *The Journal of Epidemiology and Community Health* **59**, 706–710 (2005)
17. Twisk, J.W.R., de Vente, W.: The analysis of randomised controlled data with more than one follow-up measurement. A comparison between different approaches. *European Journal of Epidemiology* **23**, 655–660 (2008)

18. Yang, M., Goldstein, H.: Multilevel models for repeated binary outcomes: attitudes and voting over the electoral cycle. *Journal of the Royal Statistical Society* **163**, 49–62 (2000)
19. Zeger, S.L., Liang, K.Y.: Longitudinal data analysis for discrete and continuous outcomes. *Biometrics* **42**, 121–130 (1986)

Chapter 11
Statistical Methods for the Assessment of Clinical Relevance

Meinhard Kieser

Abstract It is commonly accepted that the results of clinical trials to be important for medical practice do not only have to be statistically significant but also clinically relevant. While an elaborated and canonical methodology exists for statistical significance tests, there is no common sense so far on how to judge the clinical relevance of a medical finding. The assessment of the clinical relevance of a study result should provide quantified information about its practical importance. For this, both statistical procedures and appropriate effect measures on which the relevance judgment is based are required. The test for relevant superiority and the relevance assessment based on the observed effect are presented as two statistical approaches for the assessment of clinical relevance. The properties of these procedures are investigated and contrasted. Furthermore, an overview of effect measures used for relevance assessment is given and their characteristics are illustrated. Application of the methods is illustrated with a clinical trial example.

11.1 Introduction

It is commonly accepted that the results of clinical trials to be important for medical practice do not only have to be statistically significant but also clinically relevant the latter being often also denoted as clinically significant. As Friedman stated "statistical significance refers to whether or not the value of a statistical test exceeds some pre-specified level. Clinical significance refers to the medical importance of a finding. The two often agree but not always." [18]. This means that specific methods for the assessment of clinical relevance are required. However, while an elaborated and canonical methodology exists for statistical significance tests, there is no common sense so far on how to judge the clinical relevance of a medical finding.

The assessment of the clinical relevance of a study result should provide quantified information about its practical importance. Ideally, the results of an evaluation

M. Kieser (✉)
Institute of Medical Biometry and Informatics, University of Heidelberg, Heidelberg, Germany
e-mail: meinhard.kieser@imbi.uni-heidelberg.de

K. van Montfort et al. (eds.), *Developments in Statistical Evaluation of Clinical Trials*, DOI 10.1007/978-3-642-55345-5_11

of clinical relevance can be used to support decision making both for the complete population at hand and for the individual patient. Furthermore, it is favorable if the results of a relevance evaluation are easy to interpret and depend as less as possible from (arbitrary) relevance criteria. Moreover, statistical procedures for sample size calculation and analysis should be available that are tailored to the effect measure that is applied for relevance assessment. Finally, methods for relevance assessment should have good statistical properties. Especially, they should not lead to a huge increase in required sample size thus making practical implementation feasible.

This contribution reviews methods for relevance assessment and judges their characteristics according to the above criteria. In Sect. 11.2, we present the test for relevant superiority and the relevance assessment based on the observed effect as two statistical approaches for the assessment of clinical relevance. The properties of these methods are investigated and contrasted. Both these procedures can be applied to any of the effect measures presented in Sect. 11.3. Here, an overview of effect measures used for relevance assessment is given and their characteristics are discussed. Application of the methods is illustrated in Sect. 11.4 with a clinical trial example. We conclude with a summary of the findings and recommendations for practical application.

11.2 Statistical Approaches to the Assessment of Clinical Relevance

In 1987, Victor pointed out that the statistical procedure of testing the classical nullhypothesis of non-superiority does not adequately reflect the demand of judging the clinical relevance of therapeutic effects within the analysis of clinical trials [38]. Instead, he proposed to test "non-zero nullhypotheses (shifted nullhypotheses) where the 'clinically relevant difference' is the shift parameter." [38]. If the effect measure is denoted by θ and the threshold for clinical relevance by ϑ (higher values indicating here and in the following more favorable effects), the approach consists of testing the nullhypothesis $H_0^{\text{relsup}} : \theta \leq \vartheta,\ \vartheta > 0$, at one-sided level $\alpha = 0.025$. Of course, a statistical significant result for the test for relevant superiority at one-sided level α is equivalent to the lower boundary of the two-sided $1-2\alpha$ confidence interval lying above the relevance threshold. Therefore, when the test of relevant superiority is statistically significant this relates to a "large clinically significant effect" according to the categorization of Jones [23].

This approach is appealing because, unlike for classical hypothesis testing, statistical significance also implies clinical relevance. A major disadvantage of this method lies in the fact that the sample size is substantially increased as compared to the usual test for superiority. As an example, let us consider the situation of a two-group comparison between a test treatment (T) and a reference (R) with a continuous outcome and the difference in expectations between the two interventions as effect measure. We express the relevance threshold ϑ as a fraction

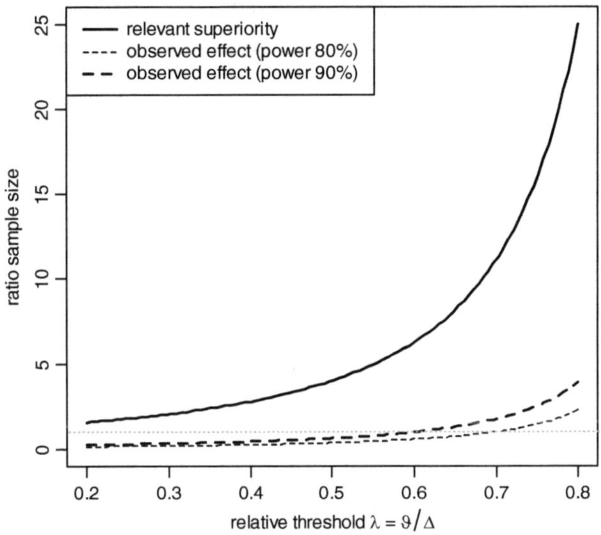

Fig. 11.1 Ratio of sample size required for the assessment of clinical relevance by testing for relevant superiority (shifted t-test, $\alpha = 0.025$, one-sided) or based on the observed treatment effect, respectively, to sample size required for significance test for superiority (t-test, $\alpha = 0.025$, one-sided) depending on fraction $\lambda = \vartheta/\Delta$. For all approaches the sample size is calculated for a treatment effect Δ to achieve the same power when applying the same relevance threshold ϑ (shifted normal distributions with common variance)

of the treatment group difference $\Delta = \mu_T - \mu_R$ used for sample size calculation, i.e., $\vartheta = \lambda \cdot \Delta,\ 0 < \lambda < 1$. It can then easily be seen that for the same desired power $1 - \beta$ the sample size required for the proof of relevant superiority is approximately by a factor $(1 - \lambda)^{-2}$ higher than when testing the common nullhypothesis of a superiority trial $H_0^{\text{sup}} : \mu_T - \mu_R \leq 0$. For example, if the relevance threshold is chosen as 0.6 or 0.7 of the assumed difference, respectively, the sample size is 6.25 or 11.1 fold for the shifted nullhypothesis approach (see Fig. 11.1). It is therefore not surprising that despite its appeal this approach has not been broadly implemented in practice. A recent literature search revealed only a single report of a clinical study where this approach was applied (but only as secondary analysis after having proven 'simple' superiority [34]). Furthermore, there is up to now only a single guideline that adopted the test for relevant superiority [20, 21].

In contrast, a number of regulatory guidelines recommend to address the judgment of clinical relevance by assessing whether the observed treatment effect lies above a pre-specified threshold (see, e.g., [5, 13]). When testing the null-hypothesis H_0^{relsup}, a one-sided p-value below 0.50 is obtained if and only if the observed treatment effect falls above the threshold. Therefore, this approach is equivalent to testing H_0^{relsup} at one-sided level 0.50. Jones [23] denoted a result where the point estimate of the treatment group difference overcomes the relevance threshold but not the lower boundary of the $1 - 2\alpha$ confidence interval as "probably clinically significant effect". For continuous outcomes and using the same relevance threshold for the test for relevant superiority and for the approach based on the observed effect, the required sample size is approximately by a factor $\left(\Phi^{-1}(1-\alpha)+\Phi^{-1}(1-\beta)\right)^2/\Phi^{-1}(1-\beta)$ higher for the former as compared to the latter. Here, $\Phi^{-1}(\gamma)$ denotes the γ-percentile of the standard normal distribution and $1-\beta$ the desired power. For the common values $\alpha = 0.025$ and $1-\beta = 0.80$ (or 0.90) this factor amounts to 11.1 (or 6.4). A similar picture can be observed in Fig. 11.1 where the sample size required for exceeding the relevance threshold with the observed effect is compared with the sample size required for testing the common superiority nullhypothesis. The same treatment group difference and power values are taken for sample size calculation, and the relevance threshold is expressed as a fraction λ of the effect. The ratio between the sample sizes is then approximately given by $\left[\Phi^{-1}(1-\beta)/\left(\Phi^{-1}(1-\alpha)+\Phi^{-1}(1-\beta)\right)\right]^2/(1-\lambda)^{-2}$. It can be seen that the relevance assessment based on the observed difference does not require a higher sample size than the common superiority test for $\lambda \leq 0.6$ (power 0.90) or $\lambda \leq 0.7$ (power 0.80) and increases only moderately for higher thresholds. Especially, the higher type I error rate inherent to the observed difference approach leads to a considerably smaller sample size as compared to testing for relevant superiority.

When testing for relevant superiority at common values for the significance level, a considerably smaller type I error rate is applied as compared to judging relevance based on the observed effect. One may therefore argue that a less restrictive (i.e., a smaller) threshold may be used for the first approach as compared to the latter in order to enable the feasibility of clinical trials implementing this method. Such an argument was implicitly used when justifying the comparatively low threshold fixed in the above cited guideline for testing shifted nullhypotheses [21]. There it was stated "The proposed 'irrelevant' quotient ($\theta_0 = 1.05$) is based on these figures, on a cautious estimation of the standard deviation and on realistic sample sizes." [21]. As an example, if the threshold for the approach that is based on the observed difference is chosen as $\lambda_{\text{obs}} = 0.7 \cdot \Delta$, the same power of $1-\beta = 0.90$ can be achieved with the same sample size by relaxing the threshold for the test for relevant superiority to $\lambda_{\text{relsup}} = 0.24 \cdot \Delta$.

Values for the significance level between 0.025 and 0.50 for the test of the shifted nullhypothesis could be chosen resulting in a practicable compromise between the two approaches described above. Nevertheless, relevance assessment is currently performed in practice throughout by inspection of the observed value of the applied effect measure. For this reason, we will focus on this approach in the following. However, methods for a relevance assessment by testing shifted nullhypotheses can also be derived for all effect measures presented in the next section. The characteristics with respect to the required sample size are very similar to those shown above for the difference in means.

11.3 Effect Measures for the Assessment of Clinical Relevance

11.3.1 Difference in Location Parameters

A frequently used measure for the judgment of clinical relevance is the between-group difference of the location parameters of the endpoints' distribution functions. For example, a CHMP guideline states for placebo-controlled superiority trials that "Establishing a clinically relevant benefit over placebo is accomplished by considering the point estimates of the difference between the test product and placebo and assessing its clinical relevance, ... using the original scale ...." [5]. This approach is easy to interpret, and assessing relevance in this way in addition to testing for statistical significance results in an acceptable increase in sample size or reduction in power, respectively, as compared to testing only the classical superiority hypothesis [25]. However, while the results provide information for the complete population, the mean difference observed for a continuous endpoint may be difficult to interpret for individual patients and may not be helpful for decision making.

For binary endpoints indicating whether some kind of treatment success has been achieved or not, the situation is more comfortable. For example, the CHMP guideline on the evaluation of medicinal products indicated for treatment of bacterial infections states that "In addition, clinical judgment should be applied to assess whether the observed difference in cure rates between the test antibacterial agent and placebo is clinically relevant." [13]. The observed cure rates and their difference give an impression about the individual benefit ('cure') in the patient population ('difference in cure rates'). Responder analyses which are described in the next section aim at creating a similar situation for continuous endpoints.

11.3.2 Responder Analysis

Responder analyses are recommended by many regulatory guidelines for the assessment of clinical relevance (see, e.g., [6–10, 12]). Here, a continuous outcome is dichotomized at a pre-defined cut point indicating a clinically important effect for the individual patient; then the observed difference in the rate of patients achieving this success criterion is judged for clinical relevance. This approach mimics the method sketched in Sect. 11.3.1 above for the case of binary endpoints and thus shows the same benefits: It provides a simple measure for the patient population based on a criterion that defines a clinically important benefit for the individual patient. This desirable property is a result of the dichotomization of the original continuous scale which, however, is criticized due to a potential loss of information and power (see e.g., [4, 16, 17, 28, 35–37]). Recently, Peacock et al. [32] proposed a distributional approach to derive a difference in proportions from continuous data together with a related confidence interval that retains the precision and power of the confidence interval for the difference in means on the original scale.

A further point of critic of responder analyses is the frequently arbitrary chosen cut point separating responders from non-responders [35–37]. It has to be noted, however, that there are also situations where internationally accepted and established thresholds for a successful outcome exist.

If in addition to a significance test on the original outcome scale the clinical relevance is assessed by responder analyses, this should already be taken into account in the planning phase when choosing the sample size (see, e.g., [7, 11] for examples of regulatory documents that adopted this requirement). Kieser et al. [26] developed methods for sample size calculation for the simultaneous assessment of statistical significance and clinical relevance based on responder analyses by exploiting the correlation between the original outcome and the dichotomized responder variable. As an example, let us consider the situation of normal distributions shifted by $\Delta/\sigma = 0.6$. We assume that dichotomization of the continuous outcome is performed at the cut point providing the maximum difference in responder rates. Figure 11.2 shows the sample size per group and the

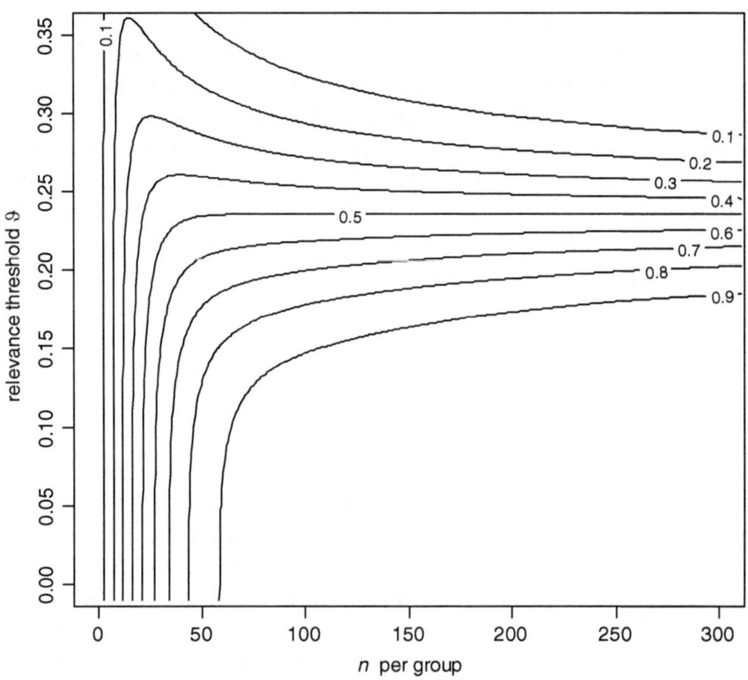

Fig. 11.2 Power for the simultaneous assessment of statistical significance (t-test, $\alpha = 0.025$, one-sided) and clinical relevance based on responder analyses depending on sample size n per group and relevance threshold ϑ for the difference in responder rates (normal distributions with common variance shifted by $\Delta/\sigma = 0.6$, dichotomization at cut point providing the maximum difference in responder rates)

power for the simultaneous assessment of statistical significance (t-test, $\alpha = 0.025$, one-sided) and clinical relevance based on a responder analysis when applying the relevance threshold ϑ to the responder rates. Note that the sample sizes per group for the threshold $\vartheta = 0$ refer to the test for statistical significance only, e.g., to sample sizes per group $n = 45$ for a power of 0.80 and $n = 60$ for a power of 0.90. When heightening the relevance threshold away from zero, the required sample size initially increases only slightly but subsequently enlarges dramatically. For the threshold $\vartheta = 2 \cdot \Phi\,[0.5 \cdot (\Delta/\sigma)] \approx 0.236$, the probability of observing a difference in responder rates above the threshold is equal to 0.5 for any sample size. If the threshold is further increased, this probability becomes smaller. This explains why for such threshold values the power for the simultaneous assessment of statistical significance and clinical relevance decreases with increasing sample size: Such relevance hurdles are too high to be achieved for the effect measure at hand. When planning a clinical trial where in addition to showing statistical significance the relevance of treatment effects shall be judged by responder analyses, considerations as those outlined above are useful for choosing an adequate relevance criterion and the correct sample size.

Shift alternatives for the original outcome were assumed for the evaluations described above. However, if one supposes that the population consists of patients that respond and others that do not respond to the investigated treatment, this situation may be modeled by mixture alternatives. Tests were proposed in the literature that are specifically tailored to this type of test problems [14,22]. However, in a recent investigation it turned out that these tests are not advantageous to standard tests such as t-test or the Wilcoxon-Mann-Whitney test [24]. This article also provides a sample size formula for the t-test in case of mixture normal distributions.

11.3.3 Probabilistic Index

As mentioned in the preceding section, regulatory guidelines strongly advocate the assessment of clinical relevance by responder analyses. It is probably due to two reasons that this approach is such attractive. Firstly, it results in a simple summary measure for the study population, namely the rate difference that can easily be transferred to other popular measures such as the number needed to treat [15]. Secondly, this approach also includes a success criterion for the patient. Thus, responder analyses combine two levels of relevance and result in measures with an easy interpretation for the physician and the patient.

The probabilistic index is an alternative measure for capturing the clinical relevance of treatment effects. This approach shows the above mentioned desirable properties of responder analyses but avoids their disadvantages, namely dichotomization (which is frequently done at arbitrarily chosen cut points) and the inherent loss of information. This effect measure is defined as $\theta = P(X_T > X_R) + 0.5 \cdot P(X_T = X_R)$, where X_i, $i = T, R$, denotes the random outcome under treatment i. For continuous outcomes the probabilistic index is thus given by

$\theta = P(X_T > X_R)$ and gives just the probability that a patient treated with the test treatment achieves a more favorable result than a patient treated with the reference. As for the dichotomization approach, this is a simple summary measure for the study population which is based on a success criterion for the individual patient. However, other than for responder analyses, the patient-based criterion is defined in terms of a direct comparison of treatments under investigation.

Another difference to responder analysis is that this approach allows a straightforward connection of the assessment of statistical and clinical relevance. This is possible due to the one-to-one relationship between the p-value of the significance test and the observed value of the probabilistic index. For example, Browne [1] pointed out for normally distributed data that $\hat{\theta} = \Phi\left[\sqrt{1/n} \cdot T^{-1}_{2(n-1)}(1-p)\right]$, where p denotes the observed one-sided p-value of the two-sample t-test performed with sample size per group n, and $T^{-1}_{df}(\gamma)$ the γ-percentile of the central t-distribution with df degrees of freedom. A rule for statistical significance ("p-value below a defined threshold") thus directly matches to a rule for clinical relevance ("observed value for probabilistic index above a defined threshold").

A further advantage of the probabilistic index as compared to responder analyses is that the former does not dichotomize the data by an (arbitrary) cut point but uses the complete shape of the underlying distribution. It can be expected that this leads to higher power or lower required sample sizes, respectively. In [24] the relationship between the observed probabilistic index and the p-value was used to derive formulae for the sample size required for the assessment of clinical relevance based on the probabilistic index in case of normally distributed data. Similar expressions can easily be derived also for other kinds of data, as, for example, censored survival data [19]. These formulae can be used to address the topic of relevance assessment already in the planning phase when defining the relevance threshold and when calculating the required sample size. An example is shown in Fig. 11.3.

We consider again the situation of a normally distributed outcome and a standardized treatment effect of $\Delta/\sigma = 0.6$ where for the responder analysis dichotomization is performed at the cut point leading to the maximum difference in responder rates. The threshold used for the probabilistic index ($\vartheta = 0.6$) relates to a threshold $\vartheta = 2 \cdot 0.6 - 1.0 = 0.2$ for the observed difference in responder rates. It can be seen from Fig. 11.3 that the sample size required for the simultaneous assessment of statistical significance and clinical relevance based on a responder analysis is mainly driven by the sample size required to fulfill the relevance criterion. Furthermore, this sample size is considerably higher than that required for statistical significance. In contrast, when assessing clinical relevance by using the probabilistic index, the required sample size is, if at all, only moderately higher than for the significance test. As a consequence, the required sample size for the simultaneous assessment of statistical significance and clinical relevance is much smaller when using the probabilistic index as an effect measure instead of the difference in responder rates.

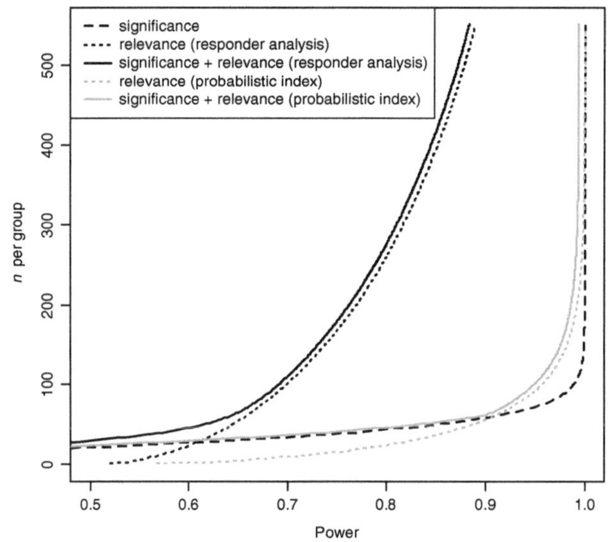

Fig. 11.3 Sample size n per group required to achieve a specified power for demonstrating statistical significance (t-test, $\alpha = 0.025$, one-sided), clinical relevance based on responder analysis (threshold $\vartheta = 0.20$), clinical relevance based on the probabilistic index (threshold $\vartheta = 0.60$) and simultaneous demonstration of statistical significance and clinical relevance (normal distributions with common variance shifted by $\Delta/\sigma = 0.6$, dichotomization at cut point providing the maximum difference in responder rates)

Proper statistical inference requires valid methods for point estimation and construction of confidence intervals for the applied effect measure. For arbitrary distributions the Mann-Whitney test statistics provides an unbiased and consistent estimator of the probabilistic index [2]. Furthermore, point estimators that share these properties can be derived for randomized trials where an adjustment for baseline covariates is performed [33] as well as for randomly censored data [27]. Newcombe [29,30] presented and evaluated various methods for the construction of confidence intervals for the probabilistic index.

Another favorable characteristic of the probabilistic index lies in the fact that it is a general concept that includes commonly used effect measures as special cases. For example, for normally distributed data θ is just a transformation of the standardized difference ($\theta = \Phi\left(\frac{1}{\sqrt{2}} \cdot \frac{\Delta}{\sigma}\right)$, where σ is the common population standard deviation), for binary endpoints it is directly related to the rate difference ($\theta = 0.5 + 0.5 \cdot (p_T - p_R)$, where p_i, $i = T, R$, denotes the rate in group i), and for survival data it has a simple relationship to the hazard ratio ($\theta = \mathrm{HR}/(1 + \mathrm{HR})$, where HR denotes the hazard ratio [3]). A common interpretation of treatment effects measured on a variety of scale levels is certainly a worthwhile characteristic facilitating the communication of trial results.

11.4 Example

In an open-label randomized controlled trial, Okun et al. [31] investigated the effects of constant-current deep brain stimulation in patients with Parkinson's disease. Primary outcome was the change from pre-implantation to 3 months after surgery in duration of time without bothersome dyskinesia. Assuming normal distribution, a common standard deviation of $\sigma = 4.9h$, a difference between treatment groups of $\Delta = 3h$ and an allocation ratio of 3 : 1, a total of 116 (= 87 + 29) patients was calculated to achieve a power of 0.80 for the test for statistical significance (t-test, $\alpha = 0.025$, one-sided). If a test for relevant superiority at the same significance level and a relevance threshold of $\vartheta_{rs} = 1.5h$ ($2h$) was employed instead, the required total sample size would be approximately fourfold (ninefold).

The secondary analysis of the trial included a responder analysis where the cut point of $c = 2h$ was chosen for the primary variable to define a treatment response. It should be mentioned that this cut point was denoted by the authors as "arbitrarily defined" [31]. For illustrative purposes, we assume that clinical relevance is demonstrated by the responder analysis if the observed difference in responder rates lies above the threshold $\vartheta_r = 0.2$. Under the above planning assumptions, the difference in responder rates in the population amounts to $\Delta_r = \Phi(c/\sigma) - \Phi\left((c - \Delta)/\sigma\right) = 0.239$. Applying the methods presented in [26], the power to observe a difference in responder rates of at least 0.2 can be calculated to be 0.649, and a total of 564 (= 423 + 141) patients are required to achieve a power of 0.80. For the simultaneous proof of statistical significance and clinical relevance based on the responder approach, the power is 0.622 for a sample size of 116 patients, and the 564 patients calculated above assure a power of 0.80. Note that there is a sharp increase in required sample size when increasing the value of the desired power as the threshold $\vartheta_r = 0.2$ is quite close to the actual difference in responder rates $\Delta_r = 0.239$.

Let us now assume that we base the assessment of clinical relevance on the observed probabilistic index. The threshold $\vartheta_r = 0.2$ for the difference in responder rates translates to a boundary for the probabilistic index of $\vartheta_{pi} = 0.5 \cdot (1 + 0.2) = 0.6$ (see, for example, [2]). The power to observe a probabilistic index of at least 0.6 for a total of 116 patients amounts to 0.882. As the power for the proof of both statistical significance and clinical relevance is the minimum of the two power values, this sample size also assures a power of 0.80 for this criterion. All computations presented in this book chapter can be performed with any statistical software that includes the probability function of the bivariate standard normal distribution.

11.5 Discussion

The purpose of this contribution was to present statistical methods and effect measures for the assessment of the clinical relevance of treatment effects and to evaluate their characteristics. When choosing whether relevance assessment should be performed based on the observed effect or on the lower bound of the confidence interval, one has to weigh the aspects of selecting an appropriate relevance threshold, the desired amount of protection against a type I error, and the required (and feasible) sample size. While the test for relevant superiority requires a considerably higher sample size when applying the same relevance threshold, it can lead to similar or only moderately higher sample sizes when the relevance threshold is relaxed. Vice versa, increasing the significance level by basing the judgment on the observed effect may be counterbalanced by applying a stricter relevance criterion. It may be worthwhile to consider approaches that apply a significance level between the two extremes $\alpha = 0.025$ and $\alpha = 0.50$. By this, a reasonable compromise between the antagonistic requirements of applying a strict relevance criterion and assuring a low type I error rate while at the same time using practicable sample sizes may be achieved.

Responder analyses are based on success rates derived from originally continuous outcomes. This provides a simple measure that is informative both for physicians and patients. However, if the cut point used for dichotomization is not widely accepted in the scientific community, its choice is somehow arbitrary. Furthermore, dichotomization may lead to a loss in power and information. The latter two disadvantages of responder analyses are overcome by the probabilistic index while saving their advantages. The probabilistic index is a general concept that provides the same interpretation for any kind of data and that allows an assessment of statistical and clinical relevance on the same scale.

In summary, in situations where a commonly accepted cut point for dichotomizing a continuous variable is available clinical relevance may be assessed by use of responder analyses. In all other situations, application of the probabilistic index seems to be beneficial due to the many advantages this effect measure offers.

References

1. Browne, R.H.: The t-test p value and its relationship to the effect size and P(X>Y). *The American Statistician* **64**, 30–33 (2010)
2. Brunner, E., Munzel, U.: *Nichtparametrische Datenanalyse*. Springer, Berlin (2002)
3. Buyse, M.: Reformulating the hazard ratio to enhance communication with clinical investigators. *Clinical Trials* **5**, 641–642 (2008)
4. Cohen, J.: The cost of dichotomization. *Applied Psychological Measurement* **7**, 249–253 (1983)
5. Committee for Medicinal Products for Human Use (CHMP): *Guideline on the choice of the non-inferiority margin* (2005). URL http://www.ema.europa.eu/docs/en_GB/document_library/Scientific_guideline/2009/09/WC500003636.pdf. Cited 14 December 2012
6. Committee for Medicinal Products for Human Use (CHMP): *Guideline on clinical investigation of medicinal products for the treatment of multiple sclerosis* (2006). URL http://www.ema.europa.eu/docs/en_GB/document_library/Scientific_guideline/2009/09/WC500003485.pdf. Cited 14 December 2012
7. Committee for Medicinal Products for Human Use (CHMP): *Guideline on clinical investigation of medicinal products indicated for the treatment of social anxiety disorder (SAD)* (2006).

URL http://www.ema.europa.eu/docs/en_GB/document_library/Scientific_guideline/2009/09/WC500003490.pdf. Cited 14 December 2012
8. Committee for Medicinal Products for Human Use (CHMP): *Guideline on clinical investigation of medicinal products in the treatment of Parkinson's disease* (2008). URL http://www.ema.europa.eu/docs/en_GB/document_library/Scientific_guideline/2009/09/WC500003540.pdf. Cited 14 December 2012
9. Committee for Medicinal Products for Human Use (CHMP): *Guideline on medicinal products for the treatment of Alzheimer's disease and other dementias* (2008). URL http://www.ema.europa.eu/docs/en_GB/document_library/Scientific_guideline/2009/09/WC500003562.pdf. Cited 14 December 2012
10. Committee for Medicinal Products for Human Use (CHMP): *Guideline on clinical investigation of medicinal products in the treatment of epileptic disorders* (2010). URL http://www.ema.europa.eu/docs/en_GB/document_library/Scientific_guideline/2010/01/WC500070043.pdf. Cited 14 December 2012
11. Committee for Medicinal Products for Human Use (CHMP): *Guideline on the clinical investigation of medicinal products for the treatment of attention deficit hyperactivity disorder (ADHD)* (2010). URL http://www.ema.europa.eu/docs/en_GB/document_library/Scientific_guideline/2010/08/WC500095686.pdf. Cited 14 December 2012
12. Committee for Medicinal Products for Human Use (CHMP): *Draft Guideline on clinical investigation of medicinal products in the treatment of depression* (2011). URL http://www.ema.europa.eu/docs/en_GB/document_library/Scientific_guideline/2011/10/WC500116160.pdf. Cited 14 December 2012
13. Committee for Medicinal Products for Human Use (CHMP): *Guideline on the evaluation of medicinal products indicated for treatment of bacterial infections* (2011). URL http://www.ema.europa.eu/docs/en_GB/document_library/Scientific_guideline/2009/09/WC500003417.pdf. Cited 14 December 2012
14. Conover, W.J., Salsburg, D.: Locally most powerful tests for detecting treatment effects when only a subset of patients can be expected to "respond" to treatment. *Biometrics* **44**, 189–196 (1988)
15. Cook, R.J., Sackett, D.L.: The number needed to treat: a clinically useful measure of treatment effect. *BMJ* **310**, 452 (1995)
16. Deyi, B.A., Kosinski, A.S., Snapinn, S.M.: Power considerations when a continuous outcome variable is dichotomized. *Journal of Biopharmaceutical Statistics* **8**, 337–352 (1998)
17. Fedorov, V., Mannino, F., Zhang, R.: Consequences of dichotomization. *Pharmaceutical Statistics* **8**, 50–61 (2009)
18. Friedman, L.: Clinical significance vs. statistical significance. In: P. Armitage, T. Coltan (eds.) *Encyclopedia of Biostatistics*, pp. 676–678. John Wiley and Sons, New York (1998)
19. Gondan, M.: How to demonstrate both statistical significance and clinical relevance in survival data. Internal Report, Institute of Medical Biometry and Informatics, University of Heidelberg (2012)
20. Heidrich, H., Cachovan, M., Kreutzig, A., Rieger, H., Trampisch, H.J.: Guidelines for therapeutic studies in Fontaine's stages II - IV peripheral arterial occlusive disease. *Vasa* **24**, 114–119 (1995)
21. Heidrich, H., Trampisch, H.J., Röhmel, J.: Comments on guidelines for therapeutic studies in Fontaine's stage II-IV peripheral arterial occlusive disease. *Vasa* **25**, 73–75 (1996)
22. Johnson, R.A., Verrill, S., Moore II, D.H.: Two-sample rank tests for detecting changes that occur in a small proportion of the treated population. *Biometrics* **43**, 641–655 (1986)
23. Jones, P.W.: Interpreting thresholds for a clinically significant change in health status in asthma and COPD. *European Respiratory Journal* **19**, 396–404 (2002)
24. Kieser, M., Friede, T., Gondan, M.: Assessment of statistical significance and clinical relevance. *Statistics in Medicine* **32**(10), 1707–1719 (2012). DOI 10.1002/sim.5634
25. Kieser, M., Hauschke, D.: Assessment of clinical relevance by considering point estimates and associated confidence intervals. *Pharmaceutical Statistics* **4**, 101–107 (2005)

26. Kieser, M., Röhmel, J., Friede, T.: Power and sample size determination when assessing the clinical relevance of trial results by 'responder analyses'. *Statistics in Medicine* **23**, 3287–3305 (2004)
27. Koziol, J., Jia, Z.: The concordance index C and the Mann-Whitney parameter Pr(X>Y) with randomly censored data. *BMJ* **51**, 467–474 (2009)
28. MacCallum, R.C., Zhang, S., Preacher, K.J., Rucker, D.D.: On the practice of dichotomization of quantitative variables. *Psychological Methods* **7**, 19–40 (2002)
29. Newcombe, R.G.: Confidence intervals for an effect size measure based on the Mann-Whitney statistic. Part 1: general issues and tail-area-based methods. *Statistics in Medicine* **25**, 543–557 (2006)
30. Newcombe, R.G.: Confidence intervals for an effect size measure based on the Mann-Whitney statistic. Part 2: asymptotic methods and evaluation. *Statistics in Medicine* **25**, 559–573 (2006)
31. Okun, M.S., Gallo, B.V., Mandybur, G., Jagid, J., Foote, K.D., Revilla, F.J., Alterman, R., Jankovic, J., Simpson, R., Junn, F., Verhagen, L., Arle, J.E., Ford, B., Goodman, R.R., Stewart, R.M., Horn, S., Baltuch, G.H., Kopell, B.H., Marshall, F., Peichel, D., Pahwa, R., Lyons, K.E., Tröster, A.I., Vitek, J.L., Tagliati, M., for the SJM DBS Study Group: Subthalamic deep brain stimulation with a constant-current device in Parkinson's disease: an open-label randomised controlled trial. *The Lancet Neurology* **11**, 140–149 (2012)
32. Peacock, J.L., Sauzet, O., Ewings, S.M., Kerry, S.M.: Dichotomising continuous data while retaining statistical power using a distributional approach. *Statistics in Medicine* **31**, 3089–3103 (2012)
33. Schacht, A., Bogaerts, K., Bluhmki, E., Lesaffre, E.: A new nonparametric approach for baseline covariate adjustment for two-group comparative studies. *Biometrics* **64**, 1110–1116 (2008)
34. Seiler, C.M., Fröhlich, B.E., Veit, J.A., Gazyakan, E., Wente, M.N., Wollermann, C., Deckert, A., Witte, S., Victor, N., Büchler, M.W., Knaebel, H.P.: Protocol design and current status of CLIVIT: a randomized controlled multicentre relevance trial comparing clips versus ligatures in thyroid surgery. *Trials* **7**, 27 (2006)
35. Senn, S.: Disappointing dichotomies. *Pharmaceutical Statistics* **2**, 239–240 (2003)
36. Snapinn, S.M., Jiang, Q.: Responder analyses and the assessment of a clinically relevant treatment effect. *Trials* **8**, 31 (2007)
37. Uryniak, T., Chan, I.S.F., Fedorov, V.V., Jiang, Q., Oppenheimer, L., Snapinn, S.M., Teng, C.H., Zhang, J.: Responder analyses - a PhRMA position paper. *Statistics in Biopharmaceutical Research* **3**, 476–487 (2011)
38. Victor, N.: Clinically relevant differences and shifted nullhypotheses. *Methods of Information in Medicine* **26**, 109–116 (1987)

Chapter 12
Statistical Considerations in the Use of Composite Endpoints in Time to Event Analyses

Richard J. Cook and Ker-Ai Lee

Abstract Many disease processes are complex and impact functional ability and quality of life of affected individuals in a multitude of ways. Diseases such as diabetes, lupus and other autoimmune disorders, for example, involve several different organ systems, which makes it challenging to select one specific endpoint. In other settings a disease puts affected individuals at risk of several different types of undesirable clinical events. This is the case in cardiovascular disease where individuals are at increased risk of myocardial infarction, angina, or stroke. In such settings it is common for clinical trialists to adopt composite endpoints on which to base treatment comparisons. We discuss issues in the use of composite endpoints and emphasize the difficulty in interpreting measures of effect.

12.1 Introduction

12.1.1 Clinical Settings Involving Multiple Endpoints

Disease processes are often complex and put individuals at risk for a wide range of clinically important events. When one type of event is of greater clinical importance than others, it can be chosen as the basis of the primary treatment comparison and hence play a central role in the trial design. Statistical analyses are then relatively straightforward and the effects of treatment on other endpoints can be assessed through secondary analyses. How best to select such a primary endpoint is, however, often not clear. We describe three settings involving multiple events.

Scenario I. Events have Similar Manifestation but Different Etiology Patients with asthma experience exacerbations of symptoms which significantly impact morbidity and quality of life, as well as incur considerable expense to the healthcare system. Trials of experimental prophylactic treatments often take the time of

R.J. Cook (✉) • K.-A. Lee
Department of Statistics and Actuarial Science, University of Waterloo, Waterloo, ON, Canada
e-mail: rjcook@uwaterloo.ca; ka2lee@uwaterloo.ca

K. van Montfort et al. (eds.), *Developments in Statistical Evaluation of Clinical Trials*, DOI 10.1007/978-3-642-55345-5_12

the first exacerbation after randomization as the endpoint of interest. Cellular analysis of sputum samples obtained during an exacerbation reveal the nature of the inflammatory process leading to the exacerbation and exacerbations can be classified as neutrophilic or eosinophilic in nature [11]. A given treatment may be more effective at preventing one type of exacerbation than another, but prevention of all exacerbations is of primary concern since they generally have the same impact on patients.

Scenario II. Events have Different Manifestation but Similar Etiology In trials of paediatric immunology, malnourished children in developing countries are at increased risk of infection, but many different organ systems can become infected [21]. Data on the time of onset and resolution of different infections can be collected over a period of observation and interest may lie in comparing the incidences of infection between one or more groups and interest may lie in comparing the incidences of infection between treatment groups.

Scenario III. Events are of Different Importance In trials of palliative therapies in cancer, interest may lie in demonstrating a new treatment is effective in preventing a non-fatal morbidity event which decreases quality of life or functional ability. In cancer metastatic to bone, patients are at risk of fractures which are associated with pain and disability, but they remain at high risk of death due to the advanced disease [18].

These three scenarios differ in the role of the various types of events. In Scenario I the different types of exacerbations have a similar consequence to patients and the exploration of the treatment effect according to the nature of the exacerbation is of secondary importance. From a patient perspective if the impact on morbidity and quality of life is the same for the different types of exacerbations, the physiological nature of the inflammatory response matters little. In Scenario II, the events are reflective of the state of an underlying condition, the strength of the immune system of the child. The occurrence of any infection is more likely with poorer immune function, and it is the immune function that the intervention is directed at improving. However, the different types of infection lead to quite different physical manifestations and risks. Infectious diarrhea can be fatal if not successfully treated and accounts for millions of deaths worldwide. Upper respiratory tract infections typically have serious but milder manifestations. Estimation of treatment effects on different types of infections are necessary in this setting to understand the consequences of any health policy decisions. Similar situations arise in diabetes trials where interventions may aim to improve glucose control but clinically important long-term endpoints may be based on measures of retinopathy and nephropathy [1]. In Scenario III, interest lies in the prevention of the non-fatal event impacting the morbidity of the patient. Death is an obviously undesirable event but treatment is not expected to impact risk of death.

12.1.2 Statistical Issues in the Analysis of Multiple Events

When different types of events are of comparable importance, but separate inferences about treatment effects are desired, co-primary endpoints can be specified. Use of co-primary endpoints, however, typically requires control of the experimental type I error rate through use of multiple comparison procedures [4, 27, 29]. Sample size requirements are often high in such designs due to the allocation of the type I error across the hypothesis tests for the individual endpoints. Moreover decision making following completion of the trial can be more complex if results are not in accord. Another strategy is to use global tests of treatment effects using methods that synthesis evidence of effect across separate analyses of the different events. Such methods are typically based on multivariate analyses [26, 32] which furnish estimates of treatment effects for the individual endpoints.

Perhaps the most common approach is to adopt a composite endpoint [7, 12]. A composite endpoint is said to have occurred when any one of a set of component endpoints occurs, and the time of the composite endpoint is the time of the first of its component endpoints. There are several reasons investigators may consider the use of composite endpoints in clinical trials. In studies involving a time-to-event analysis, the use of a composite endpoint will mean that more events will be observed than would be for any particular component. If the same clinically important effect is specified for the composite endpoint and one of its components, this increased event rate will translate into greater power for tests of treatment effects; at the design stage a reduction in the required number of subjects or duration of follow-up [7, 16, 24]. This rationale presumes that the same minimal clinically important effect applies for the composite endpoint and the component endpoint of interest. Composite endpoints are routinely adopted through the introduction of one or more less serious events, which presumably warrants changing the clinically important effect of interest. Moreover we show later that with models featuring a high degree of structure, model assumptions may not even be compatible for the composite endpoint and one of its components.

Recommendations are available in the literature on how to design trials, analyse resultant data, and report findings when composite endpoints are to be used [8, 16, 24, 25]. The main recommendations include that (1) individual components should have similar frequency of occurrence, (2) the treatment should have a similar effect on all components, (3) individual components should have similar importance to patients, (4) data from all components should be collected until the end of trial, and (5) individual components should be analysed and reported separately as secondary endpoints. The first three recommendations have face validity and seem geared towards helping ensure that conclusions regarding treatment effects on the composite endpoint have some relation to treatment effects on the component endpoints, thus helping in the interpretation of results. The collection of data on the occurrence of the component endpoints until the end of the trial facilitates separate assessment of treatment effects on each of the component endpoints. This means the consistency of findings across components can be empirically assessed.

The aforementioned issues have been actively debated in the medical literature [5, 14, 23–25]. In this chapter we discuss statistical considerations related to composite endpoint analyses and use the recommendations to guide the investigation. Since proportional hazards regression models are routinely adopted for the analysis of composite endpoints in clinical trials [8], we consider them here and point out important issues regarding model specification and interpretation. We formulate multivariate failure time models with proportional hazards for the marginal distributions which may be used to reflect the settings where composite endpoints are most reasonable according to the current guidelines. We study the asymptotic and empirical properties of estimators arising from a composite endpoint analysis. We also explore the utility of marginal methods based on multivariate failure time data [32]. We argue that the belief that composite endpoints provide an overall measure of the effect of treatment is overly simplistic, and a thoughtful interpretation of intervention effects based on composite endpoints alone is difficult. Their use as a primary basis for treatment comparison in clinical trials therefore warrants careful consideration.

12.2 Composite Endpoints in the Absence of Competing Risks

12.2.1 Notation and Modeling Issues

We first consider the case of two types of events in the context of a parallel group randomized trial. Let $Z = 1$ for patients in the experimental arm and $Z = 0$ otherwise, T_1 denote the time of an event of type 1, and T_2 denote the time of an event of type 2. Figure 12.1 contains timeline diagrams for six individuals indicating the occurrence of a type 1 (closed circle) event, a type 2 (open circle) event, and a censoring time (vertical dash). A time to event analysis for the type 1 event would based on four observed and two censored event times, and for the type 2 event there would be two observed and four censored event times. A composite endpoint analysis would involve five observed failure times and only one censored time corresponding to individual 3.

Let

$$h_k(t|z;\theta_k) = \lim_{\Delta t \downarrow 0} \frac{P(t \le T_k < t + \Delta t | t \le T_k, z)}{\Delta t} \tag{12.1}$$

denote the hazard for a type k event given $Z = z$. Under a proportional hazards model

$$h_k(t|z;\theta_k) = h_{0k}(t;\alpha_k)\exp(\beta_k z) \, , \tag{12.2}$$

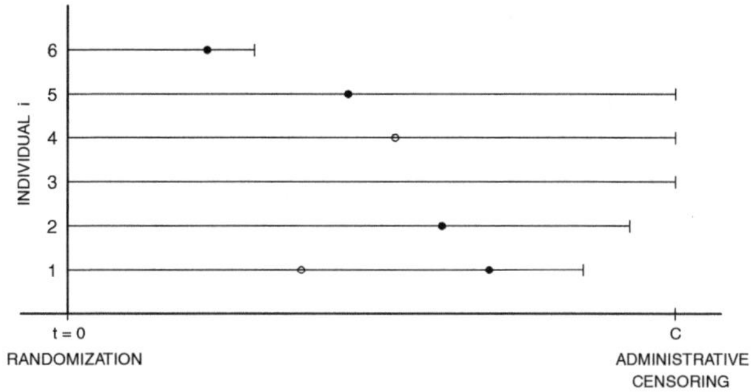

Fig. 12.1 Timeline diagrams indicating the occurrence of a type 1 (*closed circle*) and type 2 (*open circle*) event and censoring for a sample of six individuals

where α_k indexes the baseline hazard function and β_k reflects the effect of treatment on the type k events; $\theta_k = (\alpha_k', \beta_k)'$ and we let $\theta = (\theta_1', \theta_2')'$. The marginal survivor function for a type k event is then given by

$$P(T_k \geq t|z;\theta_k) = \mathscr{F}_k(t|z;\theta_k) = \exp\left(-\int_0^t h_k(u|z;\theta_k)\,du\right),$$

for $k = 1, 2$.

For preliminary reflections we suppose that the times are independent given an assigned treatment. If $T_1 \perp T_2|Z$, then the failure time of the composite endpoint is $T = \min(T_1, T_2)$, and the corresponding survivor function is

$$P(T \geq t|z;\theta) = \mathscr{F}(t|z;\theta) = \exp\left(-\int_0^t (h_1(u|z;\theta_1) + h_2(u|z;\theta_2))\,du\right).$$

At time u, the hazard ratio for T is then

$$w(u)\exp(\beta_1) + (1 - w(u))\exp(\beta_2), \tag{12.3}$$

where $w(u) = h_{01}(u)/(h_{01}(u) + h_{02}(u))$, which is in general a function of time. The marginal event times satisfy the proportional hazards assumption, the composite endpoint does not satisfy the proportional hazards assumption unless one or both of the following conditions are satisfied.

Condition I The treatment effects are common (i.e. $\beta_1 = \beta_2 = \beta$);
Condition II The baseline hazards are proportional (i.e. $h_{02}(t) = c \cdot h_{01}(t), c \geq 0$).

Under Condition I, (12.3) reduces to $\exp(\beta)$, the common hazard ratio, and under Condition II, $w(u) = (c+1)^{-1}$ and (12.3) becomes $(\exp(\beta_1) + c \cdot \exp(\beta_2))/(1+c)$.

It is of course more realistic to assume there exists a dependence between the event times given treatment. There are several ways of formulating joint models for multivariate failure times, but models based on copula functions [20] are most appealing since they enable one to link two marginal failure time distributions of any form to create a joint survival function. If $U_k \sim \text{UNIF}(0, 1)$, $k = 1, 2$, any bivariate cumulative distribution function for (U_1, U_2), denoted $C(u_1, u_2; \phi) = P(U_1 \leq u_1, U_2 \leq u_2; \phi)$, is a copula function. The association between the two components can be characterized by Kendall's τ. If we let $U_1 = \mathscr{F}_1(T_1|z; \theta_1)$ and $U_2 = \mathscr{F}_2(T_2|z; \theta_2)$, then $U_k \sim \text{UNIF}(0, 1)$, $k = 1, 2$. A joint survivor function for $T_1, T_2|Z$ is obtained as

$$\begin{aligned} P(T_1 \geq t_1, T_2 \geq t_2|z; \Omega) &= \mathscr{F}_{12}(t_1, t_2|z; \Omega) \\ &= C(\mathscr{F}_1(t_1|z; \theta_1), \mathscr{F}_2(t_2|z; \theta_2); \phi) , \end{aligned} \tag{12.4}$$

where $\Omega = (\theta', \phi)'$. Since Kendall's τ is invariant to monotonic increasing or decreasing transformations [17], it can also be interpreted as a measure of association of the transformed variables $(T_1, T_2)'$ given Z.

In this model, the random variable $T = \min(T_1, T_2)$ has survival, density and hazard function conditional on z, given by

$$P(T \geq t|z; \Omega) = \mathscr{F}(t|z; \Omega) = \mathscr{F}_{12}(t, t|z; \Omega) , \tag{12.5}$$

$f(t|z; \Omega) = -d\mathscr{F}(t|z; \Omega)/dt$ and $h(t|z; \Omega) = f(t|z; \Omega)/\mathscr{F}(t|z; \Omega)$, respectively. A key point here is that the hazard ratio $h(t|z = 1; \Omega)/h(t|z = 0; \Omega)$ is not independent of time, in general, even when Conditions I and II are satisfied. As a result, even if the marginal distributions feature the proportional hazards assumption, the model for the composite endpoint will typically not.

Wu and Cook [33] found that there is generally an incompatibility between the marginal models and the composite endpoint model even if $h_{02}(t) \propto h_{01}(t)$ or the marginal effects of treatment are the same (i.e. $\beta_1 = \beta_2$). That is, if the component endpoints feature the proportional hazards structure, the Cox model for the composite endpoint is typically misspecified because the proportional hazards assumption does not in general hold. The estimator of treatment effect under such a misspecified Cox model for the composite endpoint typically may have a conservative or anti-conservative limiting value. The factors that influence the limiting value include the specific copula function linking the component events, the strength of the association between the individual component events, the stochastic ordering of the individual components, and the degree and nature of the censoring process. These factors are relevant when the treatment effect is common across the component endpoints. When the treatment effect varies across component endpoints it becomes even more difficult to interpret estimates.

The marginal approach of Wei, Lin, and Weissfeld [32] for analysing multivariate failure time data has considerable appeal in this setting. This approach is based on formulating ordinary Cox models for each component event to obtain component-specific estimates of treatment effect, so it is compatible with the way we have formulated the joint distributions using copula functions. Estimation proceeds under a working independence assumption, as often adopted for analyses based on generalized estimating equations [22].

In what follows we again assume there are two events of interest. We suppose analysis is to be based on a sample of m independent individuals labelled $i = 1, \ldots, m$. We let $dN_{ik}(s) = I(T_{ik} = s)$ indicate that a type k event experienced by individual i at time s, and let $\{N_{ik}(s), 0 < s\}$, $k = 1, 2$ and $\{N_i(s) = (N_{i1}(s), N_{i2}(s)), 0 < s\}$ denote the univariate and bivariate counting process for individual i, $i = 1, \ldots, m$, respectively. If C_i is a right censoring time, let $Y_i(s) = I(s \leq C_i)$, $Y_{ik}(s) = I(s \leq T_{ik})$, and $\bar{Y}_{ik}(s) = Y_i(s)Y_{ik}(s)$, $k = 1, 2$, $i = 1, \ldots, m$. Under a Wei-Lin-Weissfeld approach for bivariate event times, the Cox model for a type k event is given by (12.2) and the corresponding score function for β_k is

$$U_k(\beta_k) = \sum_{i=1}^{m} \int_0^{\infty} \bar{Y}_{ik}(u) \left(Z_i - \frac{S_k^{(1)}(\beta_k, u)}{S_k^{(0)}(\beta_k, u)} \right) dN_{ik}(u) \ , \tag{12.6}$$

where $S_k^{(1)}(\beta, u) = \sum_{i=1}^{m} \bar{Y}_{ik}(u) Z_i^r \exp\{\beta_k Z_i\}$, $r = 0, 1$.

Under the copula model (12.6) the marginal distributions retain the proportional hazards structure, and so the solution to the score equation (12.6), $\hat{\beta}_k$, is consistent for the true marginal treatment effect β_k, $k = 1, 2$. If $\hat{\boldsymbol{\beta}} = (\hat{\beta}_1, \hat{\beta}_2)'$ is the estimate of $\boldsymbol{\beta} = (\beta_1, \beta_2)'$ obtained under the working independence assumption, Wei, Lin, and Weissfeld [32] show that $\sqrt{m}(\hat{\boldsymbol{\beta}} - \boldsymbol{\beta})$ converges in distribution to a multivariate normal distribution with a zero-mean vector and covariance matrix $\mathbb{V}(\boldsymbol{\beta})$ and they provide the form of a consistent sandwich-type estimate of $\mathbb{V}(\boldsymbol{\beta})$. A global estimate of the treatment effect is justified under the assumption $\beta_1 = \beta_2 = \beta$ and is a weighted combination of the component-specific estimates $\hat{\beta}_1$ and $\hat{\beta}_2$. If $\hat{\mathbb{V}}(\hat{\boldsymbol{\beta}})$ is the empirical estimate of the covariance matrix of $\hat{\boldsymbol{\beta}}$ and $\mathbf{J} = (1, 1)'$, then we take $\hat{\hat{\beta}} = \mathbf{C}'\hat{\boldsymbol{\beta}}$, where the weight $\mathbf{C} = [\hat{\mathbb{V}}(\hat{\boldsymbol{\beta}})]^{-1}\hat{\mathbf{J}}[\hat{\mathbf{J}}'[\hat{\mathbb{V}}(\hat{\boldsymbol{\beta}})]^{-1}\hat{\mathbf{J}}]^{-1}$ is chosen to correspond to the estimator with the minimize variance among all linear estimators.

A key distinction between the global approach of Wei, Lin, and Weissfeld [32] and the composite endpoint approach is that the former makes use of all observed events whereas the composite endpoint uses only information on the first event. There may also be gains in power as a result of this if model assumptions are correct.

12.2.2 Application to an Asthma Trial

We now apply both the composite endpoint analysis and the global approach to an asthma management study [19]. This is a two-phase, multicenter, randomized, parallel group effectiveness trial for comparing two treatment strategies for asthma management over a 2-year period. The control strategy is a "clinical strategy" (CS), in which the treatment was guided based on patient symptoms and spirometry readings. The experimental strategy is a so-called "sputum strategy" (SS), whereby a cellular analysis of sputum samples was used to guide cortricosteroid therapy use to keep eosinophils cell counts less than 2 %. In phase I, a total of 107 patients

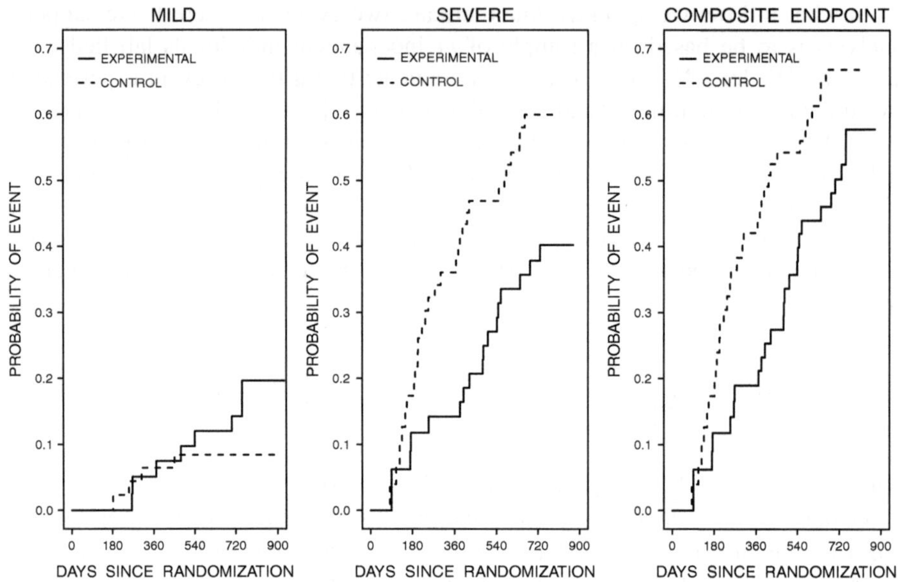

Fig. 12.2 Empirical distribution functions for mild exacerbations, severe exacerbations and the composite endpoint in asthma trial

were identified through the minimum treatment to maintain control. The aim of this asthma study was to investigate whether SS is more effective than CS on reducing the number and severity of exacerbations in phase II.

In our analysis we focus on two types of exacerbations: mild exacerbations defined as requiring a daily maintenance dose of fluticasone of $<250\,\mu$g, and severe exacerbations defined here as requiring a minimum daily maintenance dose of $\geq 250\,\mu$g. The composite endpoint is defined as the time to the first of the two type of exacerbations. Figure 12.2 displays the empirical distribution function plots for the two component types of exacerbations and for the composite endpoint. It is apparent that the severe exacerbations occur much more frequently than mild exacerbations, and thus represent the majority of the events contributing to the composite endpoint.

Table 12.1 presents the results of the proportional hazards regression analysis in which the single binary covariate is the treatment indicator taking the value one for patients in the experimental (SS) group and zero otherwise. From these results it is clear that the experimental SS strategy leads to a significantly lower hazard of severe exacerbations with a relative risk reduction (1-RR) of 0.47 (95 % CI: 0.01, 0.71; $p = 0.047$), but has little effect on the occurrence of mild exacerbations ($p = 0.247$). The result from the composite endpoint analysis is not statistically significant with $p = 0.137$. The last column of Table 12.1 gives the p-values for testing the proportional hazards assumption using univariate tests based on Schoenfeld residuals. There is insufficient evidence to reject the null hypothesis of

Table 12.1 Analysis results of the asthma management study [19]; *RR* denotes relative risk defined by the ratio of hazards

Endpoint/analysis	RR	95 % CI	p-value[a]	p-value[b]
Mild	2.07	(0.60, 7.06)	0.247	0.114
Severe	0.53	(0.29, 0.99)	0.047	0.220
Composite	0.66	(0.39, 1.14)	0.137	0.063
Global (WLW)	0.70	(0.40, 1.22)	0.209	

[a] Wald test of the null hypothesis that regression coefficients are zero
[b] p-value for test of the proportional hazards assumption [31]

proportional hazards for each component, and the test yields a $p-$value just shy of statistical significance for the composite endpoint analysis at 0.063. Thus, while we have demonstrated that, in principle, if the proportional hazards assumption holds for the components of a composite endpoint, it generally does not hold for composite endpoint itself, the tests do not suggest problems with model fit for this particular data.

12.3 Composite Endpoints with Semi-competing Risks

12.3.1 Notation and Description of the Setting

In palliative trials in oncology, patients with bone metastases are at risk of skeletal complications including vertebral and non-vertebral fractures, bone pain, and need for surgery to repair bone [18]. The study population is at high risk of death given presence of metastatic disease. We consider trials directed at assessing therapeutic interventions to prevent clinical or disease-related non-fatal events in populations at high risk of death. In this, and many similar settings, while interest lies in assessing whether a new treatment is effective in preventing non-fatal events, death is an important outcome which affects the information we can collect on the event of interest.

Individuals are randomized in state 0 and are at risk of a transition to state 1 (corresponding to event occurrence) or state 2 (corresponding to death); following event occurrence individuals can of course make a transition to the death state. In this setting individuals are at risk of a non-fatal event but they may die without experiencing this event, or after experiencing the event. The term "semi-competing risk" is used to describe this setting because one event (usually death) precludes the future occurrence of the other event, but the reverse is not true [34].

There are two broad settings where this multistate figure applies. In one scenario, interest lies primarily in evaluating the effect of a treatment on the prevention of a non-fatal event. This is the case in studies of palliative interventions in patients with advanced cancer where treatments are directed at reducing pain-related events, maintaining functional ability by preventing debilitating events. Here treatment may be envisioned to have an effect on the non-fatal event and a traditional competing risk analysis is warranted as we describe in the next subsection. In the second scenario, primary interest may lie in overall survival (i.e. time to entry into state 2). This arises in cancer and cardiovascular trials, for example, when evaluating new treatments for improvement in survival. In such settings if the mortality rate is low, a composite endpoint of event-free survival is often adopted. In oncology, for example, it is common to adopt progression-free survival as a composite endpoint. In cardiovascular disease trials similar composites such as hospitalization-free survival may be used.

12.3.2 *Assessing Treatments on Non-fatal Events*

Here we consider a trial aiming to evaluate the effect of an experimental treatment on the prevention of a single non-fatal event in a population at high risk of death. Figure 12.3 contains a multistate figure for an illness-death process reflecting the possible occurrence of the event and death. If T_{i1} denote the time of transition from state 0 to state 1 and let T_{i2} denote the time of transition from state 0 to state 2 (death). We let $q_k(t|z_i)$ denote the cause-specific hazard defined by

$$q_k(t|z_i) = \lim_{\Delta t \downarrow 0} \frac{P(t \le T_i < t + \Delta t, R_{ik} = 1 | t \le T_i, z_i)}{\Delta t}, \tag{12.7}$$

where $T_i = \min(T_{i1}, T_{i2})$ and $R_{ik} = I(T_{ik} < T_{i,3-k})$ indicates which event occurred first. It is customary to adopt a cause-specific proportional hazards model and so we often set $q_k(t|z) = q_{0k}(t) \exp(\beta_k z)$, $k = 1, 2$, and let $Q_k(t|z) = \int_0^t q_k(u|z)du$ and $Q_{0k}(t) = \int_0^t q_{0k}(u)du$, and denote the full vector of parameters for type k events by $\theta_k = (q_{0k}(\cdot), \beta_k')'$ and let $\theta = (\theta_1', \theta_2')'$.

Let $T_i = \min(T_{i1}, T_{i2})$ and $Y_i^\dagger(s) = I(s \le T_i)$ indicate that individual i is at risk of transition out of state 0 at time s. If C_i is the censoring time for individual i, $Y_i(s) = I(s \le C_i)$, and we let $\bar{Y}_i(s) = Y_i(s) Y_i^\dagger(s)$ indicate that individual i is both under observation and at risk of a transition out of state 0. The likelihood function based on a competing risk model for the first event is then

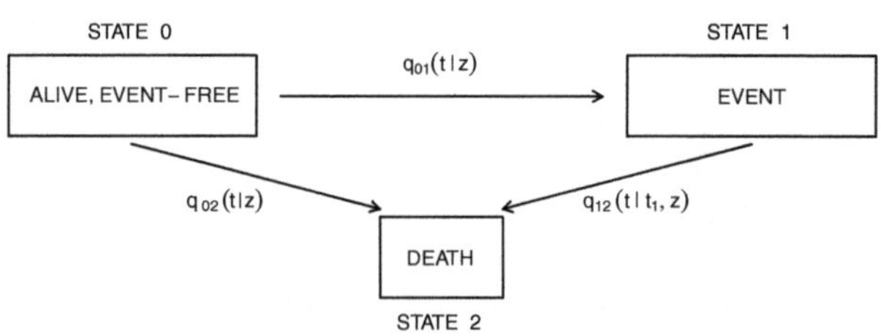

Fig. 12.3 Illness-death diagram characterizing the joint occurrence of a non-fatal and fatal event

$$L(\theta) \propto L_1(\theta_1)L_2(\theta_2) \,. \tag{12.8}$$

We maximize $L(\theta)$ by separately maximizing $L_k(\theta_k)$ for $k = 1, 2$, where

$$\log L_k(\theta_k) = \sum_{i=1}^{m} \int_0^{\infty} \left\{ \bar{Y}_i(u) dN_{ik}(u) \log dQ_k(u|z_i) - \int_0^{\infty} \bar{Y}_i(u) dQ_k(u|z_i) \right\} , \tag{12.9}$$

$k = 1, 2$ and obtain consistent estimates [9].

The fact that the likelihood factors in (12.8) means that estimation of β_1 is carried out by effectively censoring individuals at the time of death if they have not experienced the event of interest. This is viewed as unappealing to many since death is the most serious event that can occur and it is generally viewed as poor practise to selectively censor individuals based on event arising post-randomization. The multistate model in Fig. 12.3 is an accurate reflection of the possible outcomes in individuals and this approach to the analysis is well-justified. Such analyses are best presented, however, along with the results of analyses directed at assessing the treatment effect on survival in order to provide a complete representation of the data.

The aforementioned reservations about competing risk analyses, however, has prompted investigators to adopt a composite endpoint based on the minimum time to the non-fatal event and death [8, 13] even though type I events are of real interest. This strategy leads to an "event-free survival" analysis which is particularly common in cancer where progression-free survival is routinely adopted as a primary endpoint [30]. A concern with this approach is that in palliative trials, treatments under study may not be expected to affect survival times, and if a non-negligible proportion of individuals die before experiencing the clinical event of interest, this analysis can lead to a serious attenuation of the estimator of treatment effect [10,16]. The extent of the attenuation depends on the probability that the first event is the non-fatal event (i.e. a $0 \to 1$ transition occurs rather than a $0 \to 2$ transition in Fig. 12.3). The cumulative incidence function of a $0 \to k$ transition in the control arm is

$$F_k(t|Z=0) = P(T < t, R_k = 1|Z=0) = \int_0^t dQ_{0k}(u) \exp(-[Q_{01}(u) + Q_{02}(u)]) \,, \tag{12.10}$$

and the cumulative probability of either event occurring by time t is

$$F(t|Z=0) = P(T < t|Z=0) = 1 - \exp(-[Q_{01}(t) + Q_{02}(t)]) \,. \tag{12.11}$$

As an illustration, suppose $q_{01}(t) = \gamma_1 \lambda_1 (\lambda_1 t)^{\gamma_1 - 1}$ and $q_{02}(t) = \lambda_2$, where $\lambda_k > 0$, $k = 1, 2$ and $\gamma_1 > 0$. Suppose we plan a trial with follow-up over the interval $[0, 1]$. We may set γ_1 to any value to reflect a trend in the hazard for type 1 events. We set $F(1|Z=0) = p$ where $p = 0.40$ or 0.80 and let $F_1(1|Z=0)/F(1|Z=0) = p_1$ where $p_1 = 0.25, 0.50$ and 0.75. Finally we set $\beta_1 = \log(0.5)$ to correspond

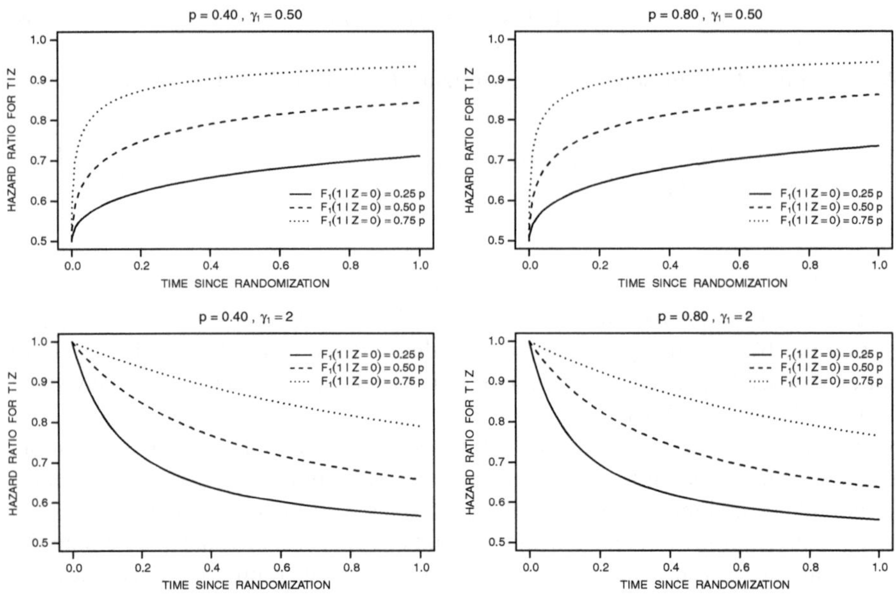

Fig. 12.4 Plots of hazard ratio for the composite event-free survival analysis based on $T|Z$

to a large treatment effect on the reduction in the intensity of type 1 events, and $\beta_2 = 0$ to reflect no effect on survival. Figure 12.4 gives plots of the hazard ratio for the composite event with $T = \min(T_1, T_2)$ for $Z = 1$ vs. $Z = 0$, for each parameter configuration with $\gamma_1 = 0.5$ and 2. Here it can be seen that the hazard ratio varies considerably with time and that at any given time the ratio is more than 0.50, the value specified for the intensity of the non-fatal event. This reveals the fact that an event-free survival analysis will tend to yield conservative estimates of the effect of treatment on intermediate events if the intervention is not expected to improve survival. This point is illustrated empirically in the example of Sect. 12.3.4.

12.3.3 Composite Endpoints When Interest Lies in Survival

We focus on the second scenario discussed earlier where interest really lies in the effect of treatment on survival but where the non-fatal event was incorporated into a composite endpoint. This is often the case in cancer trials when interest lies in showing that a treatment can prolong survival but progression is incorporated as a component in a progression-free survival endpoint. There has been much discussion in the literature about the interpretation of findings based on survival, progression-free survival and progression endpoints, and in particular the challenges in reconciling the findings when the conclusions about an experimental intervention differ [2, 6, 15].

To address this scenario we must more completely define the process in Fig. 12.3. We let $V(s)$ be the state occupied at time s and let

$$q_{12}(t|t_1, z) = \lim_{\Delta t \downarrow 0} \frac{P(V(t + \Delta t^-) = 2|V(t^-) = 1, t_1, z)}{\Delta t},$$

denote the $1 \to 2$ transition intensity. Under a Markov model this intensity depends only on the state occupied and the time, so $q_{12}(t|t_1, z) = q_{12}(t|z)$. In contrast, under a semi-Markov model, $q_{12}(t|t_1, z) = q_{12}(B(t)|z)$ where $B(t) = t - t_1$ is the time since entry to state 1. In the context of cancer trials time might measured as time since randomization and we adopt a Markov model with $q_{12}(t|z) = q_{0,12}(t)\exp(\beta_{12}z)$.

In this scenario interest lies in examining the effect of an intervention on the survival distribution. In fact in cancer trials, primary interest is in survival (entry time to state 2, regardless of path) but progression is often added to form a composite endpoint $T = \min(T_1, T_2)$ to increase the number of events and hence increase power (this is predicated on the assumption that the treatment effect is the same for the composite endpoint progression-free survival as it is for survival). The hazard for death can be obtained from the three-state model in special circumstances. Under the assumption of constant transition intensities we can compute the transition probability matrix $\mathbb{P}(0, t|z)$ with (j, k) entry.

The transition intensity matrix for the three state process under a Markov assumption is

$$d\mathbb{Q}(t|z) = \begin{bmatrix} -dQ_{01}(t|z) - dQ_{02}(t|z) & dQ_{01}(t|z) & dQ_{02}(t|z) \\ 0 & -dQ_{12}(t|z) & dQ_{12}(t|z) \\ 0 & 0 & 0 \end{bmatrix}.$$

If we let $\prod_{(s,t]}$ denote product integration over the interval $(s, t]$, then

$$\mathbb{P}(s, t|z) = \prod_{(s,t]} \{\mathbb{I} + d\mathbb{Q}(u|z)\},$$

is the transition probability matrix [3] where $\mathbb{I}$ is a 3×3 identity matrix. This matrix has $[j, k]$ entry of $\mathbb{P}(s, t|z)$ is $P_{jk}(s, t|z) = P(V(t) = k|V(s) = j, z)$ where s and t $(s < t)$ are two specified times; we write $\mathbb{P}(0, t|z) = \mathbb{P}(t|z)$ and $P_{jk}(0, t|z) = P_{jk}(t|z)$. The functions $P_{02}(t|z = 0)$ and $P_{02}(t|z = 1)$ of $\mathbb{P}(t|z)$ are the cumulative distribution functions for death for control and treated individuals, respectively. From this we can obtain the hazard functions $h^D(t|z) = -d\log(1 - P_{02}(t|z))/dt$, $z = 0, 1$. The hazard ratio is then

$$\frac{h^D(t|z=1)}{h^D(t|z=0)} = \exp\left(-d\log(1 - P_{02}(t|z=1))/dt + d\log(1 - P_{02}(t|z=0))/dt\right),$$

which again does not satisfy the proportional hazards form in general. The point is that the models we may examine in the analysis and secondary analysis of composite events are often incompatible.

Here we can consider a setting where there is a treatment effect on the risk of death reflected by models with $q_1(t|z) = q_{01}(t)\exp(\beta_1 z)$ and $q_{12}(t|z) = q_{0,12}(t)\exp(\beta_{12}z)$ with $\beta_2 = \beta_{12} = \log 0.5$ so there is a 50 % reduction in the rate of death for individuals who have and have not experienced progression. The association between progression and death is reflected by the relation between $q_{01}(t)$ and $q_{0,12}(t)$. We consider time-homogeneous transition rates with $q_{0,12}/q_{01} = 2.0$ to reflect a doubling of the risk of death following the intermediate event, and $q_{01}/q_{02} = 0.25$ and 0.75 so the odds of progression are 25 % and 75 %, respectively. Then we set $\beta_1 = \log 0.5$ and 0 to correspond to a comparable effect of treatment on progression and no treatment effect on progression. We consider a trial designed with follow-up over [0, 1] and set q_{01} so that the proportion of patients dying by the administrative end of study is $E_z[P(V(1) = 2|V(0) = 0, Z = z)] = p$ where $p = 0.50$ and 0.80; we also set $P(Z = 1) = 0.5$. Figure 12.5 displays the hazard ratio for a survival analysis in this setting. Here we can see considerable variation in the hazard ratio where this variation depends critically on the relative odds of the possible paths as well as the effect of treatment on the cause-specific hazard of the intermediate event.

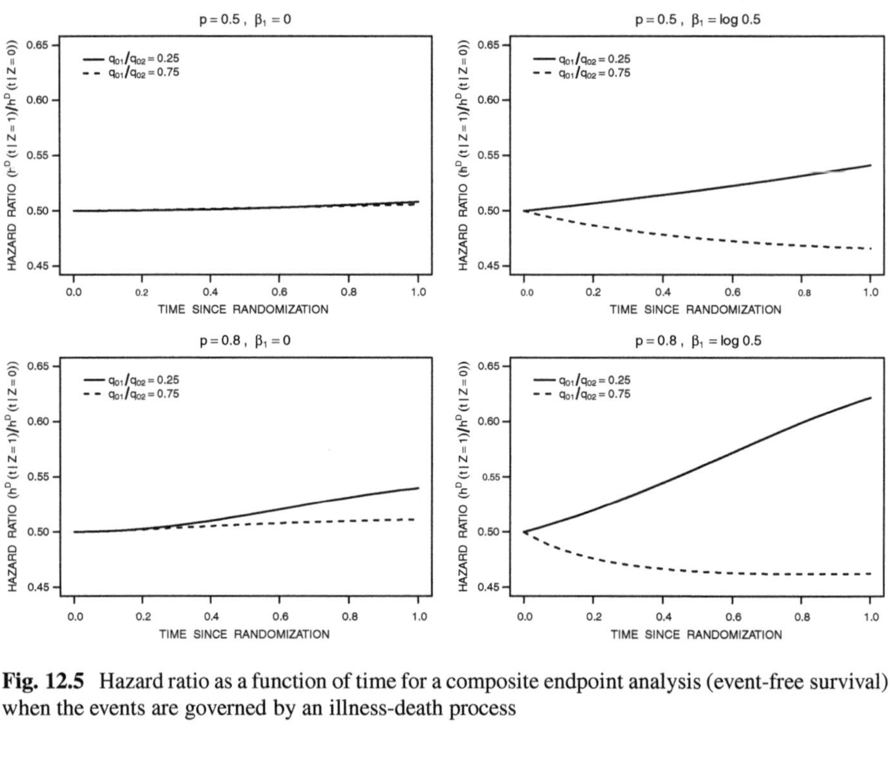

Fig. 12.5 Hazard ratio as a function of time for a composite endpoint analysis (event-free survival) when the events are governed by an illness-death process

Table 12.2 Results of several time to event analyses based on the fracture and survival times [18]

Event	$\hat{\beta}$	se($\hat{\beta}$)	RR	95 % CI	p−value
Fracture	−0.411	0.160	0.66	(0.48, 0.91)	0.010
Fracture-free	−0.256	0.142	0.77	(0.58, 1.02)	0.072
Survival	0.084	0.256	1.09	(0.66, 1.80)	0.742

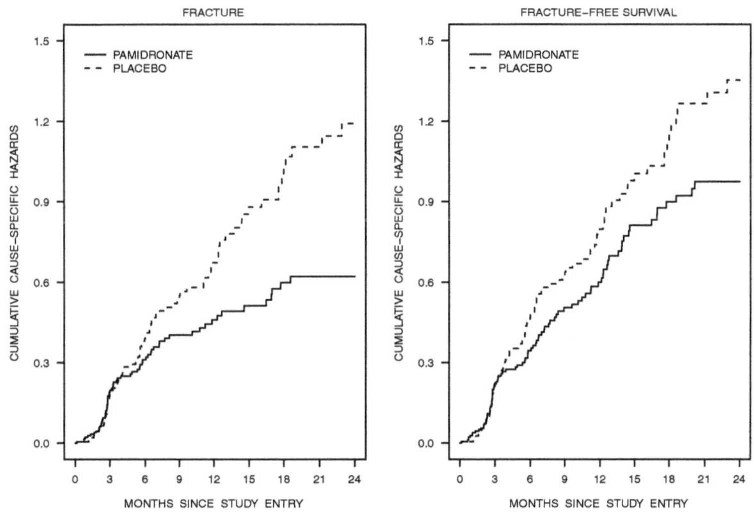

Fig. 12.6 Plot of cumulative cause-specific hazard for fracture and cumulative hazard for the fracture-free survival analysis

12.3.4 Application to a Cancer Trial for Event-Free Survival

Consider data from an international multicenter randomized clinical trial of 380 breast cancer patients with skeletal metastases. The skeletal metastases weaken bone and put patients at increased risk of fractures and hence there is a need to treat patients for the prevention of fractures. In this trial patients were randomized to receive pamidronate or placebo medication by monthly infusions and followed over time for the occurrence of fractures. A substantial number of patients could die before experiencing a fracture and hence Fig. 12.3 characterizes this setting well with the non-fatal event being fracture.

We present the results of several analyses related to these data in Table 12.2 and Fig. 12.6. The time from randomization to fracture is the time of interest in the first row, the time from randomization to the first fracture or death (whichever is first) is the event in the second row, and the time from randomization to death is the time of interest in the third row; all times are subject to right censoring. Analyses are carried out using R version 2.14.0 [28]. The estimated hazard ratio from the cause-specific Cox regression model for fracture reveals a significant reduction in risk of fracture associated with pamidronate (RR = 0.66; 95 % CI: 0.48, 0.91; p = 0.010). A fracture-free survival analysis gives a more conservative estimated relative risk of 0.77 (95 % CI: 0.58, 1.02; p = 0.072). A cause-specific survival analysis directed at 0 → 2 transitions gives a relative risk 1.45 (95 % CI: 0.76, 2.75; p = 0.258) and an overall survival analysis gives a relative risk of 1.09 (95 % CI: 0.66, 1.80; p = 0.742).

Figure 12.6 gives a plot of the cumulative cause-specific hazard for fracture (left panel) arising from a competing risk analysis, as well as the estimated cumulative hazard for the fracture-free survival analysis (right panel). The fact that the estimated cumulative hazard functions on the right are closer together reflects the attenuation that can arise from including death in a composite endpoint when treatment is not expected to (nor does) have an effect on it.

12.4 Discussion

We have focussed on the setting of composite endpoints which are used in the setting of time to event data. Separate consideration was given to the setting where mortality rates are negligible and the setting where patients are at non-negligible risk of death where competing or semi-competing risks arise. For the latter situation a competing risk analysis seems natural but there seems to be some resistance to this among clinical trialists. Concerns regarding dependent censoring are unfounded when interest lies in the effect of treatment on a non-fatal event unless one takes the latent variable view of the competing risk problem, by which the time of the non-fatal event is conceptualized to occur after death. The multistate diagram is more appealing, in our view, in that it makes it clear that this is not the case. As a result, the association between the non-fatal event time and the time of death is not appropriate to model in terms of a correlation but rather through the intensity functions $q_{02}(t)$ and $q_3(t|t_1, Z)$ and notions of dependent censoring are moot vis a vis the component endpoints.

Acknowledgements This work was supported by grants to Richard Cook from the Natural Sciences and Engineering Research Council of Canada (Grant No. 101093) and the Canadian Institutes for Health Research (Grant No. 105099). Richard Cook is a Canada Research Chair in Statistical Methods for Health Research. The authors thank Dr. Parameswaran Nair (McMaster University) for the data from the asthma trial and Novartis Pharmaceuticals for permission to use the data from the bisphosphonate trial.

References

1. Al-Kateb, H., Boright, A.P., Mirea, L., Xie, X., Sutradhar, R., Mowjoodi, A., Bharaj, B., Liu, M., Bucksa, J.M., Arends, V.L., Steffes, M.W., Cleary, P.A., Sun, W., Lachin, J.M., Thorner, P.S., Ho, M., McKnight, A.J., Maxwell, A.P., Savage, D.A., Kidd, K.K., Kidd,

J.R., Speed, W.C., Orchard, T.J., Miller, R.G., Sun, L., Bull, S.B., Paterson, A.D., Diabetes Control and Complications Trial/Epidemiology of Diabetes Interventions and Complications Research Group: Multiple superoxide dismutase 1/splicing factor serine alanine 15 variants are associated with the development and progression of diabetic nephropathy: the diabetes control and complications trial/epidemiology of diabetes interventions and complications genetics study. *Diabetes* **57**(1), 218–228 (2008)

2. Amir, E., Seruga, B., Kwong, R., Tannock, I.F., Ocaña, A.: Poor correlation between progression-free and overall survival in modern clinical trials: Are composite endpoints the answer? *European Journal of Cancer* **48**(3), 385–388 (2012)
3. Andersen, P.K., Borgan, O., Gill, R.D., Keiding, N.: *Statistical Models Based on Counting Processes*. Springer Verlag, New York (1993)
4. Benjamini, Y., Hochberg, Y.: Controlling the false discovery rate: A practical and powerful approach to multiple testing. *Journal of the Royal Statistical Society, Series B (Methodological)* **57**, 125–133 (1995)
5. Bethel, M.A., Holman, R., Haffner, S.M., Califf, R.M., Huntsman-Labed, A., Hua, T.A., McMurray, J.: Determining the most appropriate components for a composite clinical trial outcome. *American Heart Journal* **156**(4), 633–640 (2008)
6. Buyse, M., Burzykowski, T., Carroll, K., Michiels, S., Sargent, D.J., Miller, L.L., Elfring, G.L., Pignon, J.P., Piedbois, P.: Progression-free survival is a surrogate for survival in advanced colorectal cancer. *Journal of Clinical Oncology* **25**(33), 5218–5224 (2007)
7. Cannon, C.P.: Clinical perspectives on the use of composite endpoints. *Controlled Clinical Trials* **18**(6), 517–529 (1997)
8. Chi, G.Y.H.: Some issues with composite endpoints in clinical trials. *Fundamental & Clinical Pharmacology* **19**(6), 609–619 (2005)
9. Crowder, M.: *Multivariate Survival Analysis and Competing Risks*. CRC Press, Boca Raton (2012)
10. DeMets, D.L., Califf, R.M.: Lessons learned from recent cardiovascular clinical trials: Part I. *Circulation* **106**(6), 746–751 (2002)
11. D'silva, L., Cook, R.J., Allen, C.J., Hargreave, F.E., Parameswaran, K.: Changing pattern of sputum cell counts during successive exacerbations of airway disease. *Respiratory Medicine* **101**(10), 2217–2220 (2007)
12. Ferreira-González, I., Permanyer-Miralda, G., Busse, J.W., Bryant, D.M., Montori, V.M., Alonso-Coello, P., Walter, S.D., Guyatt, G.H.: Composite endpoints in clinical trials: the trees and the forest. *Journal of Clinical Epidemiology* **60**(7), 660–661 (2007)
13. Ferreira-González, I., Permanyer-Miralda, G., Busse, J.W., Bryant, D.M., Montori, V.M., Alonso-Coello, P., Walter, S.D., Guyatt, G.H.: Methodologic discussions for using and interpreting composite endpoints are limited, but still identify major concerns. *Journal of Clinical Epidemiology* **60**(7), 651–657 (2007)
14. Ferreira-González, I., Permanyer-Miralda, G., Domingo-Salvany, A., Busse, J.W., Heels-Ansdell, D., Montori, V.M., Akl, E.A., Bryant, D.M., Alonso-Coello, P., Alonso, J., Worster, A., Upadhye, S., Jaeschke, R., Schünemann, H.J., Pacheco-Huergo, V., Wu, P., Mills, E.J., Guyatt, G.H.: Problems with use of composite end points in cardiovascular trials: systematic review of randomised controlled trials. *British Medical Journal* **334**, 786–792 (2007)
15. Fleming, T.R., Rothmann, M.D., Lu, H.L.: Issues in using progression-free survival when evaluating oncology products. *Journal of Clinical Oncology* **27**(17), 2874–2880 (2009)
16. Freemantle, N., Calvert, M., Wood, J., Eastaugh, J., Griffin, C.: Composite outcomes in randomized trials: greater precision but with greater uncertainty? *Journal of the American Medical Association* **289**(19), 2554–2559 (2003)
17. Genest, C., MacKay, J.: The joy of copulas: bivariate distributions with uniform marginals. *The American Statistician* **40**(4), 280–283 (1986)

18. Hortobagyi, G.N., Theriault, R.L., Porter, L., Blayney, D., Lipton, A., Sinoff, C., Wheeler, H., Simeone, J.F., Seaman, J., Knight, R.D., for the Protocol 19 Aredia Breast Cancer Study Group: Efficacy of pamidronate in reducing skeletal complications in patients with breast cancer and lytic bone metastases. *The New England Journal of Medicine* **335**(24), 1785–1792 (1996)
19. Jayaram, L., Pizzichini, M.M., Cook, R.J., Boulet, L.P., Lemiére, C., Pizzichini, E., Cartier, A., Hussack, P., Goldsmith, C.H., Laviolette, M., Parameswaran, K., Hargreave, F.E.: Determining asthma treatment by monitoring sputum cell counts: effect on exacerbations. *European Journal of Respirology* **27**(3), 483–494 (2006)
20. Joe, H.: *Multivariate Models and Dependence Concepts*. Chapman and Hall, London (1997)
21. Lemaire, M., Islam, Q.S., Shen, H., Khan, M.A., Parveen, M., Abedin, F., Haseen, F., Hyder, Z., Cook, R.J., Zlotkin, S.H.: Iron-containing micronutrient powder provided to children with moderate-to-severe malnutrition increases hemoglobin concentrations but not the risk of infectious morbidity: a randomized, double-blind, placebo-controlled, noninferiority safety trial. *The American Journal of Clinical Nutrition* **94**(2), 585–593 (2011)
22. Liang, K.Y., Zeger, S.L.: Longitudinal data analysis using generalized linear models. *Biometrika* **73**(1), 13–22 (1986)
23. Lim, E., Brown, A., Helmy, A., Mussa, S., Altman, D.G.: Composite outcomes in cardiovascular research: a survey of randomized trials. *Annals of Internal Medicine* **149**(9), 612–617 (2008)
24. Montori, V.M., Permanyer-Miralda, G., Ferreira-González, I., Busse, J.W., Pacheco-Huergo, V., Bryant, D., Alonso, J., Akl, E.A., Domingo-Salvany, A., Mills, E., Wu, P., Schünemann, H.J., Jaeschke, R., Guyatt, G.H.: Validity of composite end points in clinical trials. *British Medical Journal* **330**(7491), 594–596 (2005)
25. Neaton, J.D., Gray, G., Zuckerman, B.D., Konstam, M.A.: Key issues in end point selection for heart failure trials: composite end points. *Journal of Cardiac Failure* **11**(8), 567–575 (2005)
26. Pocock, S.J., Geller, N.L., Tsiatis, A.A.: The analysis of multiple endpoints in clinical trials. *Biometrics* **43**(3), 487–498 (1987)
27. Proschan, M.A., Waclawiw, M.A.: Practical guidelines for multiplicity adjustment in clinical trials. *Controlled Clinical Trials* **21**(6), 527–539 (2000)
28. R Development Core Team: *R: A Language and Environment for Statistical Computing*. R Foundation for Statistical Computing, Vienna, Austria (2011). URL http://www.R-project.org/. ISBN 3-900051-07-0
29. Sankoh, A.J., B., S.D.R., Huque, M.F.: Efficacy endpoint selection and multiplicity adjustment methods in clinical trials with inherent multiple endpoint issues. *Statistics in Medicine* **22**(20), 3133–3150 (2003)
30. Soria, J.C., Massard, C., Chevalier, T.L.: Should progression-free survival be the primary measure of efficacy for advanced NSCLC therapy? *Annals of Oncology* **21**(12), 2324–2332 (2010)
31. Therneau, T.M., Grambsch, P.M.: *Modeling Survival Data: Extending the Cox Model*. Springer, New York (2000)
32. Wei, L.J., Lin, D.Y., Weissfeld, L.: Regression analysis of multivariate incomplete failure time data by modeling marginal distributions. *Journal of the American Statistical Association* **84**(408), 1065–1073 (1989)
33. Wu, L., Cook, R.J.: Misspecification of Cox regression models with composite endpoints. *Statistics in Medicine* **31**(28), 3545–3562 (2012)
34. Xu, J., Kalbfleisch, J.D., Tai, B.: Statistical analysis of illness-death processes and semicompeting risks data. *Biometrics* **66**(3), 716–725 (2010)

Chapter 13
Statistical Validation of Surrogate Markers in Clinical Trials

Ariel Alonso, Geert Molenberghs, and Gerard van Breukelen

Abstract The increasing cost of drug development has raised the demand on the use of biomarkers as surrogate endpoints for the evaluation of new drugs in clinical trials. However, failed past attempts to use surrogate endpoints made it clear that, before deciding on the use of a candidate surrogate endpoint, it is of the utmost importance to investigate its validity. Such validation process has proven challenging for conceptual and practical reasons. In the present chapter, some of the statistical methods introduced for the evaluation of surrogate markers will be discussed. Emphasis will be made on the so-called meta-analytic approach and its information-theoretic version, where information from several units is combined to carry out the validation exercise. The methods will be illustrated using a case study in ophthalmology.

13.1 Motivations and Antecedents

Recent discoveries in medicine and biology are opening an entire range of possibilities for the development of new treatments. However, these unquestionable achievements are also facing us with the challenge of having to evaluate a large number of promising therapies, using increasingly complex and costly clinical trials [2].

One of the most important factors influencing the duration and complexity of modern clinical trials is the choice of the endpoint used to assess drug efficacy. Actually, the most sensitive and relevant clinical endpoint, the so-called "true" endpoint, might often be difficult to use. This can happen, for instance, if measurement of the true endpoint is costly (e.g., to diagnose "cachexia", a condition associated with malnutrition and involving loss of muscle and fat tissue, expensive

A. Alonso (✉) • G. van Breukelen
Department of Methodology and Statistics, Maastricht University, Maastricht, The Netherlands
e-mail: ariel.alonso@maastrichtuniversity.nl; gerard.vbreukelen@maastrichtuniversity.nl

G. Molenberghs
I-BioStat, Universiteit Hasselt, Diepenbeek, Belgium

KU Leuven - University of Leuven, Leuven, Belgium
e-mail: geert.molenberghs@uhasselt.be

K. van Montfort et al. (eds.), *Developments in Statistical Evaluation of Clinical Trials*, DOI 10.1007/978-3-642-55345-5_13

equipment measuring content of nitrogen, potassium and water in the patient's body is required); requires a long follow-up time (e.g., survival in early stage cancers); or requires a large sample size due to a low incidence of the event (e.g., short-term mortality in patients with suspected acute myocardial infarction). A plausible strategy in these circumstances is the use of biomarkers for efficacy. The pursue of this strategy has been further encouraged by recent developments in many medical and biological fields that have considerably increased the number of promising biomarkers for the assessment of efficacy. In addition, a growing number of new drugs have a well-defined mechanism of action at the molecular level, allowing drug developers to measure the effect of these drugs on the relevant biomarkers [15].

Basically, one would like to replace the problematic true endpoint by a biomarker, which is measured earlier, more conveniently, or more frequently. From a regulatory perspective, a biomarker is not considered an acceptable endpoint for a determination of efficacy of new drugs, unless it has been shown to function as a valid indicator of clinical benefit, i.e., unless it is a valid surrogate marker [5].

Because of the possible benefits for the duration and cost of clinical trials, surrogate markers have been used in medical research for a long time [12, 14]. However, in spite of all its potential advantages, the use of surrogate endpoints in the development of new therapies has always been controversial. This may be due to a number of unfortunate historical instances where treatments showing a highly positive effect on a surrogate endpoint, were ultimately shown to be detrimental to the subjects' clinical outcome. One of such unfortunate events was the approval by the Food and Drug Administration (FDA) in the United States of three antiarrhythmic drugs: encainide, flecainide and moricizine, based on their efficacy to effectively suppress arrhythmias. It was believed that, since arrhythmia is associated with an almost fourfold increase in the rate of cardiac-complication-related death, the drugs would also reduce the death rate. Nonetheless, a clinical trial conducted after the drugs had been approved and introduced into clinical practice showed that, in fact, the death rate among patients treated with encainide and flecainide was more than twice the one among patients treated with placebo [8]. An increase of the risk was also detected for moricizine.

Behind many of these failures in the initial use of surrogate endpoints, was the logical but naive perception that surrogacy could be established by only evaluating the association between the biomarker on the one hand and the corresponding true endpoint on the other hand. Nevertheless, these failed past attempts made clear that the mere existence of an association between a biomarker and the true endpoint is not sufficient for using the former as a surrogate, i.e., a good correlate is not automatically a good surrogate [14]. The recognition of this fact opened an exciting and fruitful debate about the properties that a good surrogate should satisfy. After more than 20 years of research, this debate is far from settled and many questions and practical issues still need to be addressed. This notwithstanding, our level of knowledge has been dramatically increased and plethora statistical methods are now available for the evaluation of surrogate markers.

In Sect. 13.2 some important definitions are given. The single-trial methods and the meta-analytic approach to the validation of surrogate markers are introduced in Sects. 13.3 and 13.4 respectively. Section 13.5 describes some of the issues that emerge when the true and/or the surrogate endpoints are not normally distributed and in Sects. 13.6 and 13.7 a unified approach based on information theory is introduced. The meta-analytic approach is illustrated using a case study in Sect. 13.8 and the implementation of this method in widely used software packages is addressed in Sect. 13.10. Eventually, some final comments are presented in Sect. 13.11.

13.2 Some General Definitions

The terms "endpoint", "biomarker", and "marker" have often been interchangeably used to refer simply to a random variable that can be measured over the course of the disease process. Variables that are measured early in the course of the disease are frequently suggested as potential surrogates for those that are measured later. The following definitions, introduced by the Biomarker Definitions Working Group, are nowadays widely accepted and adopted in the biomedical literature [4]:

- Clinical endpoint: a characteristic or variable that reflects how a patient feels, functions, or survives;
- Biomarker: a characteristic that is objectively measured and evaluated as an indicator of normal biological processes, pathogenic processes, or pharmacologic responses to a therapeutic intervention;
- Surrogate endpoint: a biomarker that is intended to substitute for a clinical endpoint. A surrogate endpoint is expected to predict clinical benefit (or harm or lack of benefit or harm).

It is important to point out that, although extremely useful, the previous definitions do not include all situations one may encounter in practice. For instance, in our case study we analyze a potential surrogate that is not a biomarker, but an intermediate endpoint that has clinical meaning of its own. This is frequently the case in medical fields like, for instance, oncology, where progression-free survival is often considered as a potential surrogate for survival.

13.3 Single-Trial Methods

All earlier approaches to the validation of surrogate markers were framed in a single-trial setting, i.e., it was assumed that information on both the surrogate (S) and the true endpoint (T) was available from a single clinical trial. Within this setting Prentice introduced in 1989 the first formal definition of surrogacy. Basically, Prentice proposed to define a surrogate endpoint as

> a response variable for which a test of the null hypothesis of no relationship to the treatment groups under comparison is also a valid test of the corresponding null hypothesis based on the true endpoint [21].

Symbolically, Prentice's definition can be written

$$f(S|Z) = f(S) \Leftrightarrow f(T|Z) = f(T)\,, \tag{13.1}$$

where $f(X)$ denotes the probability distribution of random variable X and $f(X|Z)$ denotes the probability distribution of X conditional on the treatment variable Z. Note that this definition involves the triplet (T, S, Z) and, consequently, the endpoint S is a surrogate for T always with respect to the effect of some specific treatment Z. This implies that, at least in principle, if a new treatment is considered, then the validation process would need to be repeated. Prentice and other authors supplemented the previous definition with the following set of operational criteria that has become known as the *Prentice's Criteria*: (1) treatment has a significant impact on the surrogate endpoint $f(S|Z) \neq f(S)$, (2) treatment has a significant impact on the true endpoint $f(T|Z) \neq f(T)$, (3) the surrogate endpoint has a significant impact on the true endpoint $f(T|S) \neq f(T)$, and (4) the full effect of treatment upon the true endpoint is captured by the surrogate $f(T|S, Z) = f(T|S)$ [5].

The latter two are Prentice's original criteria and it has been proven that the definition and criteria are only equivalent when both the surrogate and the true endpoints are binary [5]. Note that the first two criteria measure the departures from the null hypothesis used in (13.1) and the third criterion implies that the surrogate has a prognostic value for the true endpoint. Finally, the fourth criterion requires S to fully capture the effect of treatment on the true endpoint, that is, there is no effect of treatment on the true endpoint after correcting for the surrogate.

Freedman et al. argued that the last criterion raises conceptual problems, since it requires the statistical test for the treatment effect on the true endpoint to be non-significant after adjustment for the surrogate [16]. In general, the nonsignificance of this test does not prove that the effect of treatment on the true endpoint is totally captured by the surrogate [5, 13]. Freedman further proposed to shift the paradigm from hypothesis testing to estimation and to calculate the so-called proportion of treatment explained (*PTE*). The *PTE* is the proportion of the treatment effect on the true endpoint captured by the surrogate and is defined as $PTE = (\beta - \beta_S)\beta$, where β denotes the effect of the treatment on the true endpoint emanating from $f(T|Z)$ and β_S is the effect of the treatment on the true endpoint after adjusting by the surrogate and can be calculated using $f(T|S, Z)$.

Note that *PTE* is large when β_S is small relative to β, Prentice fourth criterion implies $\beta_S = 0$ and therefore, if this criterion holds, $PTE = 1$. Freedman suggested that a good surrogate is one for which *PTE* is close to one. However, some conceptual problems also surround *PTE*, the most paradoxical one is that it is not a proportion. In fact, *PTE* can take any value on the real line, making its interpretation problematic [5]. Freedman himself acknowledged that the confidence limits for *PTE* will tend to be rather wide or even unbounded if Fieller's confidence intervals are used.

Frangakis and Rubin strongly criticized the conceptual foundation of Prentice's fourth criterion and the *PTE* [13]. They pointed out that the treatment effect on the true endpoint used in these two procedures is obtained after conditioning on the surrogate, i.e., a post-randomization variable and, consequently, is not a causal effect. Further, they proposed to assess surrogacy using the so-called *principal stratification* which is based on the potential outcomes model often used in causal inference. It has been argued that this method suffers from a similar drawback as the Prentice's definition and criteria, in that it is too stringent and difficult to implement in practice [27]. In addition, the intrinsically unobserved nature of the vector of potential outcomes implies that untestable assumptions are unavoidable.

In a separate line of research, Buyse et al. showed that, for continuous and normally distributed endpoints, *PTE* can be decomposed in three different quantities: the first one merely is the ratio of the surrogate and true endpoint variances and, therefore, it only represents a scale factor, the other two are the so-called relative effect *RE* and the adjusted association ρ_Z [7]. The relative effect is defined as $RE = \beta/\alpha$, where α is the treatment effect on the surrogate emanating from $f(S|Z)$ and β is defined as before. Notice that, unlike Prentice's fourth criterion and the *PTE*, the treatment effects involved in *RE* are not adjusted by post-randomization variables and, hence, have a direct causal interpretation. Indeed, α and β are simply the average causal effects of the treatment on the surrogate and the true endpoint respectively. The adjusted association is the correlation between the surrogate and the true endpoint after adjusting by treatment and is defined as $\rho_Z = \text{Corr}(S, T|Z)$.

The relative effect tries to enable prediction of the treatment effect on the true endpoint based on the treatment effect on the surrogate, but to do so strong and untestable assumptions have to be made. Essentially, in a single trial setting one is confronted with the problem of estimating the relationship between both average causal effects using a single observation, namely the vector of treatment effects (α, β). A way out of the problem is to assume that $E(\beta|\alpha) = RE \times \alpha$, i.e., the average causal effects satisfy the regression through the origin equation $\beta = RE \times \alpha + \varepsilon$. Regression through the origin has often been surrounded by controversy due to the paradoxical results it can produce, like negative coefficients of determination and negative F ratios. Even when there are theoretical reasons to believe that the function relating the two variables of interest does pass through the origin, regression through the origin may be problematic if the relationship between the variables of interest is not linear in a neighborhood of zero. Moreover, if the data at hand lie far from zero, then the assumption of linearity at this point becomes impossible to evaluate. This lack of replication is a fundamental problem of all the previously discussed approaches and it can only be overcome when more than one pair (α, β) is available for the analysis.

13.4 Data from Several Trials: The Meta-analytic Approach

Over the years, it has become clear that the single trial setting is too restrictive for the evaluation of surrogate markers and a general agreement has been growing regarding the need of replication at the trial level as well. A first formal proposal along these lines, using Bayesian methods, was given by Daniels and Hughes [11]. Buyse et al. extended these ideas using the theory of linear mixed-effects models and Gail et al. extended it further using generalized estimating equations methodology [7, 17]. In what follows, we describe the approach as proposed by Buyse et al. under the assumption that both endpoints are normally distributed and in Sects. 13.5–13.7 other types of endpoints will be addressed. To that end let us assume that data from $i = 1, \ldots, N$ trials are available, in the ith of which $j = 1, \ldots, n_i$ subjects are enrolled. Further, let us denote the true and surrogate endpoints for patient j in trial i by T_{ij} and S_{ij}, respectively, and the indicator variable for the new treatment by Z_{ij}. The random treatment allocation in a clinical trial context naturally leads to the following bivariate model

$$\begin{cases} T_{ij} = \mu_{Ti} + \beta_i Z_{ij} + \varepsilon_{Tij} \ , \\ S_{ij} = \mu_{si} + \alpha_i Z_{ij} + \varepsilon_{sij} \ , \end{cases} \tag{13.2}$$

where μ_{Ti} and μ_{si} are trial-specific intercepts quantifying the average response in the control group, β_i and α_i are trial-specific average causal effects and ε_{Tij} and ε_{sij} are correlated error terms, assumed to be zero-mean normally distributed with covariance matrix

$$\boldsymbol{\Sigma} = \begin{pmatrix} \sigma_{TT} & \sigma_{TS} \\ \sigma_{TS} & \sigma_{SS} \end{pmatrix} , \tag{13.3}$$

i.e., (13.3) denotes the within-trial covariance matrix of T and S after adjusting by treatment and considering the patient the level of analysis. Furthermore, due to replication at the trial level, one can decompose the trial-specific parameters in the following way

$$\begin{pmatrix} \mu_{si} \\ \mu_{Ti} \\ \alpha_i \\ \beta_i \end{pmatrix} = \begin{pmatrix} \mu_S \\ \mu_T \\ \alpha \\ \beta \end{pmatrix} + \begin{pmatrix} m_{si} \\ m_{Ti} \\ a_i \\ b_i \end{pmatrix} , \tag{13.4}$$

where the second term on the right hand side of (13.4) is assumed to follow a zero-mean normal distribution with covariance matrix

$$\mathbf{D} = \begin{pmatrix} d_{SS} & d_{ST} & d_{Sa} & d_{Sb} \\ d_{ST} & d_{TT} & d_{Ta} & d_{Tb} \\ d_{Sa} & d_{Ta} & d_{aa} & d_{ab} \\ d_{Sb} & d_{Tb} & d_{ab} & d_{bb} \end{pmatrix} . \tag{13.5}$$

Essentially, (13.5) denotes the between-trial covariance matrix of intercepts and treatment effects on T and S, considering now trial the level of analysis. Buyse et al. investigated how the treatment effect on the true endpoint can be predicted by the treatment effect on the surrogate [7]. The main idea is to predict the treatment effect on T in a new trial $i = 0$ based on: (a) information obtained in the validation process using trials $i = 1, \ldots, N$, and (b) the estimate of the treatment effect on S in the new trial $i = 0$. To this end, these authors notice that $(\beta + b_0 | m_{S0}, a_0)$ follows a normal distribution with mean and variance

$$E(\beta + b_0 | m_{S0}, a_0) = \beta + \begin{pmatrix} d_{Sb} \\ d_{ab} \end{pmatrix}^T \begin{pmatrix} d_{SS} & d_{Sa} \\ d_{Sa} & d_{aa} \end{pmatrix}^{-1} \begin{pmatrix} \mu_{S0} - \mu_S \\ \alpha_0 - \alpha \end{pmatrix}, \tag{13.6}$$

$$\text{Var}(\beta + b_0 | m_{S0}, a_0) = d_{bb} - \begin{pmatrix} d_{Sb} \\ d_{ab} \end{pmatrix}^T \begin{pmatrix} d_{SS} & d_{Sa} \\ d_{Sa} & d_{aa} \end{pmatrix}^{-1} \begin{pmatrix} d_{Sb} \\ d_{ab} \end{pmatrix}. \tag{13.7}$$

If the treatment effect on the surrogate conveys a lot of information about the treatment effect on the true endpoint, then the conditional variance (13.7) will be close to zero. In that case, there would be an almost deterministic relationship between the treatment effects on the true and surrogate endpoint, and a very accurate prediction of the first one would be possible if the second one has been observed. Based on these ideas Buyse et al. proposed to assess surrogacy at the trial level using the coefficient of determination

$$R^2_{\text{trial}} = R^2_{b_i | m_{Si}, a_i} = \begin{pmatrix} d_{Sb} \\ d_{ab} \end{pmatrix}^T \begin{pmatrix} d_{SS} & d_{Sa} \\ d_{Sa} & d_{aa} \end{pmatrix}^{-1} \begin{pmatrix} d_{Sb} \\ d_{ab} \end{pmatrix}. \tag{13.8}$$

This coefficient measures how precisely the treatment effect on the true endpoint can be predicted, provided that the treatment effect on the surrogate endpoint has been observed in a new trial ($i = 0$). It is unitless and ranges in the unit interval if the corresponding covariance matrix $\mathbf{D}$ is positive-definite, two desirable features for its interpretation.

One special case of the model given in (13.2) is the so-called reduced model, which assumes that the intercepts, i.e. the average responses in the control group, are constant across trials. Under this assumption, expressions (13.6) and (13.7) reduce to

$$E(\beta + b_0 | a_0) = \beta + \frac{d_{ab}}{d_{aa}}(\alpha_0 - \alpha),$$

$$\text{Var}(\beta + b_0 | a_0) = d_{bb} - \frac{d_{ab}^2}{d_{aa}},$$

with corresponding

$$R^2_{\text{trial}} = R^2_{b_i | a_i} = \frac{d_{ab}^2}{d_{aa} d_{bb}}. \tag{13.9}$$

Similar to the logic in (13.6) and (13.7), the conditional model for β_i given μ_{Si} and α_i can be written as

$$\beta_i = \theta_0 + \theta_1 \mu_{Si} + \theta_2 \alpha_i + \varepsilon_i \ , \tag{13.10}$$

where expressions for the coefficients $(\theta_0, \theta_1, \theta_2)$ follow from (13.4) and (13.5). In case the surrogate is perfect at the trial level ($R^2_{\text{trial}} = 1$), the error term in (13.10) vanishes and the linear relationship becomes deterministic, implying that β_i *equals* the systematic component of (13.10).

Notice first that, unlike for the *RE*, the regression line (13.10) does not necessarily pass through the origin. Secondly, this new approach avoids the conceptual problems surrounding the *RE*, since the relationship between β_i and α_i is studied across a family of units, rather than in a single unit. By virtue of replication, it is possible to *check* the stated relationship for the treatment effects and, if the posited linear relation does not hold, alternative regression functions can be considered. Nevertheless, one has to be aware of a potentially low power to discriminate between candidate regression functions.

At the individual level, one tries to assess how an individual's surrogate outcome is predictive for the true endpoint outcome. To this end, one needs to construct the conditional distribution of T, given S and Z. From (13.2) we obtain

$$T_{ij}|Z_{ij}, S_{ij} \sim N\left\{\mu_{Ti} - \sigma_{TS}\sigma_{SS}^{-1}\mu_{Si} + (\beta_i - \sigma_{TS}\sigma_{SS}^{-1}\alpha_i)Z_{ij} \right.$$
$$\left. + \sigma_{TS}\sigma_{SS}^{-1}S_{ij}; \sigma_{TT} - \sigma_{TS}^2\sigma_{SS}^{-1}\right\}.$$

The association between both endpoints after adjustment by treatment is captured by the coefficient of determination

$$R^2_{\text{ind}} = \frac{\sigma^2_{ST}}{\sigma_{SS}\sigma_{TT}}. \tag{13.11}$$

Basically, the R^2_{ind} is the squared correlation between both endpoints once we have adjusted for treatment and trial and, therefore, it is a natural extension of the adjusted association. Unlike the trial level surrogacy, the individual level does not depend on the treatment and it can be interpreted as a quantification of the biological plausibility of the surrogate. An endpoint producing a high individual level surrogacy is always a potential surrogate, however, it may fail to be predictive at the trial level for a specific treatment that follows a causal path that completely avoids it.

Although elegant, the above hierarchical model often poses a considerable computational challenge [5]. To address this problem, Tibaldi et al. suggested several simplifications, like treating the trial-specific parameters in (13.2) as fixed effects in a two-stage approach [25]. The first-stage model will take the form (13.2) and at the second stage, the estimated treatment effect on the true endpoint is regressed on the estimated treatment effect on the surrogate and the intercept associated with the surrogate endpoint as

$$\hat{\beta}_i = \lambda_0 + \lambda_1 \hat{\mu}_{si} + \lambda_2 \hat{\alpha}_i + \varepsilon_i \ . \tag{13.12}$$

Essentially, the trial-level surrogacy R^2_{trial} is assessed by regressing $\hat{\beta}_i$ on $(\hat{\mu}_{si}, \hat{\alpha}_i)$ and the individual-level value is calculated as before, using the estimates from (13.3). Notice that, when the fixed-effects approach is chosen, there is a need to adjust for the heterogeneity in information content between trial-specific contributions. One way of doing so is weighting the contributions according to trial size. This gives rise to a weighted linear regression model (13.12) in the second stage.

Another cornerstone of the meta-analytic method is the choice of unit of analysis such as, for example, trial, center, or country. This choice may depend on practical considerations, such as the information available in the data, experts' considerations about the most suitable unit for a specific problem, the amount of replication at a potential unit's level, and the number of patients per unit. From a technical point of view, the most desirable situation is where the number of units and the number of patients per unit is sufficiently large. Of course, after choosing a specific unit for the analysis, one always has to reflect carefully on the status of the results obtained. Arguably, they may not be as reliable as one might hope for, and one should undertake every effort possible to increase the amount of information available. This issue has been covered at large by Cortiñas et al. and we refer the interested reader to this work for more details [9].

13.5 Other Types of Endpoints

In the previous section, the formalism developed by Buyse et al. was introduced using the *simplest* setting where both endpoints are Gaussian random variables measured cross-sectionally. However, this is not always the case, for example, one can encounter:

- Binary (dichotomous): the surrogate and/or true endpoints are binary, for instance, biomarker value below or above a certain threshold (e.g., viral load in HIV+ patients below detection limit) or clinical "success" (e.g., tumor shrinkage).
- Categorical (polychotomous): the surrogate and/or true endpoints are categorical, for instance, biomarker value falling in successive, ordered classes (e.g., cholesterol levels <200, 200–299, 300+ mg/dl) or clinical response (e.g., complete response, partial response, stable disease, progressive disease).

- Longitudinal or repeated measures: the surrogate and/or true endpoints are longitudinally measured, for instance, biomarker (e.g., CD4+ counts over time) or clinical outcome (e.g., blood pressure over time).
- Multivariate longitudinal: the surrogate and/or true endpoints are multivariate outcomes measured longitudinally, for instance, several biomarkers (e.g., CD4+ and viral load over time) or several clinical measurements (e.g., dimensions of quality of life over time).
- Time to event: the surrogate and/or true endpoints are failure-time random variables, for instance, time to cancer recurrence as a surrogate marker for survival.

Assessing surrogacy in these more complex scenarios raises a number of difficult challenges. Firstly, one now needs to deal with highly complicated hierarchical models. These models frequently bring severe numerical issues and the use of alternative, simplified approaches like the ones proposed by Tibaldi et al., becomes unavoidable. Secondly, based on the outputs of these models, one needs to define meaningful measures to quantify surrogacy at both the trial and individual level.

If one is ready to only consider linear models to study the relationship between the treatment effect on the surrogate and the true endpoint, then the methodology previously described can be applied in a straightforward fashion to quantify trial level surrogacy. At the individual level, however, abandoning the realm of normality has much deeper implications. Indeed, based on this meta-analytic paradigm, several individual-level measures have been proposed. For instance, in the binary-binary setting Renard et al. assumed that the observed dichotomic outcomes emerge from two latent and normally distributed variables $(\tilde{S}, \tilde{T})$. Essentially, it is assumed that the surrogate (true endpoint) takes value one when corresponding latent variable exceeds a threshold value, i.e., when $\tilde{S} > \eta_S$ $(\tilde{T} > \eta_T)$ and zero otherwise. In this framework, using a bivariate probit model, these authors defined individual-level surrogacy as $R^2_{\text{ind}} = \rho^2_{\tilde{S}\tilde{T}}$, which is the correlation at the latent level. Alternatively, they also defined $R^2_{\text{ind}} = \psi$, the global odds ratio between both binary endpoints estimated from a so-called bivariate Plackett-Dale model [22].

When the true endpoint is a survival time and the surrogate is a longitudinal sequence, Renard et al., using Henderson's model, proposed to study the individual level based on a time function defined as $R^2_{\text{ind}}(t) = \text{corr}[W_1(t), W_2(t)]^2$, where $(W_1(t), W_2(t))$ is a latent bivariate Gaussian process [23]. Burzykowski et al. approached the case of two failure-time endpoints based on copula models and quantified the individual level surrogacy using Kendall's τ [6].

Using multivariate ideas, the so-called R^2_Λ has been proposed to evaluate surrogacy when both responses are measured longitudinally [1]. The R^2_Λ coefficient quantifies the association between both longitudinal sequences and is defined using the covariance matrices emanating from a hierarchical model that characterized the joint distribution of both endpoints. Furthermore, the R^2_Λ can be incorporated into a more general framework allowing for interpretation in terms of canonical correlations of the error vectors, based on which, one can define a family of individual-level parameters [1].

All these examples underscore a limitation of the meta-analytic methodology so far: different settings require different definitions and in some of these settings, the association is measured at a latent level, hampering interpretation. Furthermore, in all cases, a joint and often non-standard model for both endpoints is needed, frequently representing a serious computational burden. In the next section, a unified approach to the validation of surrogate markers based on information theory will be introduced. Furthermore, it will be argued that this approach may help to overcome some of the aforementioned problems.

13.6 An Information-Theoretic Unification

Information theory, originated as a rigorous science in the 1940s, deals with the study of problems concerning complex systems, and has been applied in a variety of fields such as modern communication theory. In spirit and concepts, information theory has its mathematical roots connected with the idea of disorder or entropy used in thermodynamics and statistical mechanics. An early attempt to formalize the theory was made by Nyquist in 1924 who recognized the logarithmic nature of information [19]. Another major contribution in this area came in 1948 when Shannon published a remarkable paper on the properties of information sources and communication channels [24].

R.A. Fisher's well-known measure of the amount of information supplied by data about an unknown parameter is the first use of information in statistics. Further, Kullback and Leibler in 1951 studied another statistical information measure, involving two probability distributions associated with the same experiment [18].

The concept of entropy lies at the center of information theory and it can be interpreted as a measure of the randomness or uncertainty associated with a random variable. If Y is a discrete random variable taking values $\{k_1, k_2, \ldots, k_m\}$ with probability function $P(Y = k_i) = p_i$, then the entropy of Y is defined as

$$H(Y) = -E[\log P(Y)] = -\sum_i p_i \log p_i \ .$$

$H(Y)$ can be interpreted as the average uncertainty associated with P. The joint and conditional entropies are defined in an analogous fashion. Entropy is always non-negative and satisfies $H(Y|X) \leq H(Y)$ for any pair of random variables (X, Y), with equality holding under independence. Basically, the previous inequality states that uncertainty about Y can only decrease if additional information (X) becomes available. Furthermore, entropy is invariant under a bijective transformation [10].

Similarly, the so-called differential entropy $h_d(Y)$ of a continuous random variable Y with density $f_Y(y)$ and support S_{f_Y} is defined as

$$h_d(Y) = -E[\log f_Y(Y)] = -\int_{S_{f_Y}} f_Y(y) \log f_Y(y)\, dy\,.$$

Differential entropy enjoys some but not all properties of entropy, it can be infinitely large, negative, or positive, and is coordinate dependent. For a bijective transformation $W = \upsilon(Y)$, it follows that $h_d(W) = h_d(Y) - \mathrm{E}_W\left(\log\left|\frac{d\upsilon^{-1}}{dw}(W)\right|\right)$.

One can now quantify the amount of uncertainty in Y, expected to be removed if the value of X were known, by $I(X,Y) = h(Y) - h(Y|X)$, the so-called *mutual information*, where $h = H$ in the discrete case and $h = h_d$ for continuous random variables. It is always non-negative, zero if and only if X and Y are independent, symmetric, invariant under bijective transformations of X and Y, and $I(X,X) = h(X)$.

Additionally, if $\boldsymbol{Y}$ is a n-dimensional random vector, then the entropy-power of $\boldsymbol{Y}$ can be defined as

$$\mathrm{EP}(\boldsymbol{Y}) = \frac{1}{(2\pi e)^n} e^{2h(\boldsymbol{Y})}\,.$$

The differential entropy of a continuous normal random variable is given by $h(Y) = \frac{1}{2}\log\left(2\pi e\sigma^2\right)$, a simple function of the variance and, therefore, on the natural logarithmic scale $\mathrm{EP}(Y) = \sigma^2$, i.e., for the normal distribution variability and information are equivalent concepts. However, this equivalence does not hold in the general case. Indeed, in general, $\mathrm{EP}(Y) \leq \mathrm{Var}(Y)$ with equality if and only if Y is normally distributed.

We can now define an information-theoretic measure of association as

$$R_h^2 = \frac{\mathrm{EP}(\boldsymbol{Y}) - \mathrm{EP}(\boldsymbol{Y}|\boldsymbol{X})}{\mathrm{EP}(\boldsymbol{Y})}\,, \tag{13.13}$$

which ranges in the unit interval, equals zero if and only if $(\boldsymbol{X},\boldsymbol{Y})$ are independent, is symmetric, is invariant under bijective transformation of $\boldsymbol{X}$ and $\boldsymbol{Y}$, and, when $R_h^2 \to 1$ for continuous models, there is usually some degeneracy appearing in the distribution of $(\boldsymbol{X},\boldsymbol{Y})$; often $\boldsymbol{Y} = \phi(\boldsymbol{X})$ with probability one for some nontrivial function ϕ. This means that there exists a deterministic relationship between $\boldsymbol{X}$ and $\boldsymbol{Y}$. There is a direct link between R_h^2 and the mutual information: $R_h^2 = 1 - e^{-2I(X,Y)}$. For $\boldsymbol{Y}$ discrete: $R_h^2 \leq 1 - e^{-2H(Y)}$, implying that R_h^2 has an upper bound smaller than 1; in this setting it is better to consider

$$R_{h\max}^2 = \frac{R_h^2}{1 - e^{-2H(Y)}}\,,$$

reaching 1 when both endpoints are deterministically related.

Surrogacy can now be redefined preserving previous proposals as special cases. It is important to point out that, although the focus will be on the individual-level surrogacy, all results apply to the trial level as well. Let $Y = T$ and $X = S$ be the true and surrogate endpoints, respectively. S would be considered a good surrogate for T at the individual (trial) level, if a "large" amount of uncertainty about T (the treatment effect on T) is reduced when S (the treatment effect on S) is known. This definition, in spite of being based on formal concepts rooted in information theory, is simple and intuitive, since the idea behind surrogacy is to reduce our lack of knowledge about a true endpoint through the use of a surrogate alternative. At the trial level, the situation is similar: we want to gain information about the unobserved treatment effect on the true endpoint using the known treatment effect on the surrogate.

The R_h^2 coefficient is a valuable tool to evaluate surrogacy in practice. $R_h^2 \approx 1$ implies that our potential surrogate is promising, and could be interpreted as follows: once the surrogate is known, almost all of our uncertainty about the true endpoint will be removed. On the other hand, $R_h^2 \approx 0$ evidences a poor surrogate, unable to reduce our uncertainty about the true endpoint.

For the cross-sectional normal-normal case, Alonso and Molenberghs have shown that $R_h^2 = R_{\text{ind}}^2$ [1]. The same holds for R_Λ^2, defined in a longitudinal context. Finally, when the true and surrogate endpoints have distributions in the exponential family, then LRF $\xrightarrow{P} R_h^2$ when the number of subjects per trial goes to infinity, where LRF denotes the likelihood reduction factor introduced by Alonso et al. [3]. These authors also showed that (13.13) can be estimated based on $f(T|Z, S)$ and $f(T|Z)$, i.e., two univariate models that can often be easily fitted using standard software packages, in contrast to the original meta-analytic approach that requires the fitting of the complex joint hierarchical model $f(T, S|Z, \alpha, \beta)$.

13.7 Fano's Inequality and the Theoretical Plausibility of Finding a Good Surrogate

Fano's inequality relates prediction accuracy with different information-theoretic concepts and, when applied to the evaluation of surrogate endpoints, this inequality sets a limit for our capacity to successfully predict the true endpoint using the surrogate [3, 10]. For continuous endpoints it can be written as

$$\mathrm{E}\left[(T - g(S))^2\right] \geq \mathrm{EP}(T)(1 - R_h^2)\ . \qquad (13.14)$$

Note that nothing has been assumed about the distribution of both the surrogate and true endpoint and no specific form has been considered for the prediction function g.

Essentially, Fano's inequality states a lower bound for the prediction error and this lower bound can be decomposed in two different elements. The second element on the right side of (13.14) depends on the surrogate through the value of R_h^2, the first element, however, is an intrinsic characteristic of the true endpoint and it is independent of the surrogate. It is clear from (13.14) that the prediction error increases with EP(T) and, consequently, if the true endpoint has a large entropy-power then a surrogate should produce a close to one R_h^2 to have some predictive value. In other words, the surrogate would need to be almost deterministically related to the true endpoint to have some predictive power. Essentially, this inequality hints on the fact that, for some true endpoints, the search for a good surrogate may be a dead end street.

13.8 An Age-Related Macular Degeneration (ARMD) Trial

In what follows, the use of the meta-analytic approach will be illustrated using a clinical trial involving patients suffering from age-related macular degeneration (ARMD), a condition in which patients progressively lose vision [20]. Overall, 240 patients from 43 centers participated in the trial. Patients' visual acuity was assessed using standardized vision charts (see Fig. 13.1) displaying lines of five letters of decreasing size, which patients had to read from top (largest letters) to bottom (smallest letters).

The visual acuity was measured by the total number of letters correctly read. In this example, the binary indicator for treatment (Z) is set to -1 for placebo and

Fig. 13.1 Visual acuity study. Visual chart

to 1 for treatment with interferon-α. The surrogate endpoint S is the change in the visual acuity at 6 months after starting treatment, while the true endpoint T is the change in the visual acuity at 1 year. In the meta-analytic approach the centers in which the patients were treated will be considered the units of analysis. Two out of 43 centers participating in the trial enrolled patients only to one of the two treatment arms. These centers were excluded from considerations. A total of 41 centers were thus available for analysis.

13.9 Analysis of the ARMD Trial

In this section, the data from the age-related macular degeneration trial, described in Sect. 13.8, are used to evaluate visual acuity at 6 months as a surrogate endpoint for visual acuity at 1 year. Primarily, one would like to assess, for a given patient, how much information his visual acuity at 6 months provides on his visual acuity at 1 year and, similarly, one would also like to assess how much information the treatment effect at 6 months conveys about the treatment effect at 1 year. These are the questions addressed by the individual- and trial-level surrogacy. Notice that the individual level may be especially relevant for a treating physician who, having observed a particular outcome for a patient with a treatment at 6 months, wants to know what this means for the status of the patient at 1 year. On the other hand, the trial level may be more relevant for a data analyst that wants to know if the follow up period of a new trial might be shorten by 6 months in order to reduce cost.

Figure 13.2 shows the scatterplot of the two endpoints for all patients included in the trial. Clearly, there is a correlation between both variables. Indeed, the estimated Pearson correlation coefficient equals 0.757 and the 95 % confidence interval is $CI_{95\,\%} = (0.688, 0.812)$. We have learned in previous sections that, although appealing, the existence of correlation does not imply that visual acuity at 6 months

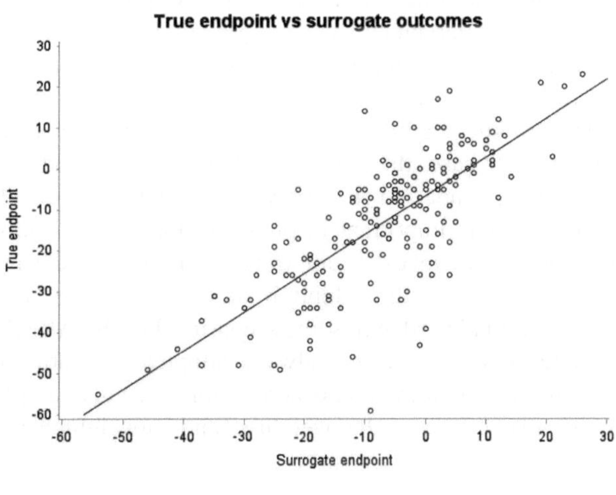

Fig. 13.2 Age-related macular degeneration trial. True endpoint (change in visual acuity at 1 year) versus surrogate endpoint (change in visual acuity at 6 months) for all individual patients, raw data

is a valid surrogate and further analyses are needed. In the present section we will follow the multi-units paradigm introduced in Sect. 13.4.

Using similar data, Buyse et al. experienced problems when fitting the full random-effects model, irrespective of whether standard statistical software or user developed alternatives were employed [7]. Similarly, our attempt to fit the complete hierarchical model given in (13.2) produced an infinite likelihood and the resulting **D** matrix was ill-conditioned with a condition number equal to 5.852−E15.

It is important to point out that when the full bivariate random-effects model is used, severe numerical issues are often encountered, especially if the surrogate and/or the true endpoint are not normally distributed. This numerical issues may have a huge impact on the assessment of surrogacy, particularly at the trial level. Indeed, the R^2_{trial} is computed based on the covariance matrix **D** and it is possible that this matrix becomes ill-conditioned and/or non-positive definite due to numerical problems. In such cases, the resulting quantities computed based on this matrix might not be trustworthy. For example, in our case study, the estimated **D** matrix produced a $R^2_{\text{trial}} = 0.972$ with a 95 % confidence interval (0.955, 0.989). Although possible, such a large value for the trial level surrogacy inevitably raises some doubts. Obviously, this result emanates from an ill-conditioned matrix and is probably misleading. One way to asses the ill-conditioning of a matrix is by reporting its condition number, i.e., the ratio of the largest over the smallest eigenvalue. A large condition number is an indication of ill-conditioning. The most pathological situation occurs when at least one eigenvalue is equal to zero. This corresponds to a positive semi-definite matrix, which occurs, for example, when the maximization procedure used to calculate the maximum likelihood estimators converges to a boundary solution. Thus, when using the full hierarchical model in the validation process, it is always necessary to check the D matrix to evaluate the presence of these issues.

Due to the numerical problems found with the ARMD data when fitting the complete hierarchical model, simplifying strategies along the lines introduced by Tibaldi et al. were called for and a two-stage approach was adopted [25]. At a first stage, the bivariate regression model given in (13.2) was fitted considering the trial-specific parameters as fixed effects. Within the two-stage approach, Tibaldi et al. explored two plausible strategies for fitting the model in (13.2), the so-called univariate and bivariate strategies, taking into account whether the surrogate and true endpoints are modeled as a bivariate outcome or rather as two univariate ones. In the latter case, the correlation between both endpoints is not incorporated into the model, rendering the study of the individual-level surrogacy more involved. However, it is important to point out that, if the trial-level surrogacy is of most interest and the investigation of the individual-level surrogacy is only of secondary importance, then the adoption of the univariate strategy can largely ease the computational burden in some scenarios. For the ARMD trial, the bivariate strategy was feasible and, hence, always adopted. In addition, the reduced model that assumes constant intercepts across units was also employed. Finally, at the second stage, one can consider weighted and unweighted versions of the model given

Table 13.1 Results of the trial and individual level surrogacy: R^2_{trial}, R_{ind} and 95 % confidence intervals (CI) obtained using the Delta method for the ARMD trial

Full model		
	Unweighted	Weighted
R^2_{trial}	0.381	0.437
R^2_{trial} CI	(0.138, 0.6234)	(0.200, 0.674)
R^2_{ind}&CI	0.512, CI= (0.422, 0.601)	
Reduced model		
	Unweighted	Weighted
R^2_{trial}	0.601	0.517
R^2_{trial} CI	(0.404, 0.797)	(0.297, 0.738)
R^2_{ind}&CI	0.581, CI= (0.499, 0.662)	

in (13.12) to estimate the trial level surrogacy. A summary of all these analyses is given in Table 13.1.

Note firstly that the individual-level surrogacy is estimated at the first stage and, consequently, it is not affected by the strategies followed to fit the second-stage model (weighted/unweighted). Secondly, the R^2_{ind} produced very similar results for both the reduced and full models. However, the AIC associated with the reduced and full model were 2668.3 and 2185.4 respectively, indicating that the assumption of equal intercepts across units produced a poorer fit to the data.

At the trial level, the results are much more variable, with the estimates R^2_{trial} varying from 0.38 to 0.60 across different settings. Because the full model seems to produce a better description of the data in what follows we will focus on the results displayed at the top panel of Table 13.1.

Taking into account that the sample size greatly varied across centers, one may consider a weighted analysis a more reliable option in this case. Nonetheless, the point estimate of R^2_{trial} was similar when the weighted or unweighted strategy was used and the confidence intervals largely overlapped in both scenarios. The general conclusion is that the trial level surrogacy seems to be rather weak, with the upper bound of the confidence intervals never exceeding 0.7.

Figure 13.3 displays the results obtained with the full-weighted model approach. Figure 13.3a shows a plot of the treatment effects on the true endpoint by the treatment effects on the surrogate endpoint and the size of the points are proportional to the sample size of each center. These effects are weakly correlated. Figure 13.3b shows a certain degree of correlation between the measurements at 6 months and at 1 year, after correction for treatment effect and center. Based on the previous findings, even with the limited data available, one may conclude that the assessment of visual acuity at 6 months seems to be a poor surrogate for the same assessment at 1 year.

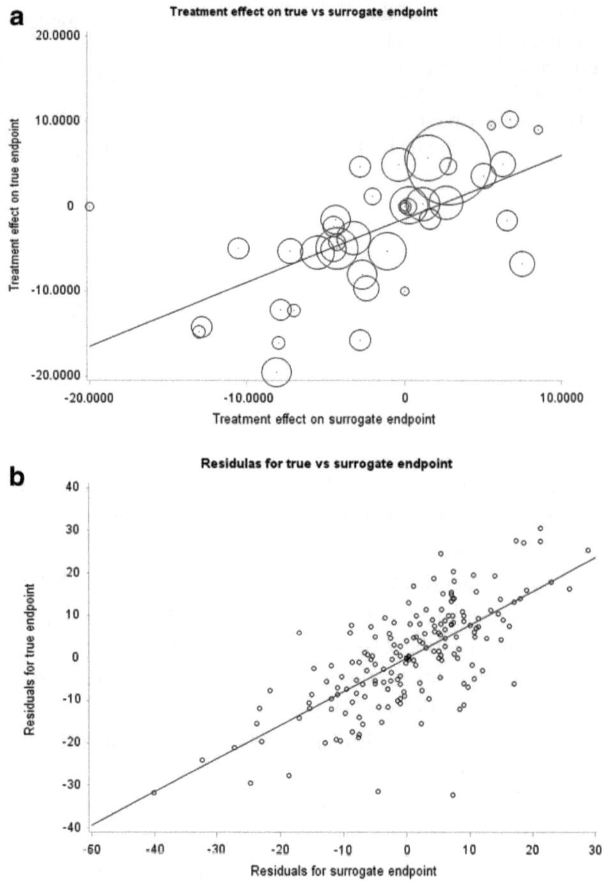

Fig. 13.3 Age-related macular degeneration trial. (**a**) Treatment effects on the true endpoint versus treatment effects on the surrogate endpoint in all centers. The size of each point is proportional to the number of patients in the corresponding center. (**b**) True endpoint versus surrogate endpoint for all individual patients, after correction for treatment effect

13.10 Software Packages

R functions and SAS macros have been developed to implement the methods discussed in the previous sections [26]. The ARMD trial was analyzed using the macro SURCONCON in SAS 9.3. The macro is a slight modification of the one that can be downloaded from http://www.ibiostat.be/software/surrogate.asp. The SAS code to carry out the analysis, the modified version of the macro and the data set will be available from the book's website. A detailed account of the macro can also be found in [26].

13.11 Conclusion

The initial enthusiasm that accompanied the use of surrogate markers, was followed by concern and skepticism after some dramatic failures. However, these failures opened a fruitful and stimulating scientific debate that has resulted in the development of different approaches and schools of thoughts for the validation of surrogate markers [2]. It is now clear that surrogate markers are a powerful tool that can play an important role in the drug development process. But it has also transpired that they need to be properly evaluated. Consequently, the initial enthusiasm and subsequent skepticism have been substituted by a more scientific and objective comprehension of their potentials and limitations.

At the same time, regulatory agencies around the globe, in particular in the United States and in Europe, have developed new policies and methods to accelerate the approval of certain types of drugs through the use of surrogate endpoints. In the United States, accelerated approval, sometimes referred as "conditional approval" or subpart H, refers to an acceleration of the overall development plan by allowing submission of an application, and if approved, marketing of a drug based on the evidence obtained, for instance, using a surrogate endpoint while further studies demonstrating direct patient benefit are underway. In the same way, the European regulatory agency has developed a set of regulations that are converging to an accelerated approval system like in the United States, perhaps with more flexibility [5].

As the previous sections illustrate, the scientific debate and research on surrogate markers, initiated more than 20 years ago, is still thriving and we believe this work together with the clear regulations established by leading regulatory agencies in the world will arguably allow, in the near future, a more rational and efficient use of this powerful tool.

References

1. Alonso, A., Geys, H., and Molenberghs, G.: A unifying approach for surrogate marker validation based on Prentice's criteria. *Statistics in Medicine* **25**, 205–221 (2006)
2. Alonso, A. and Molenberghs, G.: Surrogate endpoints: Hopes and Perils. *Pharmacoeconomics and Outcomes Research*, **8(3)** 255–259 (2008)
3. Alonso, A. and Molenberghs, G.: Surrogate marker evaluation from an information theoretic perspective. *Biometrics*, **63**, 180–186 (2007)
4. Biomarkers Definition Working Group: Biomarkers and surrogate endpoints: preferred definitions and conceptual framework. *Clin Pharmacol Ther*, **69** 89–95 (2001)
5. Burzykowski, T., Molenberghs, G. and Buyse, M. (Eds).: The Evaluation of Surrogate Endpoints. New York: Springer-Verlag (2005)
6. Burzykowski, T., Molenberghs, G., Buyse, M., Renard, D., and Geys, H. Validation of surrogate endpoints in multiple randomized clinical trials with failure-time endpoints. *Applied Statistics* **50**, 405–422 (2001).

7. Buyse, M., Molenberghs, G., Burzykowski, T., Renard, D., and Geys, H. (2000). The validation of surrogate endpoints in meta-analyses of randomized experiments. *Biostatistics* **1**, 49–67 (2000)
8. Cardiac Arrhythmia Suppression Trial (CAST) Investigators: Preliminary Report: effect of encainide and flecainide on mortality in a randomized trial of arrhythmia suppression after myocardial infraction. *New England Journal of Medicine* **321**, 406–412 (1989).
9. Cortiñas Abrahantes, J., Molenberghs, G., Burzykowski, T., Shkedy, Z., and Renard, D.: Choice of units of analysis and modeling strategies in multilevel hierarchical models. *Computational Statistics and Data Analysis*, **47**, 537–563 (2004).
10. Cover, T. and Tomas, J.: *Elements of Information Theory.* New York: Wiley (1991)
11. Daniels, M.J. and Hughes, M.D.: Meta-analysis for the evaluation of potential surrogate markers. *Stat Med* **16**, 1515–1527 (1997)
12. Ellenberg SS and Hamilton JM.: Surrogate endpoints in clinical trials: cancer. *Stat Med* **8**, 405–413 (1989)
13. Frangakis CE, Rubin DB.: Principal stratification in causal inference. *Biometrics* **58**, 21–29 (2002)
14. Fleming TR and DeMets DL.: Surrogate endpoints in clinical trials: are we being misled? *Annals of Internal Medicine* **125**, 605–613 (1996)
15. Ferentz AE.: Integrating pharmacogenomics into drug development. *Pharmacogenomics* **3**, 453–467 (2002)
16. Freedman, L., Graubard, B. and Schatzkin, A.: Statistical validation of intermediate endpoints for chronic diseases. *Stat Med* **11**, 167–178 (1992)
17. Gail, M.H., Pfeiffer, R., van Houwelingen, H.C., and Carroll, R.J.: On meta-analytic assessment of surrogate outcomes. *Biostatistics* **1**, 231–246 (2000)
18. Kullback, S. and Leibler, R.: On information and sufficiency. *Annals of Mathematical Statistics*, **22**, 79–86 (1951)
19. Nyquist, H.: Certain factors affecting telegraph speed. *Bell System Technical Journal*, **3**, 324–346 (1924)
20. Pharmacological Therapy for Macular Degeneration Study Group. Interferon α-IIA is ineffective for patients with choroidal neovascularization secondary to age-related macular degeneration. Results of a prospective randomized placebo-controlled clinical trial. *Archives of Ophthalmology* **115**, 865–872 (1997)
21. Prentice, R.L.: Surrogate endpoints in clinical trials: definitions and operational criteria. *Stat Med* **8**, 431–440 (1989)
22. Renard, Geys, H., Molenberghs, G., Burzykowski, T., Buyse, M., Bijnens, L. and Vangeneugden, T.: Validation of a longitudinally measured surrogate marker for time-to-event endpoint. *Journal of Applied Statistics* **29**, 000–000 (2002)
23. Renard, D., Geys, H., Molenberghs, G., Burzykowski, T., and Buyse, M.: Validation of surrogate endpoints in randomized trials with discrete outcomes. *Biometrical Journal* **44**, 1–15 (2002).
24. Shannon, C.: A mathematical theory of communication. *Bell System Technical Journal* **27**, 379–423 and 623–656 (1948)
25. Tibaldi, F.S, Cortiñas Abrahantes, J., Molenberghs, G., Renard, D., Burzykowski, T., Buyse, M., Parmar, M., Stijnen, T., and Wolfinger, R.: Simplified hierarchical linear models for the evaluation of surrogate endpoints. *Journal of Statistical Computation and Simulation*, **73**, 643–658 (2003).
26. Tilahun, A., Pryseley, A., Alonso, A., and Molenberghs, G.: Flexible Surrogate Marker Evaluation from Several Randomized Clinical Trials with Continuous Endpoints, Using R and SAS. *Computational Statistics and Data Analysis.* **51**, 4152–4163 (2007).
27. Weir, C.J., Walley, R.J.: Statistical evaluation of biomarkers as surrogate endpoints: a literature review. *Stat Med* **25** 183–203 (2006)

Chapter 14
Biomarker-Based Designs of Phase III Clinical Trials for Personalized Medicine

Shigeyuki Matsui, Takahiro Nonaka, and Yuki Choai

Abstract Advances in biotechnology and genomics have accelerated development of molecularly targeted treatments and prognostic and predictive biomarkers, particularly, in oncology. This chapter provides an overview of various biomarker-based designs for phase III randomized clinical trials to evaluate clinical utility of a biomarker or biomarker-based treatment, including biomarker-strategy, enrichment, and randomize-all designs. We also provide a simulation comparison of the randomize-all designs in terms of their ability to assert treatment efficacy for the correct patient population. Complex adaptive designs with development and validation of predictive biomarkers are also discussed.

14.1 Introduction

A key component to realize personalized medicine is the development of biomarkers for treatment selection. Biomarkers that are particularly important for personalized medicine can be broadly categorized as prognostic or predictive biomarkers. Prognostic biomarkers are pretreatment or baseline measurements that predict the long-term risk for untreated patients or those receiving the standard treatment, and thus can aid in the decision of whether a patient needs a more aggressive treatment (when diagnosed with high-risk) or no additional treatment (when diagnosed with low-risk). Predictive biomarkers are baseline measurements that provide information about which patients are likely or unlikely to benefit from a specific treatment.

S. Matsui (✉)
Department of Biostatistics, Graduate School of Medicine, Nagoya University, Showa-ku, Nagoya, Japan
e-mail: smatsui@med.nagoya-u.ac.jp

T. Nonaka
Pharmaceuticals and Medical Devices Agency, Chiyoda-ku, Tokyo, Japan
e-mail: nonaka-takahiro@pmda.go.jp

Y. Choai
Department of Statistical Science, School of Multidisciplinary Sciences, The Graduate University for Advanced Studies, Tachikawa, Tokyo, Japan
e-mail: choai@ism.ac.jp

K. van Montfort et al. (eds.), *Developments in Statistical Evaluation of Clinical Trials*, DOI 10.1007/978-3-642-55345-5_14

A predictive biomarker is often designated for the use of a particular new treatment, as a companion biomarker in the development of the new treatment. For example, a biomarker that captures overexpression of the growth factor receptor protein *HER-2*, which transmits growth signals to breast cancer cells, can be a companion biomarker in developing a molecularly-targeted drug for breast cancer patients, trastuzumab (Herceptin ®), which blocks the effects of *HER-2* [24].

A biomarker needs to be validated before its clinical application. Analytical validation refers to establishment of robustness and reproducibility of the assay and accuracy of measurement, such as sensitivity and specificity, relative to a gold standard assay if one is available [3, 22]. Clinical validity refers to establishment of the ability of the biomarker in predicting clinical outcomes in individual patients [22]. For a prognostic biomarker, correlation between biomarker status and a clinical endpoint, such as disease-free or overall survival, may indicate clinical validity. For reliable clinical validation of a predictive biomarker for a survival endpoint, a randomized clinical trial would be required to estimate treatment effects (of a new treatment relative to a control treatment) unbiasedly and to assess whether the treatment effects vary depending on the status of the biomarker, i.e., a treatment-by-biomarker interaction.

The establishment of clinical utility of a biomarker or a new treatment based on a biomarker is finally required as a phase III study before their clinical applications [22]. Randomized clinical trials serve as a gold standard in this phase [2, 7, 9, 13, 16, 17, 20, 23]. One category of biomarker-based designs is to establish clinical utility for the developed biomarker *itself*. The *biomarker-strategy designs* have such an objective. Another category is to establish clinical utility of a new treatment with the aid of a biomarker. The *enrichment designs* and *randomize-all designs* have such an objective. The former is to randomize a biomarker-dcfincd subpopulation of patients, while the latter is to randomize the entire patient population, but entail a *prospective* analysis plan based on the biomarker.

In this chapter, we provide an overview of various biomarker-based designs of phase III clinical trials for personalized medicine. We emphasize again that the two categories of the biomarker-based designs hold distinct objectives, although they have often been discussed as if all of them can be options of biomarker-based designs for a particular situation. We first outline the first category, i.e., the biomarker-strategy designs, in Sect. 14.2. We then focus on the second category; we outline the enrichment designs in Sect. 14.3 and the randomize-all designs in Sect. 14.4. The randomize-all designs can be more complex, reflecting the fact that the development and clinical validation of predictive biomarkers is generally difficult before initiating a phase III clinical trial. Typically, they involve some form of *adaptive* analysis that can demonstrate treatment efficacy for either the overall population or a biomarker-defined subpopulation of patients based on the observed performance of the biomarker. We provide a simulation study to assess their ability to assert treatment efficacy for the right patient population in Sect. 14.5. More complex adaptive designs with both developing and validating a predictive biomarker or genomic signature are outlined in Sect. 14.6. We present concluding remarks in Sect. 14.7.

14.2 Biomarker-Strategy Designs

With a biomarker-strategy design, patients are randomized either to a strategy of using the biomarker in determining their treatment or to a strategy of not using the biomarker in determining treatment. The primary objective is thus to compare two strategies with and without use of the biomarker in determining treatment. An example is a randomized trial for recurrent ovarian cancer that compares the strategy of determining treatment based on tumor chemosensitivity (predictive) assays with a strategy of using physician's choice of chemotherapy based on standard practice [5] (see Fig. 14.1a). Another example is a randomized trial for non-small cell lung cancer that compares a strategy of using a standard treatment (cisplatin+docetaxel) exclusively with a biomarker-based strategy in which patients diagnosed to be resistant to the standard treatment based on the biomarker are treated with an experimental treatment (gemcitabine+docetaxel) and the rest are treated with the standard treatment [4]. In these designs, the biomarker is evaluated only for the patients assigned to the biomarker-based strategy arm.

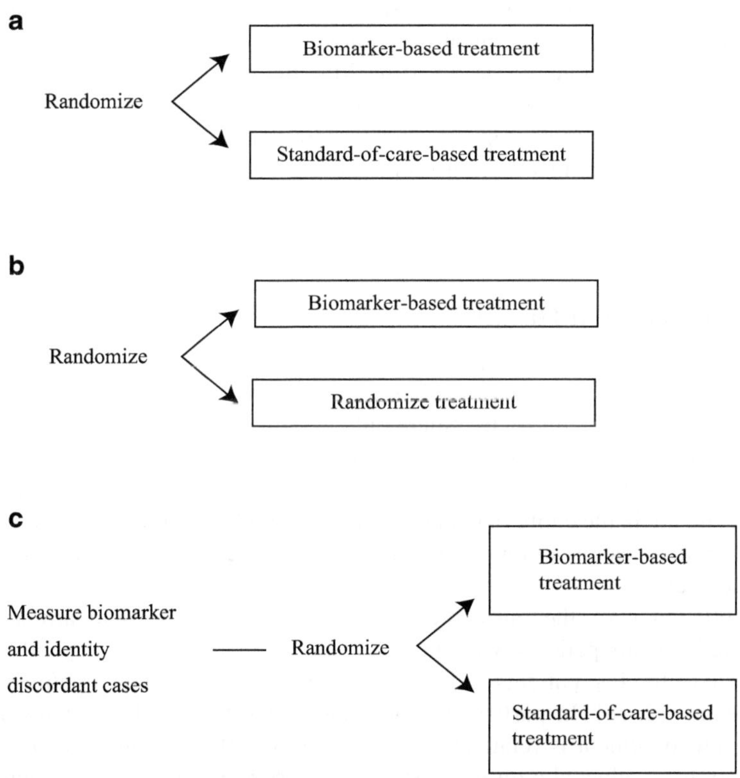

Fig. 14.1 Biomarker strategy designs

For the latter type of design with an experimental treatment, the biomarker-based arm can perform better if the experimental treatment is efficacious, regardless of whether the biomarker is predictive or not. Some authors proposed a modification in which patients in the non-biomarker-based arm undergo a second randomization to receive one of the same two treatments being used in the biomarker-based arm, i.e., the control and experimental treatments [13, 17] (see Fig. 14.1b). By measuring the biomarker status in all of the patients, the modified design would allow clinical validation of the biomarker as a predictive biomarker, through comparing treatment effects across the biomarker-based subsets of patients.

The strategy-based designs fundamentally include patients treated with the same treatment in both the biomarker-based and the non-biomarker-based arms, resulting in a large overlap in the number of patients receiving the same treatment within the two strategies being compared. Thus, a very large number of patients are required to be randomized to detect a diluted, small overall difference in the endpoint between the two arms. One modification is to randomize the two strategies to only the patients for whom the two treatments guided by the two strategies differ (see Fig. 14.1c). This modification requires measurement of the biomarker in all of the patients before randomization. The modified design is generally much more efficient than the original biomarker-strategy design. The modified design was employed in a randomized clinical trial, called the MINDACT study. In this trial, a biomarker-based strategy based on the MammaPrint prognostic signature was compared to that based on standard clinical prognostic factors for determining whether to utilize chemotherapy in women with node-negative estrogen receptor-positive breast cancer, in which discordant cases between the two strategies were subject to randomization [1].

14.3 Enrichment Designs

An enrichment or targeted design is based on a predictive biomarker and compares a new treatment and a control treatment only in biomarker-"positive" (B+) patients who are expected to be responsive to the new treatment based on the biomarker (see Fig. 14.2). Thus, the enrichment design assesses treatment efficacy only in the B+ patients, and not in the entire patient population, including biomarker-negative (B−) patients. In this design, all enrolled patients need to be screened for evaluating the biomarker status.

The efficiency of the enrichment design relative to the standard approach of randomizing all patients without using the biomarker at all depends on the prevalence of the B+ patients and on the effectiveness of the new treatment in the B− patients [12, 18]. In particular, when fewer than half of the patients are B+ and the new treatment is relatively ineffective in the B− patients, the enrichment design can be conducted with much smaller numbers of randomized patients. The enrichment design was employed in the development of trastuzumab; metastatic

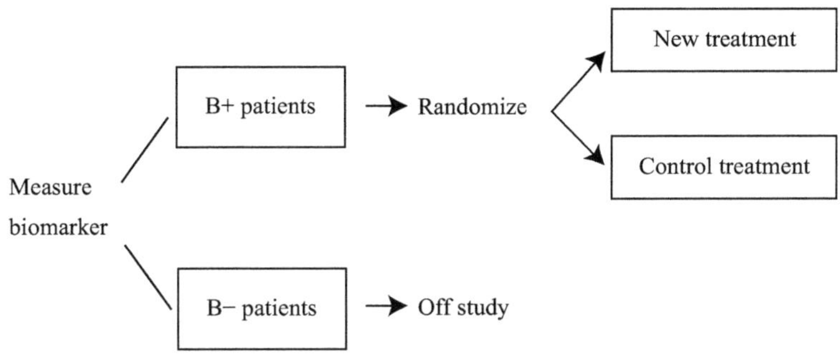

Fig. 14.2 Enrichment design

breast cancer patients whose tumors expressed *HER-2* in an immunohistochemistry test were eligible for randomization [24].

The enrichment design is appropriate for contexts where there is compelling biological evidence for believing that the B− patients will not benefit from the new treatment and that including them would raise ethical concerns [20, 23]. In addition, before initiating the trial, the biomarker used for enrichment must be analytically validated with established assay accuracy, reproducibility, and robustness.

When the biological basis is not compelling and/or assay accuracy is incomplete, assessment of clinical validity of the biomarker as a predictive biomarker would be needed. As the enrichment design does not allow it because of the absence of comparison of the new treatment with the control in the B− patients, the following designs with randomization of both B+ and B− patients, i.e., randomize-all or all-comers designs, are an alternative choice for such situations.

14.4 Randomize-All Designs

Randomization can be either unstratified or stratified on the basis of the predictive biomarker. Unstratified randomization does not diminish the validity of inference regarding treatment effects within the B+ or B− subsets of patients with moderate-to-large sizes. Under unstratified randomization, biomarker can be measured at the time of analysis. This strategy may permit such situations where an analytically validated biomarker is not available at the start of the trial but will be available by the time of analysis [20, 23]. However, careful consideration for missing biomarker data is needed for ensuring collection of sufficient numbers of patients with observed status of biomarker. On the other hand, stratified randomization requires determination and measurement of biomarker at the start of the trial, but ensures that all randomly assigned patients have biomarker status observed (see Fig. 14.3). For other practical considerations in randomized trials with biomarkers, see the references [2, 7, 9, 13, 17, 19, 20, 25, 26, 28].

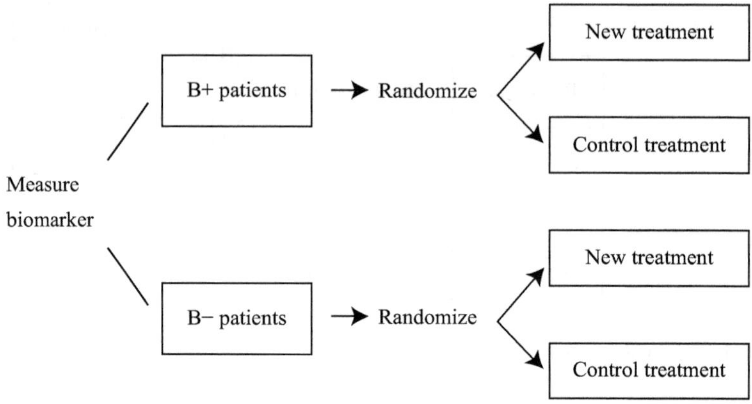

Fig. 14.3 Randomized-all design with prestratification based on the biomarker

The randomize-all designs can demonstrate the efficacy of the treatment for either the overall population or a biomarker-based subset of patients, through inspecting the predictive capability of the biomarker candidate based on the observed trial data. Various designs with a single biomarker candidate have been proposed, including fixed-sequence (FS), fallback (FB), and treatment-by-biomarker-interaction (TBBI) designs.

In what follows, we specifically consider these designs to compare a new treatment and a control treatment based on survival outcomes using a log-rank test. For a particular patient population, we assume proportional hazards between treatment arms and use the asymptotic distribution of a log-rank test statistic S under equal treatment assignment and follow-up, $S \sim N(\theta, 4/E)$ [27]. Here θ is the logarithm of the ratio of the hazard function under the new treatment relative to that under the control treatment, and E is the total number of events observed.

For a clinical trial with a given number of events, we express a standardized test statistic for testing treatment efficacy for the B+ subset of patients as

$$Z_+ = \hat{\theta}_+ / \sqrt{V_+} \,,$$

where $\hat{\theta}_+$ is an estimate of θ_+, such as a log-rank statistic S_+, and $V_+ = 4/E_+$. Similarly, we have a test statistic $Z_- = \hat{\theta}_- / \sqrt{V_-}$ for testing treatment efficacy for the B− subset, where $V_- = 4/E_-$. We consider the following standardized test statistic for testing treatment efficacy for the overall population,

$$Z_{\text{overall}} = \hat{\theta}_{\text{overall}} / \sqrt{V_{\text{overall}}} \,,$$

where $\hat{\theta}_{\text{overall}} = (E_+\hat{\theta}_+ + E_-\hat{\theta}_-)/(E_+ + E_-)$ and $V_{\text{overall}} = 4/E_{\text{overall}} = 4/(E_+ + E_-)$. We assume that the aforementioned standardized statistics follow asymptotically normal distributions with variance 1, where the means of Z_+, Z_-, and Z_{overall} are $\theta_+/\sqrt{V_+}, \theta_-/\sqrt{V_-}$, and $\sqrt{V_{\text{overall}}}(\theta_+/V_+ + \theta_-/V_-)$, respectively.

14.4.1 FS (Fixed-Sequence) Designs

If evidence from biological or early trial data suggests the predictive ability of the biomarker, it is reasonable to consider first testing treatment efficacy for the B+ subset of patients. In such a situation, one would not expect the treatment to be effective in the B− patients unless it is effective in the B+ patients. As such, the following FS design is derived [20, 23]. In the first stage, we compare the treatment versus control in the B+ patients using the test statistic Z_+ at a significance level of 5 %. If this test is significant, we proceed to the second stage; otherwise, the analysis is stopped. In the second stage, we compare the treatment versus control in the B− patients using the test statistic Z_- at a significance level of 5 %. All tests are two-sided. This sequential approach controls the experiment-wise Type I error at 5 %. When both the first test for the B+ patients and the second test for the B− patients are significant, one may assert treatment efficacy for the overall patient population. When only the first test for the B+ patients is significant, one may assert treatment efficacy only for future patients who are biomarker positive. We refer to this method as the FS-1 design.

A simple way for determining sample size in this design is to ensure the prespecified level of power, such as 90 %, for the first test, and calculate the required number of events for the B+ patients, E_+. This coincides with the required number of events for randomized patients in the enrichment designs. In this calculation, the number of events for the B− patients, E_-, is not determined at the design stage. The B− patients are enrolled concurrently until sufficient numbers of the B+ patients with E_+ are enrolled. As such, E_- can depend on the prevalence of B+, p_+, and the event rates λ_+ and λ_- in the B+ and B− control groups, respectively, at the time that there are E_+ events in the B+ subset. Specifically,

$$E_- = E_+ \left(\frac{\lambda_-}{\lambda_+}\right)\left(\frac{1-p_+}{p_+}\right)$$

is held approximately [20]. We expect a small (large) E_-, especially when p_+ is large (small). A small E_- can lead to a lack of power for detecting clinically important treatment effects in the B− patients at the second stage. On the other hand, a large E_- can yield ethical and practical concerns about enrolling a large number of the B− patients who are unlikely to benefit from the treatment [23]. Hence, sample size determination and/or planning of an interim futility analysis for the B− patients would be warranted.

In another variation of the FS design, the second stage involves testing treatment efficacy for the overall population rather than for the subset of B− patients [13]. With this approach, when only the test for the B+ subset in the first stage is significant, one may assert treatment efficacy for the B+ subset. When the second overall test is significant (following a significant result in the first stage), one may assert treatment efficacy for the overall population. We refer to this method as the FS-2 design.

14.4.2 FB (Fallback) Designs

When there is limited confidence in the predictive biomarker, it is generally reasonable to assess treatment efficacy for the overall patient population and prepare the subset analysis as a fallback option. Specifically, in the first stage, the treatment is compared with the control overall at a reduced significance level α_1, such as 3 %. If this test is significant, the analysis is stopped. Otherwise, in the second stage, the treatment is compared with the control for the B+ patients at a reduced significance level α_2, such as 2 %, in order to control the experiment-wise type I error rate within 5 % in testing treatment efficacy for the overall population or B+ subset [19,28]. All tests are two-sided. The significance level α_2 can be specified by taking into account the correlation between the first test in the overall population and the second test in the subset of B+ patients [25, 26, 28]. Specifically, the covariance (or correlation) between Z_+ and Z_{overall} reduces to $\sqrt{p_+}$. As the test on treatment efficacy for the overall patient population precedes the fallback test for the B+ patients, it is reasonable to set the significance values such that $\alpha_1 \geq \alpha_2$. When the first test is significant, one may assert treatment efficacy in the overall population. On the other hand, when only the second test for the B+ patients is significant (following a non-significant result of the first test for the overall population), one may assert treatment efficacy only in future B+ patients.

Sample size determination will be based on the first test on treatment efficacy for the overall population, like in the traditional randomized trials, apart from the use of the significance level $\alpha_1 (<0.05)$. Because of possible treatment effects that are clinically important in the B+ patients, it is advisable to perform sample size calculation for the second test for the B+ patients and plan for the option of delaying the second stage analysis until collection of the required number of events for the B+ patients when it is needed [23].

14.4.3 TBBI (Treatment-by-Biomarker Interaction) Designs

TBBI designs, like FB designs, are used when there is limited confidence in the predictive biomarker. This approach involves deciding whether to compare treatments overall or within the biomarker-based subsets based on a preliminary test of interaction of treatment and biomarker [17, 20, 23]. Here the test of interaction is to assess whether there is no difference in treatment effects (in term of the relative hazards ratio between the two treatment arms) between the B+ and B− subsets of patients. Specifically, we use the following standardized statistic for testing the interaction:

$$Z_{\text{int}} = \frac{\hat{\theta}_+ - \hat{\theta}_-}{\sqrt{V_+ + V_-}} .$$

It is reasonable to consider a one-sided interaction test to detect larger treatment effects in the B+ subset [20, 23]. To be specific, we propose the following design: a preliminary test of interaction is performed as the first stage using Z_{int} at a one-sided significance level of α_{int}. If this test is not significant, the treatment is compared with the control overall using Z_{overall} at a two-sided significance level α_3. Otherwise, the treatment is compared with the control in the B+ patients using Z_+ at a two-sided significance level α_4. Here the significance levels, α_3 and α_4, are chosen such that the experiment-wise type I error rate in testing treatment efficacy for the overall population or B+ subset is less than or equal to 5 % based on an asymptotic distribution of Z_{int}, Z_{overall}, and Z_+, where the covariances between Z_{int} and Z_{overall} or Z_+ may reduce to $\text{cov}(Z_{\text{int}}, Z_{\text{overall}}) = 0$ or $\text{cov}(Z_{\text{int}}, Z_+) = \sqrt{V_+/(V_+ + V_-)} = \sqrt{E_-/(E_+ + E_-)}$. Under the null hypothesis of no treatment efficacy for the B+ and B− patients (and thus indicating no effects for the overall population), for which we will search for the significance level, α_4, for Z_+, given α_{int} for Z_{int} and α_3 for Z_{overall}, to control the experiment-wise type I error rate within 5 %, we propose to set $\text{cov}(Z_{\text{int}}, Z_+) = \sqrt{1 - p_+}$ if the hazard rate in the B+ subset can be considered to be the same as that in the B− subset. When the predictive biomarker is prognostic, a larger number of events is expected for the B+ patients, resulting in an overestimation of the correlation. This would lead to use of a stringent significance level of α_4 and thus a conservative design.

When the test for the B+ patients is significant (following a significant result of the preliminary interaction test), one may assert treatment efficacy only for B+ patients. When the overall test is significant (following a non-significant result of the preliminary interaction test), one may assert treatment efficacy for the overall population.

The TBBI designs have been discussed in the literature as a design for clinical validation of the predictive biomarker based on a test on treatment-by-biomarker interaction [17, 20, 23]. However, sizing the trial to have high power for the interaction test may require a substantially large sample size, compared to sizing trials with the other randomize-all designs. This cannot generally be justified as it requires exposing an excessive number of B− patients to a treatment from which they are unlikely to benefit [23].

On the other hand, the proposed TBBI design with strict control of the experiment-wise type I error rate described above aims to assess the clinical utility of a new treatment with the aid of the biomarker. As our simulation study indicated (see Sect. 14.5.1), it could be more efficient compared with the other randomize-all designs. An additional advantage of the proposed TBBI design is that even if the interaction test is regarded as a preliminary test, a significant interaction could be regarded as relatively firm evidence for the clinical validity of the biomarker. Further studies on the proposed TBBI design, including determination of optimal levels of α_{int} and α_3 and sample size determination, would be worthwhile.

14.5 Probability of Asserting Treatment Efficacy

The randomize-all designs described in Sect. 14.4 can make either of two kinds of assertions regarding treatment efficacy, one for the overall population and the other for the B+ subset of patients. Which of the two assertions is considered to be valid may depend on the underlying treatment effects in the biomarker-based subsets. Specifically, let HR_+ and HR_- denote the hazard ratios of the treatment relative to the control in the B+ and B− subsets of patients, respectively. If the treatment truly has clinically meaningful effects in all of the patients, e.g., $HR_+ = HR_- = 0.7$, the assertion of treatment efficacy for the overall population would be more valid than that for the B+ patients only because the latter assertion would deprive the remaining B− patients of the chance of receiving the effective treatment. On the other hand, if the treatment can exert a clinically important effect only in the B+ patients, e.g., $HR_+ = 0.5$, and no effect in the remaining B− patients, e.g., $HR_- = 1.0$ (indicating a qualitative interaction between treatment and biomarker), the assertion of treatment efficacy for the B+ patients would be more valid than that for the overall population because the latter assertion would yield overtreatment for the remaining B− patients using the ineffective, even toxic treatment. Let P_{overall} and P_{subset} denote the probability of asserting treatment efficacy for the overall population and for the subset of B+ patients, respectively.

However, there can be other scenarios in which it is not clear which of the two assertions is valid. For example, the treatment can exert a clinically important effect for the B+ patients, e.g., $HR_+ = 0.5$, but some moderate or small effects for the remaining B− patients, e.g., $HR_- = 0.8$ (indicating a quantitative interaction between treatment and biomarker). Such a treatment effect profile could be explained by the treatment having multiple mechanisms of action, the misclassification of responsive patients into the B− subset (low sensitivity of the biomarker), and so on. Which of the two assertions is considered to be valid will be determined on a case-by-case basis incorporating many factors, including the size of the prevalence p_+, possible adverse effects, treatment costs, prognosis of the disease, availability of other treatment choices, and so on. In such situations, the probability of asserting treatment efficacy for either the overall population or the subset of B+ patients could be another meaningful criterion. From the point of view of treatment developers (e.g., pharmaceutical companies), this probability would be always important, because it can be interpreted as the *probability of success* in treatment development. Let P_{success} denote this probability. Apparently, $P_{\text{overall}} + P_{\text{subset}} = P_{\text{success}}$ for the randomize-all designs described in Sect. 14.4. As such, there is a trade-off between the two probabilities P_{overall} and P_{subset} for a given value of P_{success}.

Table 14.1 Empirical probabilities of P_{overall}, P_{subset}, and P_{success} under null effects

								TBBI	
HR_+	HR_-	p_+	Prob.	Traditional	FS-1	FS-2	FB	$\alpha_{\text{int}} = 5\,\%$	$\alpha_{\text{int}} = 10\,\%$
1.0	1.0	0.1	P_{overall}	0.051	0.003	0.005	0.032	0.030	0.029
(null effect)			P_{subset}	0.000	0.041	0.039	0.016	0.019	0.021
			P_{success}	0.051	0.044	0.044	0.047	0.049	0.050
		0.3	P_{overall}	0.050	0.002	0.010	0.031	0.029	0.028
			P_{subset}	0.000	0.048	0.040	0.020	0.020	0.022
			P_{success}	0.050	0.050	0.050	0.051	0.049	0.050
		0.5	P_{overall}	0.052	0.002	0.018	0.031	0.029	0.028
			P_{subset}	0.000	0.047	0.032	0.019	0.021	0.020
			P_{success}	0.052	0.050	0.050	0.050	0.050	0.048

14.5.1 Simulations

We provide a comparison of the randomize-all designs in Sect. 14.4 in terms of P_{overall}, P_{subset}, and P_{success}. We considered the prevalence of B+, $p_+ = 0.1, 0.3$, or 0.5. As to the underlying treatment effects within biomarker-based subsets, we considered the following scenarios: $(HR_+, HR_-) = (1.0, 1.0), (0.7, 0.7), (0.5, 1.0)$, or $(0.5, 0.8)$, i.e., null effects, constant effects, qualitative interaction, and quantitative interaction. In the FB and TBBI designs, we specified the same significance levels for the overall test, $\alpha_1 = \alpha_3 = 3\,\%$, for a fair comparison of these designs. The significance level for the one-sided interaction test, α_{int}, in the TBBI designs was specified as 5 or 10 %. The significance levels for the B+ subset tests, α_2 and α_4, in the FB and TBBI designs were determined such that the experiment-wise type I error rates were equal to 5 %. We also evaluated the traditional design without use of a biomarker as a reference, with $P_{\text{overall}} = P_{\text{success}}$ and $P_{\text{subset}} = 0$ (because there is no option for asserting treatment efficacy for the B+ subset in this design). We conducted 10,000 simulations (clinical trials) for each configuration to obtain empirical values of the probabilities. We provide the results when 400 patients with a baseline event rate of 0.2 (per year) are randomized and followed up for 5 years in each clinical trial. For larger sample sizes, P_{overall}, P_{subset}, and P_{success} became large, but similar conclusions in terms of the relative sizes of these probabilities across the designs under comparison were obtained. R codes for conducting simulations are available from author upon request. A web-based simulation program that provides estimates of required sample size for biomarker-based analysis plans for time to event or binary endpoints is also available [15].

We first confirmed control of the experiment-wise type I error rate, i.e., $P_{\text{success}} \leq 5\,\%$, for all of the designs in Table 14.1. We also confirmed control of P_{overall} as the specified significance levels for the overall tests, $\alpha_1 = \alpha_3 = 3\,\%$, for the FB and TBBI designs.

Table 14.2 summarizes the empirical values of P_{overall}, P_{subset}, and P_{success} for scenarios with non-null treatment effects. For the scenarios with constant treatment

Table 14.2 Empirical probabilities of P_{overall}, P_{subset}, and P_{success} under non-null treatment effects

HR_+	HR_-	p_+	Prob.	Traditional	FS-1	FS-2	FB	TBBI $\alpha_{\text{int}} = 5\,\%$	TBBI $\alpha_{\text{int}} = 10\,\%$
0.7	0.7	0.1	P_{overall}	0.758	0.083	0.106	0.690	0.658	0.623
(constant			P_{subset}	0.000	0.036	0.013	0.007	0.045	0.079
effect)			P_{success}	0.758	0.120	0.120	0.698	0.703	0.702
		0.3	P_{overall}	0.774	0.198	0.300	0.703	0.669	0.634
			P_{subset}	0.000	0.124	0.022	0.020	0.052	0.098
			P_{success}	0.774	0.322	0.322	0.723	0.721	0.732
		0.5	P_{overall}	0.764	0.222	0.450	0.691	0.659	0.623
			P_{subset}	0.000	0.252	0.025	0.027	0.049	0.097
			P_{success}	0.764	0.474	0.474	0.717	0.708	0.720
0.5	1.0	0.1	P_{overall}	0.074	0.016	0.039	0.048	0.027	0.019
(qualitative			P_{subset}	0.000	0.301	0.279	0.178	0.296	0.304
interaction)			P_{success}	0.074	0.317	0.317	0.225	0.323	0.323
		0.3	P_{overall}	0.301	0.038	0.281	0.230	0.053	0.031
			P_{subset}	0.000	0.719	0.476	0.449	0.706	0.743
			P_{success}	0.301	0.757	0.757	0.680	0.759	0.774
		0.5	P_{overall}	0.688	0.047	0.682	0.607	0.102	0.052
			P_{subset}	0.000	0.891	0.256	0.305	0.825	0.893
			P_{success}	0.688	0.938	0.938	0.913	0.927	0.945
0.5	0.8	0.1	P_{overall}	0.519	0.115	0.205	0.432	0.326	0.266
(quantitative			P_{subset}	0.000	0.192	0.102	0.077	0.222	0.278
interaction)			P_{success}	0.519	0.307	0.307	0.509	0.548	0.544
		0.3	P_{overall}	0.762	0.232	0.644	0.692	0.369	0.270
			P_{subset}	0.000	0.532	0.119	0.124	0.455	0.584
			P_{success}	0.762	0.764	0.764	0.816	0.824	0.854
		0.5	P_{overall}	0.914	0.214	0.882	0.873	0.403	0.288
			P_{subset}	0.000	0.722	0.054	0.073	0.533	0.666
			P_{success}	0.914	0.936	0.936	0.946	0.937	0.954

effects, $(HR_+, HR_-) = (0.7, 0.7)$, where P_{overall} would be a relevant criterion, the traditional design provided the greatest values of P_{overall}, as was expected. The FB and TBBI designs provided slightly reduced values of P_{overall} than those of the traditional design. On the other hand, the FS designs, especially FS-1, provided much smaller values of P_{overall}. Similar trends were observed for P_{success}.

For the scenarios with a qualitative interaction, $(HR_+, HR_-) = (0.5, 1.0)$, where P_{subset} would be relevant, the FS-1 and TBBI designs performed best. The FS-2 and FB designs provided much smaller values of P_{subset} when $p_+ \geq 0.3$. With respect to P_{success}, all biomarker-based designs, except the FB design, generally provided comparable P_{success} values, while the traditional design provided much smaller values of P_{success}.

Lastly, for the scenarios with a quantitative interaction, $(HR_+, HR_-) = (0.5, 0.8)$, the characteristics of the respective designs became clearer. The FS-2 and FB designs tended to provide larger P_{overall}, while the FS-1 and TBBI designs tended to provide larger P_{subset} values. With respect to P_{success}, the TBBI designs provided the largest P_{success} values, followed by the FB design with slight reductions in P_{success}.

In summary, the FS-1 design would be suitable for cases with qualitative interactions between treatment and biomarker and large treatment effects in the B+ patients, but could suffer from a serious lack of power for nearly constant treatment effects in the overall population. Interestingly, the FS-2 design has quite different properties, but was not shown to be so efficient for various profiles of treatment effects. In contrast, a FB design would be suitable for cases with nearly constant treatment effects in the overall population, but could suffer from a serious lack of power for qualitative interactions between treatment and biomarker. The TBBI designs generally performed well for various patterns of treatment effects within biomarker-based subsets in terms of all the probabilities, P_{overall}, P_{subset} and P_{success}. This can be explained by the effectiveness of the preliminary interaction test in selecting the appropriate population for testing treatment efficacy.

14.6 More Complex Adaptive Designs

When the biology of the target of a new treatment is not well understood because of the complexity of disease biology, it is quite common that a completely specified predictive biomarker is not available before initiating the definitive phase III trial. One approach in such situations is to design and analyze the randomized phase III trial in such a way that both developing a predictive biomarker and testing treatment efficacy based on the developed biomarker are possible and conducted validly. Apparently, this approach works with randomize-all designs without prestratification based on any biomarkers, and careful prespecification of the analysis plan is mandatory.

Jiang et al. [10] developed the *adaptive threshold design* for settings where a single predictive biomarker candidate is available but no threshold of positivity for the biomarker is predefined. The basic idea is, for a set of candidate threshold values $(b_1, \ldots, b_K)$, to search for an optimal threshold value through maximizing a log likelihood ratio of treatment effect for the patients with biomarker value $\geq b_k$ over possible threshold values $(k = 1, \ldots, K)$. The maximum log likelihood ratio at the optimal threshold value is used as the test statistic. Its null distribution is approximated by repeating the whole analysis after randomly permuting treatment levels several thousand times. This approach can be applied to searching for a subset determined by a positive value of any single biomarker when there is a set of candidate binary biomarkers [23]. This approach can be used as the second stage analysis of the FB designs or as a stand-alone basis by incorporating the log likelihood statistic for testing the overall treatment effects in obtaining a maximum test statistic [10].

Another adaptive design, called *adaptive signature design*, is to develop a predictor or signature using a set of covariates x, possibly high-dimensional genomic data [6, 8]. As the second stage of the FB designs, the full set of patients in the clinical trial is partitioned into a training set and a validation set. A prespecified algorithmic analysis plan is applied to the training set to generate a predictor. This is a function of x and to predict, for a given patient with a particular value of x, to be responsive or not responsive to the new treatment. The predictor is used to make a prediction for each patient in the validation set. Then, the treatment efficacy is tested for the patient subset predicted as "responsive" to the treatment in the validation set.

This modified second stage analysis of the FB designs can be based on split-sample [6] or cross-validation [8]. In the latter approach, at the end of the prediction process, each of all the patients in the clinical trial is predicted as either responsive or not. Again, the treatment efficacy is tested for the patient subset predicted as "responsive" to the treatment. However, because this subset is determined by the cross-validation using the all patient data, the standard asymptotic theory does not apply. To address this issue, a permutation method that repeats the whole processes of the cross-validated prediction analysis after randomly permuting treatment levels is employed [8].

Recently, Matsui et al. [14] developed another framework designed to estimate treatment effects quantitatively as a function of a continuous cross-validated predictive score for the entire patient population, rather than qualitatively classifying patients as in or not in a responsive subset. Average absolute treatment effects for the entire population or a responsive subset of patients can be estimated based on the estimated treatment effects function and tested using a permutation method. In this framework, patient-level survival curves can be developed to predict survival distributions of individual future patients as a function of the cross-validated predictive score and a cross-validated prognostic score that is developed independently from the development of the predictive score, through correlating genomic data with survival outcomes without reference to treatment assignment.

14.7 Concluding Remarks

In this chapter, we have discussed a wide variety of biomarker-based designs of phase III clinical trials to establish the clinical utility of a biomarker or a new treatment with the aid of a biomarker. In biomarker-strategy, enrichment, and prestratified randomize-all designs, collection of specimens and biomarker assays are conducted prospectively for newly accruing patients. As these prospective designs are highly resource-intensive and time-consuming, a study using archived specimens is sometimes used as an alternative. This type of study is retrospective with regard to using archived specimens, but should prospectively specify a protocol. An unstratified randomize-all trial, possibly with the adaptive designs in

Sect. 14.6, could be categorized to this type of study because specimens archived at the beginning of the trial are analyzed. Simon et al. [21] proposed several conditions for appropriately conducting such a study with archived specimens. In summary,

1. Archived specimen, adequate for a successful assay, must be available from a sufficient large number of patients to permit appropriately powered analyses in the pivotal trial and to ensure that the patients included in the biomarker evaluation are representative of the patients in the trial.
2. Substantial data on the analytical validity of the biomarker must exist to ensure that results obtained from the archived specimens will closely resemble those that would have been obtained from analysis of specimens collected in real time. Assays should be conducted blinded to the clinical data.
3. The analysis plan for the biomarker evaluation must be completely developed before the performance of the biomarker assays. The analysis should focus on a single diagnostic biomarker that is completely defined and specified. The analysis should not be exploratory, and practices that might lead to a false-positive conclusion (e.g., multiple analyses of different candidate biomarkers based on archived specimens from the same trial) should be avoided.
4. The results must be validated in at least one or more similarly designed studies using the same assay techniques.

These conditions are also applicable to previously conducted clinical trials (with archived specimens) that evaluated the efficacy of the treatment of interest. When substantial preliminary evidence that a new biomarker predicts treatment responsiveness has been accumulated by the middle or completion of a phase III trial of the treatment, one may consider assay of the biomarker in archived specimens from this trial. As an example, an analysis based on a *KRAS* mutation in a randomized trial for the *anti-EGFR* antibody, cetuximab, which was approved for the treatment of advanced colorectal cancer, demonstrated that the treatment was not effective for patients with *KRAS* mutations [11]. Another possibility is to analyze archived specimens from a failed pivotal trial that showed no treatment effect for the entire patient population using the methods for biomarker development described in Sect. 14.6. The developed biomarker from such an analysis can provide useful information for designing a second confirmatory trial of the same treatment, possibly with an enrichment design with small sample sizes.

The recent advances in biotechnology and genomics have posed biostatisticians further important roles and challenges in various phases of biomarker development and validation, including systematic collection of specimens and measurement of biomarker/clinical data, development of an analytically and clinically-validated biomarker, and establishment of the clinical utility of the biomarker or biomarker-based treatment, through utilizing archived or prospectively-collected specimens in the context of clinical trials. Further biostatistical researches are required indeed in this important field for accelerating modern clinical studies toward personalized medicine.

Acknowledgements This research was partly supported by a Grant-in-Aid for Scientific Research (24240042) from the Ministry of Education, Culture, Sports, Science and Technology of Japan. The views expressed in this chapter are the result of independent work and do not necessarily represent the views of the Pharmaceuticals and Medical Devices Agency.

References

1. Bogaerts, J., Cardoso, F., Buyse, M., Braga, S., Loi, S., et al.: Gene signature evaluation as a prognostic tool: challenges in the design of the MINDACT trial. *Nature Clinical Practice Oncology* **3**(10), 540–551 (2006). doi:10.1038/ncponc0591
2. Buyse, M., Michiels, S., Sargent, D.J., Grothey, A., Matheson, A., et al.: Integrating biomarkers in clinical trials. *Expert Review of Molecular Diagnostics* **11**(2), 171–182 (2011). doi:10.1586/erm.10.120
3. Chau, C.H., Rixe, O., McLeod, H., Figg, W.D.: Validation of analytic methods for biomarkers used in drug development. *Clinical Cancer Research* **14**(19), 5967–5976 (2008). doi:10.1158/1078-0432.CCR-07-4535
4. Cobo, M., Isla, D., Massuti, B., Montes, A., Sanchez, J.M., et al.: Customizing cisplatin based on quantitative excision repair cross-complementing 1 mRNA expression: a phase III trial in non-small-cell lung cancer. *Journal of Clinical Oncology* **25**(19), 2747–2754 (2007). doi:10.1200/JCO.2006.09.7915
5. Cree, I.A., Kurbacher, C.M., Lamont, A., Hindley, A.C., Love, S.: A prospective randomized controlled trial of tumour chemosensitivity assay directed chemotherapy versus physician's choice in patients with recurrent platinum-resistant ovarian cancer. *Anticancer Drugs* **18**(9), 1093–1101 (2007). doi:10.1097/CAD.0b013e3281de727e
6. Freidlin, B., Simon, R.: Adaptive signature design: an adaptive clinical trial design for generating and prospectively testing a gene expression signature for sensitive patients. *Clinical Cancer Research* **11**(21), 7872–7878 (2005). doi:10.1158/1078-0432.CCR-05-0605
7. Freidlin, B., McShane, L.M., Korn, E.L.: Randomized clinical trials with biomarkers: design issues. *Journal of the National Cancer Insititute* **102**(3), 152–160 (2010). doi:10.1093/jnci/djp477
8. Freidlin, B., Jiang, W., Simon, R.: The cross-validated adaptive signature design. *Clinical Cancer Research* **16**(2), 691–698 (2010). doi:10.1158/1078-0432.CCR-09-1357
9. Hoering, A., Leblanc, M., Crowley, J.J.: Randomized phase III clinical trial designs for targeted agents. *Clinical Cancer Research* **14**(14), 4358–4367 (2008). doi:10.1158/1078-0432.CCR-08-0288
10. Jiang, W., Freidlin, B., Simon, R.: Biomarker-adaptive threshold design: a procedure for evaluating treatment with possible biomarker-defined subset effect. *Journal of the National Cancer Insititute* **99**(13), 1036–1043 (2007). doi:10.1093/jnci/djm022
11. Karapetis, C.S., Khambata-Ford, S., Jonker, D.J., O'Callaghan C.J., Tu D., et al.: K-ras mutations and benefit from cetuximab in advanced colorectal cancer. *New England Journal of Medicine* **359**(17), 1757–1765 (2008). doi:10.1056/NEJMoa0804385.
12. Maitournam, A., Simon, R.: On the efficiency of targeted clinical trials. *Statistics in Medicine* **24**(3), 329–339 (2005). doi:10.1002/sim.1975
13. Mandrekar, S.J., Sargent, D.J.: Clinical trial designs for predictive biomarker validation: theoretical considerations and practical challenges. *Journal of Clinical Oncology* **27**(24), 4027–4034 (2009). doi:10.1200/JCO.2009.22.3701
14. Matsui, S., Simon, R., Qu, P., Shaughnessy, J.D. Jr, Barlogie, B., et al.: Developing and validating continuous genomic signatures in randomized clinical trials for predictive medicine. *Clinical Cancer Research* **18**(21), 6065–6073 (2012). doi:10.1158/1078-0432.CCR-12-1206
15. National Institutes of Health: Sample Size Calculation for Randomized Clinical Trials. http://linus.nci.nih.gov/brb/samplesize/sdpap.html

16. Pusztai, L., Hess, K.R.: Clinical trial design for microarray predictive marker discovery and assessment. *Annals of Oncology* **15**(12), 1731–1737 (2004). doi:10.1093/annonc/mdh466
17. Sargent, D.J., Conley, B.A., Allegra, C., Collette, L.: Clinical trial designs for predictive marker validation in cancer treatment trials. *Journal of Clinical Oncology* **23**(9), 2020–2027 (2005). doi:10.1200/JCO.2005.01.112
18. Simon, R., Maitournam A.: Evaluating the efficiency of targeted designs for randomized clinical trials. *Clinical Cancer Research* **10**(20), 6759–6763 (2004). doi:10.1158/1078-0432.CCR-04-0496
19. Simon, R., Wang, S.J.: Use of genomic signatures in therapeutics development in oncology and other diseases. *Pharmacogenomics Journal* **6**(3), 166–173 (2006). doi:10.1038/sj.tpj.6500349
20. Simon, R.: The use of genomics in clinical trial design. *Clinical Cancer Research* **14**(19), 5984–5993 (2008). doi:10.1158/1078-0432.CCR-07-4531
21. Simon, R., Paik, S., Hayes, D.F.: Use of archived specimens in evaluation of prognostic and predictive biomarkers. *Journal of the National Cancer Insititute* **101**(21), 1446–1452 (2009). doi:10.1093/jnci/djp335
22. Simon, R.: Clinical trial designs for evaluating the medical utility of prognostic and predictive biomarkers in oncology. *Personalized Medicine* **7**(1), 33–47 (2010). doi:10.2217/pme.09.49
23. Simon, R.: Clinical trials for predictive medicine. *Statistics in Medicine* **31**(25), 3031–3040 (2012) doi:10.1002/sim.5401
24. Slamon, D.J., Leyland-Jones, B., Shak, S., Fuchs, H., Paton, V., et al.: Use of chemotherapy plus a monoclonal antibody against HER2 for metastatic breast cancer that overexpresses HER2. *New England Journal of Medicine* **344**(11), 783–792 (2001). doi:10.1056/NEJM200103153441101
25. Song, Y., Chi, G.Y.: A method for testing a prespecified subgroup in clinical trials. *Statistics in Medicine* **26**(19), 3535–3549 (2007). doi:10.1002/sim.2825
26. Spiessens, B., Debois, M.: Adjusted significance levels for subgroup analyses in clinical trials. *Contemporary Clinical Trials* **31**(6), 647–656 (2010). doi:10.1016/j.cct.2010.08.011
27. Tsiatis. A.A.: The asymptotic joint distribution of the efficient score test for the proportional hazards model calculated over time. *Biometrika* **68**(1), 311–315 (1981). doi:10.1093/biomet/68.1.311
28. Wang, S.J., O'Neill, R.T., Hung, H.M.: Approaches to evaluation of treatment effect in randomized clinical trials with genomic subset. *Pharmaceutical Statistics* **6**(3), 227–244 (2007). doi:10.1002/pst.300

Chapter 15
Dose-Finding Methods for Two-Agent Combination Phase I Trials

Akihiro Hirakawa and Shigeyuki Matsui

Abstract In this chapter, we discuss the toxicity-based dose-finding methods for two-agent combinations in phase I oncology trials. The model-based approaches, such as the continual reassessment method (CRM), have been gradually applied to single-agent trials to determine the maximum tolerated dose (MTD). By contrast, the rule-based approaches have commonly been applied to two-agent combination trials, probably due to the absence of well-understood model-based methods for two-agent combination trials. In developing a dose-finding method for two-agent combination trials, we require a reasonable model that can adequately capture joint toxicity probabilities for two agents, taking into consideration of possible interactions of the two agents on toxicity probability (such as synergistic or antagonistic effects). We provide an overview of two useful dose-finding approaches based on Bayesian copula-type models and partial orderings across dose levels for two-agent combination trials. We also supply examples of successful software implementations and discuss the operating characteristics of these approaches.

15.1 Introduction

The purpose of many phase I trials in oncology is to determine the maximum tolerated dose (MTD), defined as the highest dose that can be administered to a population of subjects that will produce the desired effect at acceptable toxicity levels. To determine MTD in phase I populations with limited sample sizes, model-based approaches are generally efficient. When evidence regarding the dose-response relationship is available from preclinical studies or previous clinical trials for similar agents, the data can be effectively incorporated as prior information in

A. Hirakawa (✉)
Center for Advanced Medicine and Clinical Research, Nagoya University Graduate School of Medicine, Showa-ku, Nagoya, Japan
e-mail: hirakawa@med.nagoya-u.ac.jp

S. Matsui
Department of Biostatistics, Nagoya University Graduate School of Medicine, Showa-ku, Nagoya, Japan
e-mail: smatsui@med.nagoya-u.ac.jp

K. van Montfort et al. (eds.), *Developments in Statistical Evaluation of Clinical Trials*, DOI 10.1007/978-3-642-55345-5_15

the Bayesian framework. O'Quigley et al. [11] developed the continual reassessment method (CRM) that functions as the prototype for such an approach. In recent years, many investigators have studied on Bayesian dose-finding methods for phase I trials under various conditions.

The rate of implementation of two-agent combination trials, involving dose combinations of two currently marketed drugs where MTDs have already been determined or of a single new investigational drug to be used in combination with an approved drug, has rapidly increased. Furthermore, the concurrent development of two novel agents intended for use in combination to treat a single disease has attracted considerable attention.

In developing dose-finding methods for two-agent combination phase I trials, an inherent difficulty exists in modeling the complex dose-toxicity relationship characterized by the main effects of, and the interaction between, two agents on the probability of toxicity. One approach toward address this issue is to apply models that assume joint toxicity probabilities when two agents are co-administrated, including an interaction term of the two agents, within the Bayesian framework. Thall et al. [14] proposed a six-parameter model for the toxicity probabilities of the dose combinations and a toxicity equivalence contour for two-agent combinations. Wang and Ivanova [17] proposed a logistic-type regression for dose combinations that used the doses of the two agents as the covariates. Yin and Yuan considered a Bayesian adaptive design based on latent 2×2 tables [18] and copula-type models [19] for two combinatorial agents.

Another dose-finding approach to two-agent combination phase I trials is the introduction of a partial ordering for combinations of the dose levels of the two agents and the application of CRM. Conaway et al. [2] distinguished the simple and partial orders of the toxicity probabilities by defining the nodal and non-nodal parameters. Wages et al. [15] proposed a two-dimensional dose-finding method that simplifies CRM when the ordering of the toxicity probabilities is fully known, known as CRM with partial ordering. Wages et al. [16] extended this method to the case where there exist pairs of dose combinations for which the ordering of the probabilities of toxicity cannot be known a priori. Wages et al. [15] also demonstrated that their two methods were competitive with the Yin and Yuan methods [18, 19].

In this chapter, we focus on examining the Bayesian method based on copula-type models by Yin and Yuan [19] and the likelihood-based CRM with partial orderings by Wages et al. [16], as one of the most effective methods from each dose-finding approach to two-agent combination phase I trials [16]. These methods are attractive because they can be simply implemented using publicly available software.

This chapter is organized as follows. Section 15.2 provides an overview of the (one-parameter Bayesian) CRM with some modifications to improve practical performance, as well as a naive application of CRM to the two-agent combination trials. Section 15.3 presents the Bayesian dose-finding method of Yin and Yuan [19] and provides some discussion on the use of copula-type models. Section 15.4 presents CRM with partial orderings by Wages et al. [16]. Sections 15.3 and 15.4 also include examples of successful software implementation. Section 15.5 examines the operating characteristics of these two methods through simulation studies. Based on the results of the simulations, we demonstrate the inherent limitations of the models in recommending unacceptable toxicity dose combinations in certain instances by applying these methods. A possible solution to address these limitations is presented under the concluding remarks in Sect. 15.6.

15.2 CRM

15.2.1 Overview of Dose-Finding Approaches for a Single Agent

CRM is based on dose-toxicity models and is used to estimate MTD. For patient i, if a predefined toxicity is observed, primarily the dose limiting toxicity (DLT), we denote $Y_i = 1$; otherwise, $Y_i = 0$. We let $Pr\{Y_i = 1\}$ be the probability that $Y_i = 1$, often modeling this probability using a one-parameter logistic regression model with a fixed intercept β_0:

$$Pr\{Y_i = 1\} = \psi(x|\beta_1) = \frac{\exp(\beta_0 + \beta_1 x_i)}{1 + \exp(\beta_0 + \beta_1 x_i)}, \tag{15.1}$$

where x_i is the dose levels of an agent for patient i, and β_1 is the regression coefficient. O'Quigley et al. [11] have also introduced alternate models, including power and hyperbolic tangent models. It should be noted that the numerical dose label x_is in CRM is not necessarily the actual dose administered, but rather is defined on a conceptual scale that represents an ordering of the risks of toxicity based on initial guesses of toxicity probabilities, for example, *skeleton* [11].

In the original CRM, the first patient is allocated to the dose level initially believed to have toxicity closest to the target toxicity probability ϕ [11]. After obtaining the data on the toxicity outcomes from the first j patients, $D_j = \{y_1, \cdots, y_j\}$, CRM updates the posterior estimates of dose toxicity probabilities through the estimation of the posterior probability distribution $p_{j+1}(\beta_1|D_j)$ in order to determine the dose level allocated to the $(j + 1)$th patient as follows:

$$p_{j+1}(\beta_1|D_j) = \frac{L_j(\beta_1|D_j)p(\beta_1)}{\int_{-\infty}^{\infty} L_j(\beta_1|D_j)p(\beta_1)d\beta_1}, \tag{15.2}$$

where $L_j(\beta_1|D_j)$ is the likelihood function of Eq. (15.1) for j patients; that is,

$$L_j(\beta_1|D_j) = \prod_{i=1}^{j} \{\psi(x_i|\beta_1)^{y_i}\}\{1 - \psi(x_i|\beta_1)\}^{(1-y_i)}, \tag{15.3}$$

and $p(\beta_1)$ is the prior probability distribution for β_1. In this simple one-parameter setting, the posterior estimate of β_1 may be most easily computed by using a standard numerical quadrature method (e.g., trapezoidal rule), but computer-intensive simulation-based methods, such as the Markov chain Monte Carlo method, have been widely applied. Using the posterior mean of β_1, the posterior estimates of dose toxicity probabilities were obtained. The dose level at which posterior toxicity probability is closest to the target value ϕ was then determined and the $(j + 1)$th patient was allocated to that dose level. Thus, dose allocation based on the posterior toxicity probability was subsequently performed until the maximum sample size $N_{\max}$ was reached. Conclusively, the dose level with a posterior toxicity probability closest to the target value ϕ at the end of trial was selected as MTD.

Practical performance of CRM can be improved by introducing a safety stopping rule, limiting each dose escalation to one level and treating patients in cohorts [6]. When treating in cohorts of three using the same dose level within the cohort, the first three patients are allocated to the lowest dose level in practice due to ethical considerations. Cheung [1] provided comprehensive reviews and extensive discussions on CRM.

15.2.2 Software Implementation

CRM can be implemented based on several dose-toxicity models using the R package "dfcrm". The function `crmsim` in this package can be used to examine the operating characteristics of CRM when planning a trial. A simulation experiment was conducted using the function `crmsim` with the following configurations: five dose levels were set with true toxicity probabilities $\{0.01, 0.15, 0.30, 0.45, 0.65\}$. A skeleton of $\{0.05, 0.10, 0.20, 0.30, 0.55\}$ was specified, and the target toxicity probability ϕ was set as 0.30. The cohort size and total sample size were specified as 3 and 30, respectively. The first three patients were allocated to the lowest dose level. A one-parameter logistic model with a fixed intercept of 3, that is, $\text{logit}(p) = 3 + \exp(\beta) \times \text{dose}$, was assumed with a normal prior of mean 0 and variance 1.34 for β. Next, 1,000 simulations were conducted to obtain an empirical selection rate for each dose level. The following is an R code for implementing the simulations, followed by average selection rates for each dose level:

```
>## Simulation experiments for examining
  the operating characteristics of CRM
>library(dfcrm)
>p<-c(0.01,0.15,0.30,0.45,0.65)
      #true toxicity probabilities
>prior<-c(0.05,0.10,0.20,0.30,0.55) #skeleton
>target<-0.3 #target toxicity probability
>init<-1 #initial dose level
>sim<-crmsim(p,prior,target,n=30,x0=init,nsim=1000,
mcohort=3,model="logistic",intcpt=3,seed=19810314)
>sim$MTD
[1] 0.000 0.121 0.577 0.295 0.007
```

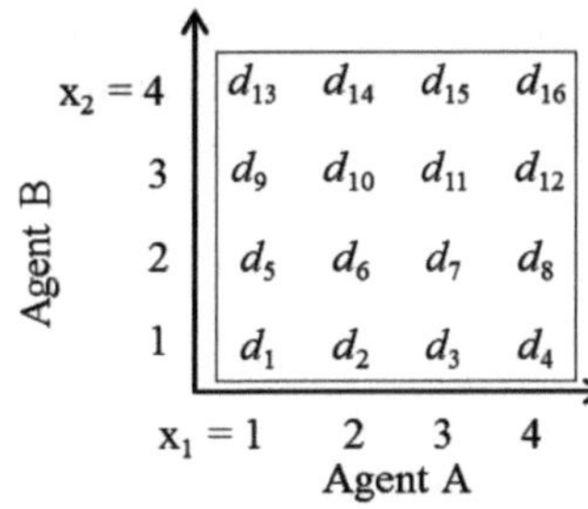

Fig. 15.1 Two-agent dose combination matrix

15.2.3 Application of CRM to Two-Agent Combination Trials

One simple way of applying CRM to a two-agent combination trial would be to fix one agent at each given dose level and vary the dose level over the second agent. For example, a two-agent combination trial with agents A and B at four dose levels, that is, $x_1 = \{1, 2, 3, 4\}$ and $x_2 = \{1, 2, 3, 4\}$, as displayed in Fig. 15.1, could be converted into four one-dimensional trials to determine the dose level of agent A for each of the four dose levels of agent B. When the total sample size is 60, one may allocate 15 patients to each of the four one-dimensional trials. Obviously, restricted CRM does not take into consideration the joint toxicity probabilities when the two agents are used simultaneously. In this instance, restricted CRM would probably not be used in practice.

15.3 Bayesian Dose-Finding Approach Using Copula Models

15.3.1 Copula-Type Regression Models

In this section, we introduce the Bayesian dose-finding approach using copula-type regressions developed in [19]. We considered a two-agent combination trial using agents A and B. In designing this trial, we first specified prior information obtained from previous studies where each agent was administrated alone. Let p_j be the pre-specified toxicity probability corresponding to A_j, the jth dose level of agent A, $p_1 < \cdots < p_J$. Similarly, we let q_k be the pre-specified toxicity probability corresponding to B_k, the kth dose level of agent B, $q_1 < \cdots < q_K$. Since the maximum dose level for each agent in the combination (i.e., A_J and B_K) is often calculated using individual MTDs that have already been determined in the singe-agent trials, the upper bounds p_J and q_K are usually known, and are typically defined as less than 30 % (or 40 %). The probabilities for the remaining dose levels $(p_1, \cdots, p_{J-1})$ for agent A and $(q_1, \cdots, q_{K-1})$ for agent B, are specified based on prior information. In order to enhance the flexibility and to accommodate the uncertainty of the prior information, Yin and Yuan [19] take p_j^{α} and q_k^{β} as the true probabilities of toxicity for agent A and agent B, respectively, where $\alpha > 0$ and $\beta > 0$ are unknown parameters with prior means centered at 1.0. In modeling the joint toxicity probabilities π_{jk}s when both agents are combined as a treatment, Yin and Yuan [19] reported that a reasonable model should satisfy the following criteria: for $j = 1, \cdots, J$ and $k = 1, \cdots, K$,

1. If $p_j^\alpha = 0$ and $q_k^\beta = 0$, then $\pi_{jk} = 0$;
2. If $p_j^\alpha = 0$, then $\pi_{jk} = q_k^\beta$, and if $q_k^\beta = 0$, then $\pi_{jk} = p_j^\alpha$; and
3. If either $p_j^\alpha = 1$ or $q_k^\beta = 1$, then $\pi_{jk} = 1$.

Motivated by the Clayton copula model [10], they proposed to use a copula-type regression model to link the joint toxicity probability π_{jk} with (p_j^α, q_k^β) in the form of

$$\pi_{jk} = 1 - \left\{(1 - p_j^\alpha)^{-\gamma} + (1 - q_k^\beta)^{-\gamma} - 1\right\}^{-1/\gamma}, \tag{15.4}$$

where $\gamma > 0$ characterizes the interaction of two agents. $\lim_{p_j \to 1}\{(1 - p_j^\alpha)^{-\gamma}\} = \infty$ and $\lim_{q_k \to 1}\{(1 - q_k^\beta)^{-\gamma}\} = \infty$, and thus $\pi_{jk} = 1$ as p_j or q_k goes to 1. If only one agent is tested, for example $p_j > 0$ and $q_k = 0$ for all k, model (15.4) reduces to CRM with $\pi_j = p_j^\alpha (j = 1, \cdots, J)$. Other copula models may also be applied, depending on mathematical convenience and computational simplicity. For example, Yin and Yuan [19] introduced the Gumbel-Hougaard copula as

$$\pi_{jk} = 1 - \exp\left[-\left\{(-\log(1 - p_j^\alpha))^{1/\gamma} + (-\log(1 - q_k^\beta))^{1/\gamma}\right\}^\gamma\right]. \tag{15.5}$$

The likelihood function can be constructed based on the binomial distribution with probabilities π_{jk}. If y_{jk} out of n_{jk} patients treated at dose levels (j, k) have experienced toxicity, the likelihood is provided by

$$L(\alpha, \beta, \gamma \mid \text{data}) \propto \prod_{j=1}^{J} \prod_{k=1}^{K} \pi_{jk}^{y_{jk}} (1 - \pi_{jk})^{(n_{jk} - y_{jk})}, \tag{15.6}$$

and correspondingly, the posterior distribution is

$$p(\alpha, \beta, \gamma \mid \text{data}) \propto L(\alpha, \beta, \gamma \mid \text{data}) p(\alpha) p(\beta) p(\gamma), \tag{15.7}$$

where $p(\alpha)$, $p(\beta)$, and $p(\gamma)$ are gamma prior distributions with a mean of 1 and suitable variance.

15.3.2 Dose-Finding Algorithm

We let c_e and c_d be the fixed probability cut-offs for dose escalation and de-escalation, respectively. Next, c_e and c_d can be calibrated through simulation studies such that the trial has desirable operating characteristics, and $c_e + c_d > 1$. Patients are treated in cohorts of three. The target toxicity probability that is clinically allowed is defined as ϕ. To be conservative, dose escalation or de-escalation were restricted to one dose level of change only, while not allowing a translation along the diagonal direction (corresponding to simultaneous escalation or de-escalation of both agents). For a trial involving two drugs, the dose-finding algorithm functions as follows:

1. Patients in the first cohort are treated at the lowest dose combination (A_1, B_1).
2. If, at the current dose combination $(j,\ k)$, $Pr(\pi_{jk} < \phi) > c_e$, the dose is escalated to an adjacent dose combination with probability of toxicity higher than the current value and closest to ϕ. If the current dose combination is (A_J, B_K), the doses remain at the same levels.
3. If, at the current dose combination $(j,\ k)$, $Pr(\pi_{jk} > \phi) > c_d$, the dose is de-escalated to an adjacent dose combination with the probability of toxicity lower than the current value and closest to ϕ. If the current dose combination is (A_1, B_1), the trial is terminated.
4. Otherwise, the next cohort of patients continues to be treated at the current dose combination (doses staying at the same levels).
5. Once the maximum sample size has been achieved, the dose combination that has the probability of toxicity that is closest to ϕ is selected as the MTD combination.

As is common to model-based clinical trial designs, the dose-finding algorithm is difficult to apply at the beginning of the trial, because very limited information (except for prior knowledge) is available. Thus, the posterior estimates of the probabilities of toxicity for dose combinations may not be reliable. To circumvent this difficulty, the following start-up rules were applied: the first patients were treated along the vertical dose escalation in the order of (A_1, B_1), (A_1, B_2), $\cdots$ until the first toxicity was observed; the patients continued to be treated by escalating doses in the horizontal direction (A_2, B_1), (A_3, B_1), $\cdots$ until the first toxicity occurs. As long as one toxicity is observed in both the vertical and the horizontal directions (e.g., if one patient experiences toxicity at (A_1, B_k) and (A_j, B_1) for some values of j and k), the Bayesian dose-finding algorithm will be seamless in effect for the remainder of the trial.

15.3.3 Discussion on the Use of Copula-Type Models

Yin and Yuan [19] have concluded that the method they proposed can fully evaluate the joint toxicity profiles of the combined drugs, in addition to preserving their single-agent properties. Furthermore, the drug-drug interactive effects are naturally modeled through a copula-type model, which reduces to CRM if only one drug is considered. However, Gasparini et al. [5] reported some concerns about the limitations of the copula-type regression model. They pointed out that during drug development of multi-agent therapies, investigators generally encounter the following three cases: (i) no interaction applies if the two agents act independently: they have no apparent effect on one another's potential toxicity; (ii) two agents are said to exhibit an antagonistic effect when one drug reduces or neutralizes the toxic potential of the other; and (iii) two agents are said to exhibit a synergistic effect if they exhibit greater toxicity when administered together than would be expected if they functioned independently.

Following these arguments, Gasparini et al. [5] provided the following formulation. They let p and q be the probabilities of toxicity when using only the first or only the second agent, respectively, and let $\pi(p, q)$ be the probability of a toxicity when both drugs are administered in combination. No interaction was observed if the two drugs function independently, that is, if the probability of no toxicity is equal to the product of the marginal probabilities of no toxicity. In terms of the probabilities of toxicity, the no-interaction model is defined as

$$\pi(p, q)^{\perp} = 1 - (1 - p)(1 - q) = p + q - pq \, . \tag{15.8}$$

Thus, the three instances of drug-drug interaction are: (a) antagonism, $\pi(p, q) < p + q - pq$; (b) no interaction, $\pi(p, q) = p + q - pq$; and (c) synergy, $\pi(p, q) > p + q - pq$. Any model that is applied to dose finding in combination studies should have the potential to allow for these situations, with the synergistic case being most plausible, based on the often toxic nature of both treatments.

The prototype of the copula-type regression model in [19] can be displayed as follows:

$$\pi_C(p, q) = 1 - \{(1 - p)^{-\gamma} + (1 - q)^{-\gamma} - 1\}^{1/\gamma} \, . \tag{15.9}$$

Gasparini et al. [5] objected to the use of copulas for modeling the joint-probability of toxicity, since the following double inequality holds

$$\max(p, q) \le \pi(p, q) \le \min(p + q, 1) \tag{15.10}$$

for any toxicity probability $\pi(p, q)$ obtained from copula arguments and for $\pi_C(p, q)$ [10]. The primary criticism by Gasparini et al. [5] is that there exists no reason for the joint-probability of toxicity to satisfy these constraints. Under extreme synergistic or extreme antagonistic effects (less likely with drug combinations), the joint-probability of toxicity should be allowed to approach 1 or 0 without restrictions. Assuming that, for two specific doses of agent A and agent B, the marginal probabilities of toxicity are $p = 0.1$ and $q = 0.2$, respectively, co-administration of both agents will cause any copula to confine the probability of toxicity to the interval (0.2, 0.3), the upper bounds being very close to the probability of toxicity under no interaction, $(0.1 + 0.2) - (0.1 \times 0.2) = 0.28$. Clearly, these restrictions are too severe. In practice, synergy could lead to a probability of toxicity that is much greater than 0.3.

Gasparini et al. [5] also argued that the Clayton copula with $\gamma > 0$ in [19] is a particularly poor choice since it can be shown that $\pi_C(p, q)$ is a strictly decreasing function of $\gamma > 0$ for fixed p and q. Since $\gamma \to 0$ represents no interaction, this implementation of the joint-probability of toxicity cannot model a synergistic effect, which is the most common effect for drug combinations. For example, in the numerical example above with $p = 0.1$ and $q = 0.2$, if $\gamma = 1$, the joint-probability of toxicity reduces to approximately 0.265. To consider the synergistic effect with the Clayton copula, Gasparini et al. [5] suggested that one could consider the extra range $-1 < \gamma < 0$, although that would not avoid the overly restrictive constraints Eq. (15.10).

15.3.4 Software Implementation

In this section, we used the software released by Yin and Yuan [19] to implement their method, downloading the .exe program from http://as.wiley.com/WileyCDA/. In this program, we entered the following configurations: the number of dose levels for two agents and their true joint toxicity probabilities for dose combinations; target toxicity probability; prior estimate of toxicity probabilities for dose levels for each agent; total number of cohorts; cohort size; and number of simulated trials.

For example, we obtain the following simulation results when two dose levels for each agent were tested:

```
-------------------------------------------------
CPU time (hour)= 0.00190944     # of trials = 10
Number of cohorts = 10; cohort size = 2
Escalate if Pr(toxicity<0.3) > 0.8
De-escalate if Pr(toxicity<0.3) < 0.45

True toxicity probabilities:
  0.20    0.50
  0.05    0.15
Selection probabilities (%):
  20.0    60.0
   0.0    20.0
Number of patients treated at each dose:
   3.4     8.0
   5.2     3.4
Number of toxicities observed at each dose:
   0.3     3.7
   0.2     0.6
```

```
Total number of observed toxicities: 4.8
Percentage of inconclusive trials: 0.0%
--------------------------------------------------
```

In the default setting, c_e and c_d were set to be 0.8 and 0.45, respectively. By utilizing the C++ program (copula.cpp), we could change the values for c_e and c_d. Furthermore, we could select a copula model (i.e., Clayton or Gumbel copula model) and the prior distributions for their model parameters.

15.4 CRM Using Partial Orderings of Toxicity Probabilities

15.4.1 Partial Orderings

The assumption that the probability of toxicity increases monotonically with dose level is generally reasonable in single-agent trials. Using the terminology of Robertson et al. [13], Wages et al. [15, 16] introduced the concept that the probabilities of toxicity with respect to dose level follow a "simple order," that is, the ordering of the toxicity probabilities between any two dose levels is known, with the higher dose corresponding to a greater probability of toxicity. While the assumption of a simple order will often appear reasonable in single-agent trials, the monotonicity assumption may not hold in two-agent combination trials since the ordering of the toxicity probabilities may be unknown for several of the dose combinations. In order to address this limitation, Wages et al. [15] focused on the fact that investigators may be able to identify the order of the toxicity probabilities for only a subset of the available dose levels, resulting in a partial ordering. The method of Wages et al. [15] relies, to some degree, upon the framework of Conaway et al. [2], which identifies all possible simple orders for the toxicity probabilities that are consistent with the known orderings among the dose combinations. Each of these simple orders consistent with a partial order can be thought of as a model. Readers are referred to [2, 15] and [16] for further details.

Wages et al. [16] introduced an example of a partially ordered trial using the dose-finding trial for combinations of topotecan and irinotecan [9]. This trial consists of eight dose combinations: $d_1, \cdots, d_8$ (Fig. 15.2). The toxicity ordering between dose combinations d_1 and d_2 is known, since the dose of irinotecan remains the same while the dose of topotecan increases. This is also the case for the ordering between dose combinations d_3 and d_4. However, the order relationships between dose combinations d_2 and d_3 and between d_4 and d_5 are unknown because the dose of topotecan decreases while the dose of irinotecan increases. If the known and unknown toxicity order relationships continue to be assessed in this manner, the following known order relationships hold: (1) $d_1 \to d_2$; (2) $d_3 \to d_4$; (3) $d_5 \to d_6$; and (4) $d_7 \to d_8$. In these diagrams, dose combinations whose orderings are known are connected by arrows, with the dose combinations to the right being more toxic (i.e., it is known that d_8 is more toxic than d_7, d_6 is more toxic than d_5, and so on).

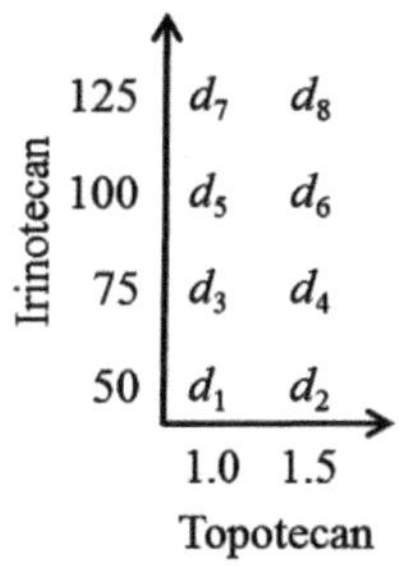

Fig. 15.2 Combination matrix of Topotecan and Irinotecan in $mg/m^2/wk$

Table 15.1 Eight possible simple orders for a combination trial of topotecan and irinotecan

Ordering (m)	Simple order							
1	$d_1 \rightarrow$	$d_2 \rightarrow$	$d_3 \rightarrow$	$d_4 \rightarrow$	$d_5 \rightarrow$	$d_6 \rightarrow$	$d_7 \rightarrow$	d_8
2	$d_1 \rightarrow$	$d_3 \rightarrow$	$d_2 \rightarrow$	$d_4 \rightarrow$	$d_5 \rightarrow$	$d_6 \rightarrow$	$d_7 \rightarrow$	d_8
3	$d_1 \rightarrow$	$d_2 \rightarrow$	$d_3 \rightarrow$	$d_5 \rightarrow$	$d_4 \rightarrow$	$d_6 \rightarrow$	$d_7 \rightarrow$	d_8
4	$d_1 \rightarrow$	$d_2 \rightarrow$	$d_3 \rightarrow$	$d_4 \rightarrow$	$d_5 \rightarrow$	$d_7 \rightarrow$	$d_6 \rightarrow$	d_8
5	$d_1 \rightarrow$	$d_3 \rightarrow$	$d_2 \rightarrow$	$d_5 \rightarrow$	$d_4 \rightarrow$	$d_6 \rightarrow$	$d_7 \rightarrow$	d_8
6	$d_1 \rightarrow$	$d_3 \rightarrow$	$d_2 \rightarrow$	$d_4 \rightarrow$	$d_5 \rightarrow$	$d_7 \rightarrow$	$d_6 \rightarrow$	d_8
7	$d_1 \rightarrow$	$d_2 \rightarrow$	$d_3 \rightarrow$	$d_5 \rightarrow$	$d_4 \rightarrow$	$d_7 \rightarrow$	$d_6 \rightarrow$	d_8
8	$d_1 \rightarrow$	$d_3 \rightarrow$	$d_2 \rightarrow$	$d_5 \rightarrow$	$d_4 \rightarrow$	$d_7 \rightarrow$	$d_6 \rightarrow$	d_8

Escalation to a previously untried dose combination depends on the prior specification of "possible escalation combinations" associated with each dose combination. For example, the possible escalation combinations for d_1 are d_2 and d_3, indicating that if d_1 were tested and found to be well tolerated, escalation could proceed to the previously untried levels d_2 or d_3.

In general, we surmise that the dose combinations follow a partial order for which there are $M(m = 1, \cdots, M)$ possible simple orders consistent with the partial order; therefore, there exists a class of M models of interest. In the context of the aforementioned example, there exist eight possible simple orders (Table 15.1). Supposing that we can account for any prior information concerning the plausibility of each model (i.e., $p(m) = \{p(1), \cdots, p(M)\}$, where $p(m) > 0$ and $\sum_{m=1}^{M} p(m) = 1$), these probabilities are determined by prior knowledge, and equal probability on each model $p(m) = 1/M$ is used when there is no available information. Given a particular ordering, the toxicity probabilities are modeled by a parametric model from the CRM class of models.

15.4.2 *Framework of Dose-Finding Based on CRM Using Partial Orderings*

Assuming that there are H dose combinations, $d_1, \cdots, d_H$. For a particular model, m $(m = 1, \cdots, M)$, the simple power model can be assumed as follows:

$$\psi(d_h, a) = \alpha_{mh}^a, \ h = 1, \cdots, H, \tag{15.11}$$

where α_{mh} for each model represents the skeleton of the model. After obtaining the data on the toxicity outcome for dose combination, we estimate the model parameter a and subsequently calculate the toxicity probability for each dose combination, $\psi(d_h, a)$. The estimates for the toxicity probabilities at each dose combination are based on the likelihood approach of the CRM of O'Quigley and Shen [12]. If we suppose that $L_m(a \mid D)$ is the log-likelihood for ordering m after obtaining data D on toxicity outcomes with the administered dose combinations after treating a certain number of patients, for each M orderings, $L_m(a \mid D)$ can be maximized in order to obtain an estimate $\hat{a}_m$ for a. As the weight of evidence in favor of model m, Wages et al. [16] introduced

$$\pi(m) = \frac{\exp\{L_m(\hat{a}_m \mid D)\}}{\sum_{m=1}^{M} \exp\{L_m(\hat{a}_m | D)\}}. \tag{15.12}$$

Furthermore, they incorporated prior probabilities $p(m)$ as follows:

$$\pi(m) = \frac{\exp\{L_m(\hat{a}_m \mid D)\} p(m)}{\sum_{m=1}^{M} \exp\{L_m(\hat{a}_m | D) p(m)\}}. \tag{15.13}$$

Notably, the inclusion of $p(m)$ is not a true Bayesian approach, as indicated in [16]. A true Bayesian approach to model selection requires both the prior $p(m)$ on the model and a prior distribution on the parameter a in model m for each model. Thus, the partial order m^* such that $m^* = \arg\max_m \pi(m), \ m = 1, \cdots, M$ was determined.

Given the partial order m^*, the working model associated with this ordering was taken and the likelihood approach of CRM was applied to obtain estimates of the toxicity probabilities at each of the H available dose combinations. Using the simple power model in Eq. (15.11), the estimated probability of toxicity at each dose combination is provided by $\psi_{m^*}(d_h, \hat{a}_{m^*})$. Thus, the dose combination that minimizes $|\psi_{m^*}(d_h, \hat{a}_{m^*}) - \phi|$ is allocated to the next patient.

15.4.3 Two-Stage Dose-Finding Algorithm

Wages et al. [16] have adopted a two-stage design for dose-finding. Supposing that toxicity increases monotonically across the rows and up the columns of the matrix of doses, we consider a two-agent combination trial using agents A and B with four dose levels (Fig. 15.3). The 16 dose combinations were divided into 7 groups (or zones) (G1, $\cdots$, G7) consisting of the diagonals of the combination matrix, that is, the first group contains the single combination d_1, the second group contains the dose combinations d_2 and d_5, and so on. Toxicity is considered to

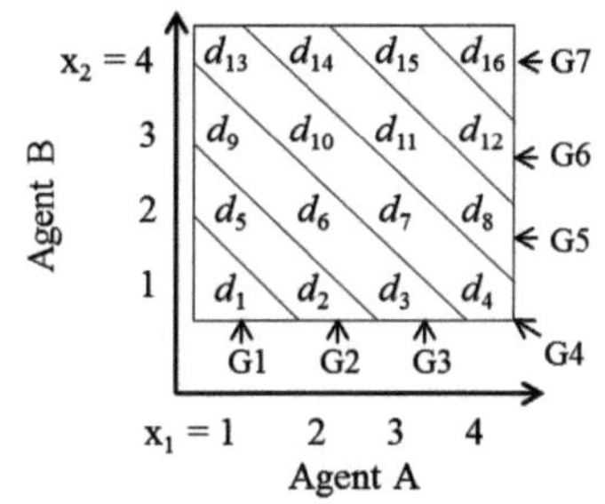

Fig. 15.3 Two-agent dose combination matrix consisting of seven groups

increase monotonically with respect to treatment during translation from group 1 to group 7. The first patient (or cohort) is allocated to the lowest dose combination, d_1. If a predefined toxicity is observed for this patient (or cohort), the first stage is closed and the second stage is opened. If toxicity is not observed, the patient (or cohort) is escalated to the second group. If more than one dose combination is contained within a particular group, the investigator can sample without replacement from the dose combinations available because the ordering of the treatments is unknown. Specifically, the next dose combination is chosen randomly from the dose combinations within the current group. In addition, the trial is not allowed to advance to the third group in the first stage until the patient (or cohort) is enrolled into both d_2 and d_3. This procedure is continued until one DLT is observed or all available groups have been exhausted.

In the second stage, the MTD combinations are found using the dose-finding approach described in the previous section. Once the maximum sample size $N_{\max}$ (or the pre-specified stopping rule is met) is achieved, the dose combination that should be assigned to the next patient (or cohort) is selected as the MTD combination

15.4.4 *Software Implementation*

In this section, we report on the implementation of the method of Wages et al. [15] using the R program downloaded from http://faculty.virginia.edu/model-based_dose-finding/POCRM-06.12.txt. The function `crm` calculates the maximum likelihood estimate (MLE) of parameter a in the simple power model and recommends a dose combination for the next patient. The functions `twostgcrm` and `pocrm.sim` return the results of a single simulated trial and multiple simulated trials, respectively.

For simplicity, two dose levels are considered for each agent: the total number of dose combinations is four. Two possible orderings are considered, $d_1 \rightarrow d_2 \rightarrow d_3 \rightarrow d_4$ and $d_1 \rightarrow d_3 \rightarrow d_2 \rightarrow d_4$. In implementing this program, we also entered the following configurations: the true joint toxicity probability for each dose combination; target toxicity probability; maximum sample size; number of patients on a combination needed to stop the trial; and skeleton (generated using the getprior function of [8]). Thus, we obtained the result of the single simulated trial as follows:

```
>d<-4 #number of dose combinations
>s<-2 #number of possible orderings
>orders<-matrix(nrow=s,ncol=d)
      #specify the possible orderings
>orders[1,]<-c(1,2,3,4)
>orders[2,]<-c(1,3,2,4)

>zones<-list(z1=c(1),z2=c(2,3),z3=c(4))
 #specify the zone for dose-finding

>ff<-function(x){
+     if(length(x)==1){
+         x
+     } else
+         sample(x)
+ }
>
> r<-c(0.05,0.15,0.20,0.50)
      #true toxicity probabilities
> theta<-0.30 #target toxicity probability
> n<-20 # Maximum sample size
> stop<-21
# number of patients on a combination
          needed to stop the trial

> library(dfcrm)
# skeleton generator by Lee and Cheung (2009)
> skeleton<-round(getprior(0.03,theta,2,d),2)
> skeleton
[1] 0.24 0.30 0.36 0.42

> alpha<-matrix(0,nrow=s,ncol=d)
# skeleton for each ordering
> for(j in 1:s){
+     alpha[j,]<-skeleton[order(orders[j,])]
+ }
>
> alpha
     [,1] [,2] [,3] [,4]
[1,] 0.24 0.30 0.36 0.42
[2,] 0.24 0.36 0.30 0.42
```

```
> fit<-twostgcrm(n,alpha,r,theta)
> fit
$trial
   patient level tox        a order
1        1     1   0 0.000000    99
2        2     2   0 0.000000    99
3        3     3   0 0.000000    99
4        4     4   1 1.361205     1
5        5     4   0 1.630720     1
.
.
.
19      19     2   0 1.084570     1
20      20     2   0 1.131424     1
21      21     3   0 0.000000     0

$MTD.rec
[1] 3
```

The R package "pocrm" for implementing the method of Wages et al. [16] was released in December 2012, and can also be used. Wages et al. [16] reported that their method was competitive with the method proposed in [15], and with the methods of Conaway et al. [2] and Yin and Yuan [18, 19].

15.5 Comparison of Operating Characteristics

15.5.1 Simulation Setting

We performed a simulation study to compare the operating characteristics of the methods of Yin and Yuan [19] and Wages et al. [16] in two-agent combination trials under the some scenarios we selected. Four dose levels were considered for both agents A and B: $A = \{1, 2, 3, 4\}$ and $B = \{1, 2, 3, 4\}$. The target toxicity probability that is clinically allowed, ϕ, was set to 0.3. The maximum sample size $N_{\max}$ was set to 60 throughout.

We also performed the method of Yin and Yuan [19] with the following specifications: the prior toxicity probabilities for 4×4 dose combinations (i.e., $A = \{1, 2, 3, 4\}$ and $B = \{1, 2, 3, 4\}$) were set to be (0.075, 0.15, 0.225, 0.30). Three patients were allocated to a single dose level at a single time. We performed the method of Wages et al. [16] with the following specifications: (1) the skeletons were generated using the `getprior` function; (2) the six possible orderings of the drug combinations were used for 4×4 dose combinations trials; and (3) the number of patients on a dose combination needed to stop the trial was 61 (i.e., we avoid

Table 15.2 Four scenarios for a two-agent combination trial with the target probability of toxicity 0.3 (MTD combinations are in boldface)

	A = 1	2	3	4	1	2	3	4
	Scenario 1				Scenario 2			
B = 4	**0.30**	0.50	0.55	0.60	**0.30**	0.50	0.60	0.70
3	0.12	**0.30**	0.50	0.55	0.15	0.35	0.50	0.55
2	0.10	0.15	**0.30**	0.45	0.08	**0.30**	0.45	0.50
1	0.08	0.12	0.16	0.18	0.05	0.10	0.20	**0.30**
	Scenario 3				Scenario 4			
4	0.20	0.50	0.55	0.70	0.08	0.55	0.60	0.75
3	0.15	**0.30**	0.50	0.60	0.05	0.50	0.55	0.65
2	0.10	0.18	**0.30**	0.50	0.03	**0.30**	0.40	0.50
1	0.06	0.08	0.10	0.15	0.01	0.10	0.15	0.45

Table 15.3 Recommendation rates for true MTD and unacceptable toxicity dose combinations in all scenarios

Scenario	1	2	3	4
	True MTD combinations (%)			
Yin and Yuan [19]	47.0	39.0	36.0	11.9
Wages et al. [16]	60.6	46.1	42.1	30.1
	Unacceptable toxicity dose combinations (%)			
Yin and Yuan [19]	32.8	42.0	37.3	62.5
Wages et al. [16]	13.7	31.8	16.9	44.1

the use of this stopping rule). We compared the operating characteristics of the two methods by simulating four scenarios as shown in Table 15.2. We conducted 1,000 simulation trials for each scenario.

15.5.2 Simulation Results

The primary simulation results are summarized in Table 15.3. In terms of recommending true MTD combinations, the method of Wages et al. [16] outperformed that of Yin and Yuan [19] under the scenarios we selected. The recommendation rates for unacceptable dose combinations (i.e., the dose combinations with toxicity probability greater than 0.3) of the two methods were greater than or equal to those for true MTD combinations under some scenarios. According to the results from our further simulation studies (data not shown), the two methods highly recommended the unacceptable dose combination levels relative to the recommendation rates for true MTD combinations in certain instances.

15.6 Discussion

In this chapter, we have provided an overview of dose-finding approaches based on toxicity for combinations of two agents. The approaches can be categorized into two groups: (1) those using a flexible Bayesian model, including an interaction term of the two agents; and (2) those that extend CRM, taking into consideration the partial ordering of toxicity probabilities for dose combinations. The methods of Yin and Yuan [19] and Wages et al. [16] represent these two categories, respectively. Although these two methods employ substantially different dose-toxicity models and dose-finding algorithms, their operating characteristics have been shown to be competitive in [16]. Regarding the method of Yin and Yuan [19], the adequacy and utility of the copula-type regression models should be further examined, as discussed in Sect. 15.3.3.

The operating characteristics of the methods of Yin and Yuan [19] and Wages et al. [16] can be varied depending on the implementation configurations (e.g., prior toxicity specifications, cohort size, assumed toxicity scenarios, etc.), although we displayed the results of simulation studies under several scenarios in Sect. 15.5. In practice, it is important to determine the suitable configurations for investigational agents through a simulation study. However, we were concerned that the two aforementioned methods tended to yield a high recommendation rate of unacceptable toxicity dose combinations in certain instances. Since the use of unacceptable toxicity dose levels is one of the primary causes for the high attrition rate of investigational drugs in confirmatory trials, as well as an increasing ethical concern, it is therefore desirable to develop a dose-finding approach that can suppress the recommendation of unacceptable dose combinations, while maintaining a high probability of selecting true MTD combinations. To address this issue, it is appropriate to tailor dose-finding methods to learn the zone of unacceptable dose combinations more effectively at an early part of the dose-finding process and then avoid unacceptable dose combinations later on. One promising approach is the application of empirical Bayes shrinkage regression [3, 4]. A dose-finding method using this approach is reported [7].

Acknowledgements This work was partially supported by JSPS KAKENHI Grant Number 25730015 (Grant-in-Aid for Young Scientists B).

References

1. Cheung, Y. K.: Dose Finding by the Continual Reassessment Method. New York, Chapman & Hall (2011)
2. Conaway, M. R., Dunbar, S., Peddada, S. D.: Designs for single- or multiple-agent phase I trials. *Biometrics* **60**, 661–669 (2004)
3. Copas, J. B.: Regression, prediction and shrinkage. *Journal of the Royal Statistical Society Series B* **45**, 311–354 (1983)

4. Copas, J. B.: Using regression models for prediction: Shrinkage and regression to the mean. *Statistical Methods in Medical Research* **6**, 167–183 (1997)
5. Gasparini, M., Bailey, S., Neuenschwander, B.: Correspondence: Bayesian dose finding in oncology for drug combinations by copula regression. *Journal of the Royal Statistical Society Series C* **59**, 543–546 (2010)
6. Goodman, S. N., Zahurak, M. L., Piantadosi, S.: Some practical improvements in the continual reassessment method for phase I studies. *Statistics in Medicine* **14**, 1149–1161 (1995)
7. Hirakawa, A., Hamada, C., Matsui, S.: A dose-finding approach based on shrunken predictive probability for combinations of two agents in phase I trials. *Statistics in Medicine* **32**, 4515–4525 (2013)
8. Lee, S. M., Cheung, Y. K.: Model calibration in the continual reassessment method. *Clinical Trials* **6**, 227–238 (2009)
9. Lokich, J.: Phase I clinical trial of weekly combined topotecan and irinotecan. *American Journal of Clinical Oncology* **24**, 336–40 (2001)
10. Nelsen, R. B.: An Introduction to Copulas, 2nd edn. New York, Springer (2006)
11. O'Quigley, J., Pepe, M., Fisher, L.: Continual reassessment method: A practical design for phase 1 clinical trials in cancer. *Biometrics* **46**, 33–48 (1990).
12. O'Quigley, J., Shen, L. Z.: Continual reassessment method: A likelihood approach. *Biometrics* **52**, 673–684 (1996)
13. Robertson, T., Wright, F. T., Dykstra, R. L., Dykstra, R.: Order Restricted Statistical Inference. New York, John Wiley & Sons (1988)
14. Thall, P. F., Millikan, R. E., Mueller, P., Lee, S. J.: Dose finding with two agents in phase I oncology trials. *Biometrics* **59**, 487–496 (2003)
15. Wages, N. A., Conaway, M. R., O'Quigley, J.: Continual reassessment method for partial ordering. *Biometrics* **67**, 1555–1563 (2011a)
16. Wages, N. A., Conaway, M. R., O'Quigley, J.: Dose-finding design for multi-drug combinations. *Clinical Trials* **8**, 380–389 (2011b)
17. Wang, K., Ivanova, A.: Two-dimensional dose finding in discrete dose space. *Biometrics* **61**, 217–222 (2005)
18. Yin, G., Yuan, Y.: A latent contingency table approach to dose finding for combinations of two agents. *Biometrics* **65**, 866–875 (2009a)
19. Yin, G., Yuan, Y.: Bayesian dose finding in oncology for drug combinations by copula regression. *Journal of the Royal Statistical Society Series C* **58**, 211–224 (2009b)

Chapter 16
Multi-state Models Used in Oncology Trials

Birgit Gaschler-Markefski, Karin Schiefele, Julia Hocke, and Frank Fleischer

Abstract Among the surrogate endpoints for overall survival (OS) in oncological trials, progression-free survival (PFS) is used as an important endpoint especially in first or second line of cancer therapies. Basic formulae for the determination of sample sizes based on time to event data can be found in the literature. Assumptions about the distributions of the survival time for OS and PFS, the accrual time and the censoring time are of key importance. Most often only uniformly distributed patient accrual and no censoring are mentioned, whereas the event time is assumed to be exponentially distributed. Considering the dependence between PFS and OS, we will investigate how a three-state model that includes states of progression/response and death can be used for a joint modelling of progression-free survival and overall survival. Sample size/power calculations are discussed for the three-state model and compared to the estimations based on exponentially distributed OS times. These sample size calculations are based on the assumption of piecewise uniformly accrual and exponentially distributed censoring time. The new three-state model approach results in a 10–30 % lower sample size and a corresponding higher power. An application to a Phase III lung cancer trial illustrates how the new approach can be successfully applied to the planning of a trial and to the monitoring of the needed events for the PFS and OS analyses.

16.1 Introduction

Oncological trials are often performed as event-driven trials, i.e. trial length and analysis time points are tied to the occurrence of a specific number of events. The most commonly used endpoint for new anticancer drug studies is overall survival

B. Gaschler-Markefski (✉) • J. Hocke • F. Fleischer
Department of Biostatistics, Boehringer Ingelheim Pharma GmbH and Co. KG, Biberach, Germany
e-mail: birgit.gaschler-markefski@boehringer-ingelheim.com; julia.hocke@boehringer-ingelheim.com; frank.fleischer@boehringer-ingelheim.com

K. Schiefele
Department of Epidemiology and Medical Biometry, University Ulm, Ulm, Germany
e-mail: karin.schiefele@uni-ulm.de

K. van Montfort et al. (eds.), *Developments in Statistical Evaluation of Clinical Trials*, DOI 10.1007/978-3-642-55345-5_16

(OS). If a patient develops progression of the tumor, then the therapy will be stopped and the patient will be switched to another (probably new) anticancer therapy. With regard to this, progression-free survival (PFS) is also used as a trial endpoint especially in early stages of cancer therapy. We will analyse OS mathematically by incorporation of the PFS information via a multi-state model. Multi-state models are probabilistic models which allow for studying transitions of a subject (in this context a person or patient) between different states over the course of time. In this chapter, an introduction to the basic concepts of multi-state modelling will be given and models commonly used in medical contexts, especially in oncology, will be presented.

In oncology as well as other indications like stroke or asthma a time to event outcome is often used as primary endpoint. For operational aspects it may be important to plan the time points of the final analysis and possible interim analyses. The time points of the interim analyses and final analysis in time to event studies are in most cases driven by the needed number of events (*landmark event number*). Therefore, a precise monitoring and prediction of the time point for the landmark event number is needed. The estimation of this time point with respect to OS can be derived based on different assumptions on the distribution of the lost-to-follow-up and the overall survival. For estimation of the time to occurrence of the landmark event number in this article an illness-death model, i.e. a three-state model for OS, is applied instead of the frequently used but oversimplifying assumption of exponentially distributed OS. An application to a phase III lung cancer trial illustrates how the new approach can be successfully applied to monitor event numbers for the OS analyses.

This chapter is structured as following: Different kinds of relevant multi-state models will be defined and their application to different contexts given. Of special interest is the three-state model for estimation of the time point of the landmark event number. Therefore, in the second part, after introducing the model assumptions as well as deriving relevant distributions, the expected number of events depending on the current time point and the planned accrual period will be derived. Based on this, the predicted landmark event time may be derived. We will compare the three-state model with an alternative one, which is restricted to exponentially distributed OS and does not account for progression. How the choice of the model influences sample size and power calculations is shown in an example. The performance of our new approach is demonstrated on real data of a non-small cell lung cancer trial.

16.2 Background Information

This section is based on models for the analysis of data with the primary endpoint being the time until occurrence of a certain event, which is also called *failure*. In the following an overview about common multi-state models will be given, whereas the kind of model is defined by the types of states it is consisting of.

16.2.1 Overview of Multi-state Models

The nonnegative random variable T corresponds to the period of time lasting from the initial time t_0 (e.g. time point of birth, randomisation, etc., mostly $t_0 = 0$) to the occurrence of the event of interest. In accordance to common terminology T is assumed to be continuous on $\mathbb{R}_+$. For analysis of discrete failure time distributions see for example [20].

Definition 16.1. A right-continuous piecewise constant stochastic process $X(t)$, $t \in [0, \infty)$ with a finite state space $S = \{1, \ldots, n\}$, $n \in \mathbb{N}$, is called a *multi-state model* (MSM).

The value of the process corresponds to the state occupied at time t and the *initial distribution* of the stochastic process is noted by $\pi_s(0) = \mathbb{P}(X(0) = s)$ for $s \in S$ (cf. [25]). The shift from one state to another is referred to as a *transition* or an *event*.

The probability for being in state $j \in S$ at time $t \in \mathbb{R}_+$ given that the process started in $i \in S$ at $u \in \mathbb{R}_+$, $u < t$, is called the *transition probability* and is noted by

$$p_{i,j}(t, u) = \mathbb{P}(X(u) = j \,|\, X(t) = i, \mathfrak{H}_t), \tag{16.1}$$

whereas $\mathfrak{H}_t$ denotes the *history* of the process $X(.)$ (a σ-algebra in mathematical terms). $\mathfrak{H}_t$ consists of all the information of the process from the initial time (mostly time point 0) until t, i.e. all of the previous states and related times of transition in the interval $[0, t]$. Based on (16.1), the *state probabilities* $\pi_j(t) = \mathbb{P}(X(t) = j)$ are

$$\pi_j(t) = \sum_{i \in S} \pi_i(0) p_{i,j}(0, t) \tag{16.2}$$

for $j \in S$ and $t \in \mathbb{R}_+$. The *transition intensity* (also called transition rate, hazard function or (age-specific) failure rate) is defined by

$$\alpha_{i,j}(t) = \lim_{\Delta t \searrow 0} \frac{p_{i,j}(t, t + \Delta t)}{\Delta t}. \tag{16.3}$$

The $\alpha_{i,j}(t)$ gives the instantaneous event (or failure) rate at time t, provided the individual has been event-free until t. Consequently, the product $\alpha_{i,j}(t)\Delta t$ corresponds to the approximate probability of an event in $[t, t + \Delta t)$, given there has been no event until t (cf. [23]). A state $i \in S$ is called *absorbing* when it is not possible to leave this state once it has been reached and therefore, it holds $\alpha_{i,j}(t) = 0$ for all $t \in \mathbb{R}_+$ and $j \in S$. The time point when the process has left state $i \in S$ and first reaches $j \in S$ is called *transition time*.

Different kinds of models are defined by dependency of the transition intensity on time (cf. [25]):

1. **Time homogeneous models** have transition rates being constant over time, i.e. $p_{i,j}(t,u)$ depends only on $u-t$ and so it holds $p_{i,j}(t,u) = p_{i,j}(0,u-t)$.
2. **Markov models** have transition intensities only depending on the current state and neither on more of the previous states nor on future states, i.e. for $i, j \in S$ and $t, u \in \mathbb{R}_+$ with $0 \leq t < u$

$$\mathbb{P}(X(u) = j \,|\, X(t) = i, \mathfrak{H}_t) = \mathbb{P}(X(u) = j \,|\, X(t) = i)\,. \qquad (16.4)$$

3. **Semi-(homogeneous) Markov models** have transition intensities depending on the current state $i \in S$ as well as on the time spent in state i.

In the following, only time-homogeneous Markov models will be analysed. For further description and examples on semi-(homogeneous) Markov models see e.g. [4] or [34].

A more detailed introduction to the theory of stochastic processes and multi-state models may be found in [3] (Chapter I). Also a good overview about multi-state models is given in [1,25] and [18].

16.2.2 *Types of Models*

Uni-directional (or progressive) models allow for forward transitions only; once a state has been left, it can not be returned to it again. On the other hand in bi-directional (or alternating) models, the process can return to each state provided that it does not enter an absorbing state. Alternating models are relevant for e.g. reversible diseases but they would not be considered in detail in this chapter.

16.2.2.1 *k*-State Model

The k-state model is characterized by $k-1$ transient but uni-directional passable states ($k \in \mathbb{N}, k \geq 2$) and one absorbing state. Commonly, the first of the transient states is the starting point and the absorbing state is reachable from each of the transient states. Each of the following kinds of k-state models is Markovian. A method for testing the Markov property for example in a three-state progressive model is presented in [30].

Mortality Model

The simplest kind of the k-state model is the *mortality model* (*two-state model*) consisting of only two states (cf. Fig. 16.1). The process starts in '0' (alive) and stops after reaching the absorbing state '1' (dead). It holds $\alpha_{1,0}(t) = 0$ for all $t \in \mathbb{R}_+$ and the initial distribution is $\pi_0(0) = 1$. For example, Birnbaumer et al. apply this model to the kinetics of an enzyme [7].

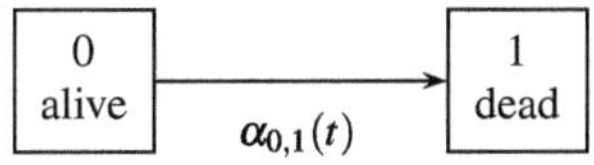

Fig. 16.1 Mortality model

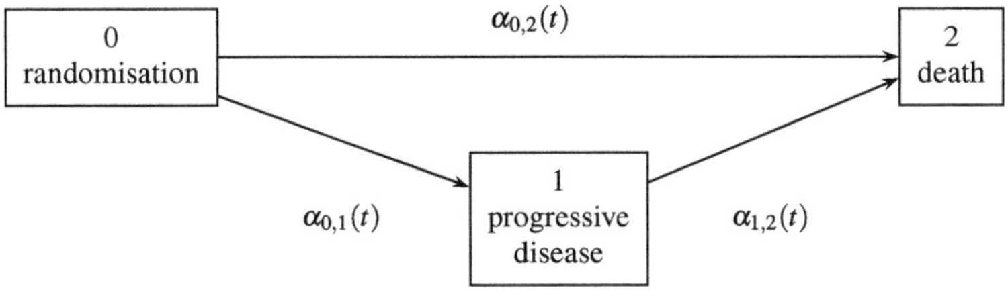

Fig. 16.2 Three-state model

Disability Model

The *disability model* (*three-state model*) is the specific multi-state model regarded in more detail in the subsequent sections. It consists of one absorbing and two transient states. Common applications of this model are state sequences like 'healthy – diseased – death' or as illustrated in Fig. 16.2 'disease – progressive disease (PD) – death'. The first mentioned setting enables inferences on the incidence of the regarded disease as well as on health rate whereas the decision if death rates of healthy subjects and patients differ may be problematic (cf. [25] p. 2). Andersen [2] applied the three-state model to the setting 'illness – comorbity – death'.

Obviously, for $\alpha_{0,1} = 0$ the disability model corresponds to the mortality model illustrated in Fig. 16.1. The transition probabilities introduced in (16.1) are for the three-state model given by (cf. [1,25])

$$p_{0,0}(s,t) = \exp\left\{-\int_s^t \alpha_{0,1}(u) + \alpha_{0,2}(u)\,\mathrm{d}u\right\}, \tag{16.5}$$

$$p_{1,1}(s,t) = \exp\left\{-\int_s^t \alpha_{1,2}(u)\,\mathrm{d}u\right\}, \tag{16.6}$$

$$p_{0,1}(s,t) = \int_s^t p_{0,0}(s,u-)\alpha_{0,1}(u)p_{1,1}(u,t)\,\mathrm{d}u\,, \tag{16.7}$$

$$p_{2,2}(s,t) = 1\,, \tag{16.8}$$

$$p_{1,2}(s,t) = \int_s^t p_{1,1}(s,u-)\alpha_{1,2}(u)\,\mathrm{d}u\,, \tag{16.9}$$

$$p_{0,2}(s,t) = \int_s^t p_{0,0}(s,u-)\underbrace{[\alpha_{0,2}(u) + \alpha_{0,1}(u)p_{1,2}(u,t)]}_{=\alpha^*_{0,2}(u,t)}\,\mathrm{d}u\,. \tag{16.10}$$

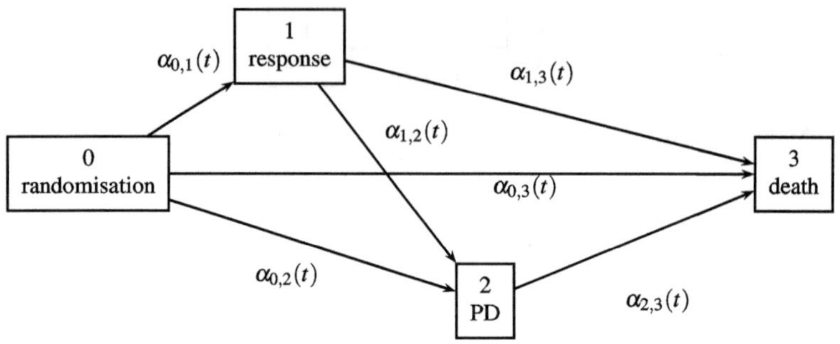

Fig. 16.3 Four-state model

The probability to stay in state 0 from time s until t is equal to the probability that the (random) time point of leaving this state is after t. It is well known that for a random variable T with hazard rate $h(.)$ it holds $\mathbb{P}(T > t) = \exp\left\{-\int_0^t h(u)\mathrm{d}u\right\}$. According to Fig. 16.2, leaving state 0 corresponds to switching into state 1 or 2 and since these are exclusive events, the hazard of the time point leaving state 0 is given by the sum of the single hazard rates. So Eq. (16.5) is verified, (16.6) can be shown analogously. Since $p_{0,1}(s,t)$ corresponds to staying in state 0 until an infinitesimal time unit before u, with u an arbitrary time between s and t, switching to state 1 at u and staying there until t, (16.7) is clear.

The overall transition rate $\alpha^*_{0,2}(u,t)$ corresponds in case of discrete time to the probability $\mathbb{P}(X(t) = 2|X(u) = 0)$ and is for continuous time equal to $\alpha_{0,2}(u) + \alpha_{0,1}(u)\, p_{1,2}(u,t)$.

In some settings it is necessary to consider also the state 'response', leading to a four-state model as shown in Fig. 16.3. Since patients having suffered progressive disease are assumed not being able to respond to the treatment without adjustment of dose/treatment, the state switches between 'progression' and 'response' are only one-directional.

In oncological trials, in particular in the metastatic setting, commonly the treatment is changed after occurrence of progressive disease in order to stop further progression. This new or adopted therapy is called second line treatment or kth line treatment in case of further previous switches. Modelling this proceeding leads to the k-state model (cf. Fig. 16.4).

16.2.2.2 Further Models

The *recurrent events model* consists of k transient states and optionally an absorbing state at the end of the line, whereas the transient ones only can be passed one after another. This model is applied if the event of interest occurs repeatedly, e.g. hospitalization, birth of a child, infections, recurrence of cancer, etc. A broad overview about the analysis of recurrent events is given in the book of Nelson [24], for further reading see also [21] or [8].

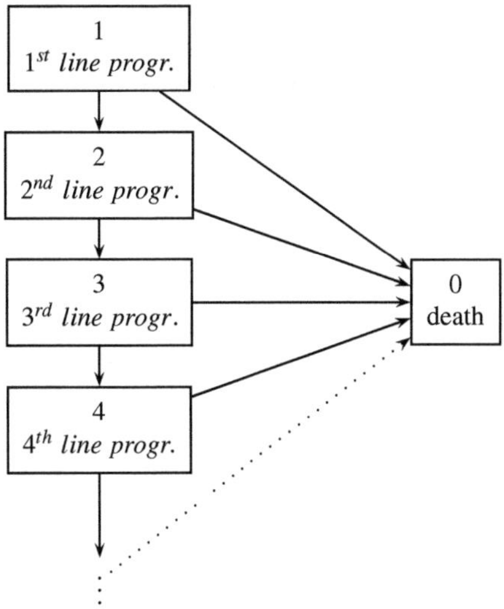

Fig. 16.4 k-state model

Adding further mutually exclusive absorbing states to the mortality model (i.e. death caused by different reasons) is called *competing risks model*. An introduction to the theory of those models is for example given in Beyersmann et al. [6] as well as in [28] and [13]. R. Chappell discusses in his manuscript two different methods for analysing competing risks models [9]. When switching to an absorbing state censors a non-terminal event, we are faced with *semi-competing risks models* which have been studied in [15] or [26]. Some authors (cf. [1], Section 3.6) call those models *bone marrow transplantion* model, since this setting is the common application. The *bivariate model* is used for modelling bivariate failure times, e.g. the survival of twins. For a more detailed description of this see for example [18], Section 5.2.

16.2.3 Recent Research in Multi-state Modelling

In recent research there are numerous applications of multi-stage modelling in the medical context given. Especially for models of chronic diseases this approach is frequently used. A three-state model for cognitive aging and suffering from dementia, with a kind of 'sub-state' (the pre-diagnosis) between 'healthy' and 'ill' and an increased transition rate after this additional state, is given by Dantan et al. [12]. They used a mixed-model approach and regarded non-informative censoring. An informative censoring mechanism is given in the model of Sweeting et al., which is a type of hidden Markov model for the analysis of disease progression in hepatitis C [31]. Lan and Datta compare a semi-Markov five-state model to a Markovian four-state model, both with assumption of log-normal as well as Weibull distributed transition times and an uniformly or rather Weibull distributed censoring mechanism, in the context of measurement of sexual development of juvenile in puberty [22]. A four-state model with Weibull distributed transition rates for survival of dental fillings was developed by Joly et al. [19].

The most prominent area for multi-state modelling is the analysis of survival time and time until non-fatal events in oncology. There are numerous extensions and adjustments of the above basic modelling approaches. Only a few examples will be given. Porta et al. [27] combine a three-state model, including the possibility of disease recurrence, with a competing risk model and apply their dynamic model to patient data on bladder cancer. In some cases, the patient history has an effect on the transition rates and consequently the Markov property is no longer given. Putter and van Houwelingen model this by introduction of frailties (i.e. unobservable random interaction of survival times of different individuals). They apply this in the context of a three-state model, a competing risks model, a recurrent event model as well as a recurrent event model combined with mutually exclusive endpoints to breast cancer patient data [29]. Different kinds of multi-state Markov models with consideration of several progression stages are given in [35] and also applied to breast cancer data.

16.2.4 Questions to Be Solved/Data to Be Collected

Patients in oncological trials will typically receive several lines of treatment because of treatment adjustment after suffering progressive disease. For the sake of simplicity, in the following only a three-state model is investigated, i.e. each of the patients considered receives at most one change of treatment regime after progression. There are two endpoints being of interest in oncological trials, the primary endpoint is progression-free survival (PFS) and the key secondary one is overall survival (OS), both visualized gray-colored in Fig. 16.5. We are primarily

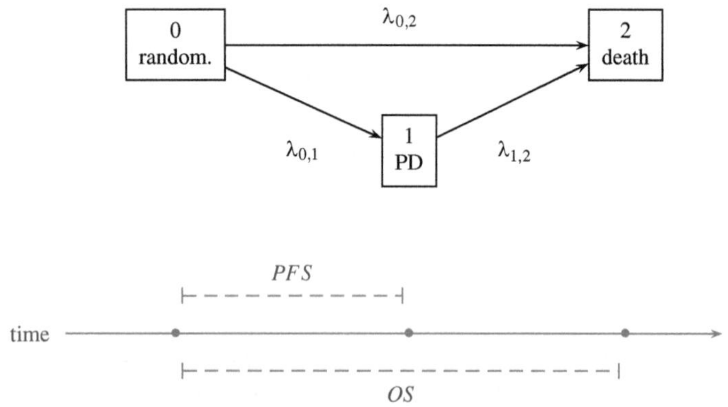

Fig. 16.5 Three-state model with constant transition rate

interested in information on overall survival. Instead of modelling OS via a single random variable, we can also incorporate the information on PFS by use of a three-state model for OS. A careful and precise definition of tumor progression is crucial [16] for accurate determination of PFS. Since there are no standard regulatory criteria, the RECIST criteria [14, 33] for solid tumours or other criteria can be used, e.g. for specific hematologic indications see [10] or [11].

Definition 16.2. The time from randomisation until death from any cause is called *overall survival* (OS).

Commonly, oncological trials are performed as event-driven trials, which means the trial length as well as the analysis time points are related to the occurrence of a specific number of events. So the study duration is a random quantity and the estimation of the time point t^* when the required number of events is observed is in question. At the begin of the study the estimated duration will be calculated and this value will be updated during the course of the trial. Furthermore, the time t^* of occurrence of the landmark event number is also relevant for planning of any interim analysis.

16.3 Statistical Methods

16.3.1 Model Assumptions

In the following, we will concentrate on the three-state model as given in Figs. 16.2 and 16.5.

16.3.1.1 Modelling of PFS and OS

For simplicity reasons, the transition rates (as defined in Eq. (16.3)) are assumed to be constant over time:

$$\begin{aligned} \alpha_{0,1}(t) &= \lambda_{0,1}\,, \\ \alpha_{0,2}(t) &= \lambda_{0,2}\,, \\ \alpha_{1,2}(t) &= \lambda_{1,2}\,, \end{aligned} \tag{16.11}$$

with $\lambda_{i,j} \in \mathbb{R}_+$ for $i = 0, 1,\ j = 1, 2$. From Eq. (16.11) follows that the random time to progression (TTP), i.e. the period between randomization and occurrence of progression, is exponentially distributed with parameter $\lambda_{0,1}$. Furthermore, the random time between progression and death as well as between randomization and dying directly is also exponentially distributed with parameter $\lambda_{1,2}$ and $\lambda_{0,2}$, respectively.

According to the definition of PFS, the PFS time corresponds to the waiting time of the stochastic process in the initial state 0, i.e. the PFS time is given by $T_0 = \min\{t \in \mathbb{R}_+ : X(t) \neq 0\}$. Based on this, the PFS time is exponentially distributed with parameter $\lambda_{0,1} + \lambda_{0,2}$.

In the regarded context, *death* is termed event. Let $f(t)$ and $g(t)$ be the density function of the event time and the lost to follow up time, respectively. The event times as well as the censoring times are assumed to be stochastically independent and identically distributed for all individuals $i = 1, \dots, N$. Because of this, the subscript i may be suppressed for the censoring and event times in order to shorten expressions. The censoring process is assumed to follow an exponential distribution with parameter θ, i.e. $g(t) = \theta\, \mathrm{e}^{-\theta t}$ for $t \in \mathbb{R}_+$. Note that the quantities derived in the following may also be given in case that no censoring is assumed. Without consideration of censoring it is $\theta = 0$.

Since OS is the event of interest, the overall survival time will be denoted by the random variable T. The distribution of T is depending on the present state of the process, so the distribution function of OS is for $t \in \mathbb{R}_+$ given by

$$\begin{aligned} F_{T,C}(t) &= \mathbb{P}(T \le t, C > T) = \mathbb{P}(C > T | T \le t) \cdot \mathbb{P}(T \le t) \\ &= \frac{\lambda_{0,1}\lambda_{1,2}}{(\lambda_{0,1} + \lambda_{0,2} - \lambda_{1,2})(\lambda_{1,2} + \theta)} \left(1 - \mathrm{e}^{-(\lambda_{1,2}+\theta)t}\right) \\ &\quad - \frac{(\lambda_{1,2} - \lambda_{0,2})(\lambda_{0,1} + \lambda_{0,2})}{(\lambda_{0,1} + \lambda_{0,2} - \lambda_{1,2})(\lambda_{0,1} + \lambda_{0,2} + \theta)} \left(1 - \mathrm{e}^{-(\lambda_{0,1}+\lambda_{0,2}+\theta)t}\right). \end{aligned} \tag{16.12}$$

In case that there is no censoring regarded, the previous distribution function simplifies to

$$\begin{aligned} F_T(t) &= \mathbb{P}(T \le t) \\ &= 1 - \frac{\lambda_{0,1}}{\lambda_{0,1} + \lambda_{0,2} - \lambda_{1,2}} \mathrm{e}^{-\lambda_{1,2}t} + \frac{\lambda_{1,2} - \lambda_{0,2}}{\lambda_{0,1} + \lambda_{0,2} - \lambda_{1,2}} \mathrm{e}^{-(\lambda_{0,1}+\lambda_{0,2})t}. \end{aligned} \tag{16.13}$$

A more detailed derivation of the above equations may be found in the appendix and in Fleischer et al. [16]. For determination of corr(*PFS*, *OS*) see Fleischer et al. [16]. Heng et al. applied these results and showed that the PFS time can be used as an intermediate endpoint for OS [17].

For a patient being already progressive, the event time is the waiting time in state 1 and therefore the distribution of overall survival in this case is

$$F_{T,C}(t) = \frac{\lambda_{1,2}}{\lambda_{1,2} + \theta} \left(1 - \mathrm{e}^{-(\lambda_{1,2}+\theta)t}\right). \tag{16.14}$$

16.3.1.2 Modelling the Accrual Process

It is assumed that all patients enrolled during the accrual period will also be randomized, i.e. screening failures are not considered. The following derivations will be done for an one-arm trial, whereas generalizations to multi-arm trials work in an analogous manner (cf. [5]).

There are two different approaches for modelling the accrual process, the common one is a Poisson-process. Especially in case of numerous randomized patients, it is possible to loose restriction on randomized accrual by assumption of a fixed accrual rate $r \in \mathbb{R}_+$ over the whole time period.

At start of the trial, we assume a linear randomization with rate $r \in \mathbb{R}_+$. Therefore, the number of patients randomized until the current calendar time t_c is given by $N(t_c) = r \cdot t_c$. In general, the observed randomization rate at time $t > t_c$ will be different from r. Henceforth, from current time t_c randomization of the remaining $N - N(t_c)$ patients is assumed with constant rate $r(t_c)$, whereas

$$r(t_c) = \begin{cases} 0, & \text{if } N(t_c) \geq N\,, \\ \frac{N - r\cdot t_c}{a(t_c) - t_c}, & \text{else}\,, \end{cases} \tag{16.15}$$

for $t_c \in \mathbb{R}_+$ and $a(t_c)$ denoting the end of the randomization period. With $u \geq t_c$ a future time-point, the density of the randomization rate for the remaining randomization time is

$$r(t_c, u) = \frac{N - r \cdot t_c}{a(t_c) - t_c}\, \mathbb{I}_{(0, a(t_c) - t_c)}(u)\,, \tag{16.16}$$

because of assumption of uniformly accrual in the remaining time interval.

16.3.2 *Prediction of OS Events*

In the following, we will derive a closed formula for the expected number of events at a future time point t, depending on the current time point t_c. Based on this, the expected time point of the landmark number of events can be calculated. Let the number of events (i.e. deaths) observed until a certain time $t \in \mathbb{R}_+$ be given by the random variable $D(t)$. Then, $\mathbb{E}[D(t)|N, \mathfrak{H}_{t_c}]$ is the conditional expectation of the number of events that will be observed by calendar time $t > t_c$, given the data up to current calendar time t_c. The value in question is the predicted calendar time t^* when the required number of events $\hat{d}$ is expected, i.e. $\mathbb{E}[D(t^*)|N, \mathfrak{H}_{t_c}] = \hat{d}$.

The expected number of events is given by

$$\mathbb{E}[D(t)|N, \mathfrak{H}_{t_c}] = d(t_c) + \mathbb{E}[D_R(t)|N, \mathfrak{H}_{t_c}] + \mathbb{E}[D_{NR}(t)|N, \mathfrak{H}_{t_c}]\,, \tag{16.17}$$

with $d(t_c) \in \mathbb{N}_0$ the number of events until the current time t_c, which is not a random variable but rather an observed quantity. $\mathbb{E}[D_R(t)|N, \mathfrak{H}_{t_c}]$ is the number of newly expected events between t_c and t of patients being already randomized at t_c given the data up to t_c and $\mathbb{E}[D_{NR}(t)|N, \mathfrak{H}_{t_c}]$ denotes the analogous quantity for patients randomized between t_c and t.

16.3.2.1 Calculation of $\mathbb{E}[D_R(t)|N, \mathfrak{H}_{t_c}]$

The conditional expectation $\mathbb{E}[D_R(t)|N, \mathfrak{H}_{t_c}]$ of the patients already randomized, alive and on study at time t_c who will have been observed to die by time t, has to be distinguished between patients who have already progressed until time t_c or not. Let $Y_i(t) = 0$ if patient i has not progressed until t and is under observation and at risk for an event at time t and $Y_i(t) = 1$ if the patient has already suffered progressive disease. Therefore, $\mathbb{E}[D_R(t)|N, \mathfrak{H}_{t_c}]$ is

$$\mathbb{E}[D_R(t)|N, \mathfrak{H}_{t_c}] = \mathbb{E}[D_R(t), Y(t_c) = 0|N, \mathfrak{H}_{t_c}] + \mathbb{E}[D_R(t), Y(t_c) = 1|N, \mathfrak{H}_{t_c}], \tag{16.18}$$

with $\mathbb{E}[D_R(t), Y(t_c) = 0|N, \mathfrak{H}_{t_c}]$ the expected number of events of patients not progressive until t_c and $\mathbb{E}[D_R(t), Y(t_c) = 1|N, \mathfrak{H}_{t_c}]$ the analogous quantity of patients already progressive at time t_c.

Let E_i, $i = 1, \ldots, N$ denote the random variable for the randomization time of the ith patient and let ϵ_i denote the observed randomization time of the ith patient. It is assumed that the randomization time E_i of every individual is stochastically independent from the associated event and censoring times. The randomization time of each individual is measured from $t = 0$, the calendar date when the first patient is randomized. The individual survival times (overall survival) and censoring times are measured from the calendar date of a patients randomization. The probability that the ith patient is at risk between $t_c - \epsilon_i$ and $t - \epsilon_i$, i.e. the probability that the ith patient has the event time within the time interval $(t_c - \epsilon_i, t - \epsilon_i)$ and does not get censored before the event, given that the patient survived uncensored at $t_c - \epsilon_i$, is denoted by $\mathbf{P}_i^{f,g}(t_c, t)$. It is

$$\begin{aligned}
\mathbf{P}_i^{f,g}(t_c, t) &= \mathbb{P}(T < C, T \in (t_c - \epsilon_i, t - \epsilon_i)|T > t_c - \epsilon_i, C > t_c - \epsilon_i) \\
&= \frac{\int_{t_c-\epsilon_i}^{t-\epsilon_i} f(u) \overbrace{\mathbb{P}(T < C|T = u)}^{=\mathbb{P}(C>u)} \, du}{(1 - F(t_c - \epsilon_i))(1 - G(t_c - \epsilon_i))} \\
&= \frac{F(t - \epsilon_i) - F(t_c - \epsilon_i) - \int_{t_c-\epsilon_i}^{t-\epsilon_i} f(u)G(u) \, du}{(1 - F(t_c - \epsilon_i))(1 - G(t_c - \epsilon_i))}.
\end{aligned}$$

The dependency of $\mathbf{P}_i^{f,g}(t_c, t)$ on the distribution of the event and censoring times is symbolized by the indexes f and g, the corresponding density functions. By the assumption of memoryless distributions for the event and the censoring time (i.e. for $F(.)$ and $G(.)$) it holds

$$\begin{aligned}
\mathbf{P}_i^{f,g}(t_c, t) &= \mathbb{P}(T < C, T \in (0, t - t_c)) \\
&= \int_0^{t-t_c} f(u) \underbrace{\mathbb{P}(C > u)}_{=1-G(u)} \, \mathrm{d}u \\
&= F(t - t_c) - \int_0^{t-t_c} f(u) G(u) \, \mathrm{d}u \, ,
\end{aligned} \tag{16.19}$$

whereas the first transformation uses the definition of memoryless distributions. Obviously, in this case the risk probability of patient i is independent of the individual randomization time ϵ_i, $i = 1, \ldots, N$. In the following, we will provide that event times and censoring times follow memoryless distributions.

For the expectation $\mathbb{E}[D_R(t)|N, \mathfrak{H}_{t_c}]$ we get

$$\mathbb{E}[D_R(t)|N, \mathfrak{H}_{t_c}] = \sum_{i=1}^{N(t_c)} (1 - Y_i(t_c)) \cdot \mathbf{P}_i^{f_0,g}(t_c, t) + \sum_{i=1}^{N(t_c)} Y_i(t_c) \cdot \mathbf{P}_i^{f_1,g}(t_c, t) \, .$$

Since $\mathbf{P}_i^{f,g}(t_c, t)$ is independent of i (cf. (16.19)), we obtain

$$\begin{aligned}
\mathbb{E}[D_R(t)|N, \mathfrak{H}_{t_c}] &= N_0(t_c) \cdot \mathbf{P}_i^{f_0,g}(t_c, t) \cdot \mathbb{I}_{Y(t_c)=0} \\
&\quad + (N(t_c) - N_0(t_c)) \cdot \mathbf{P}_i^{f_1,g}(t_c, t) \cdot \mathbb{I}_{Y(t_c)=1} \, ,
\end{aligned} \tag{16.20}$$

with $N_0(t_c)$ the number of patients not yet progressive and $N(t_c) - N_0(t_c)$ the number of randomized patients suffering progression until t_c. In the above equations, the index of the density f symbolises the corresponding distribution function of the event time, i.e. $f_0(t)$ denotes the density function of OS of a randomized patient (cf. (16.12)) and $f_1(t)$ is the density of OS time for a patient already progressive (cf. (16.14)).

The probability of dying between $t_c - \epsilon_i$ and $t - \epsilon_i$ given that PFS $> t_c - \epsilon_i$ equals the probability of dying before $t - t_c$, irrespective of the randomization time. By plugging $\mathbf{P}_i^{f_0,g}(t_c, t)$ into formula (16.20) we get

$$\begin{aligned}
\mathbb{E}[D_R(t)|Y(t_c) = 0, N, \mathfrak{H}_{t_c}] &= \frac{N_0(t_c)}{\lambda_{0,1} + \lambda_{0,2} - \lambda_{1,2}} \left[\frac{\lambda_{0,1}\lambda_{1,2}}{\lambda_{1,2} + \theta} \left(1 - \mathrm{e}^{-(\lambda_{1,2}+\theta)(t-t_c)}\right) \right. \\
&\quad \left. - \frac{(\lambda_{1,2} - \lambda_{0,2})(\lambda_{0,1} + \lambda_{0,2})}{\lambda_{0,1} + \lambda_{0,2} + \theta} \left(1 - \mathrm{e}^{-(\lambda_{0,1}+\lambda_{0,2}+\theta)(t-t_c)}\right) \right] .
\end{aligned} \tag{16.21}$$

If the patient has already progressed his further survival follows the distribution given in (16.14) and therefore

$$\mathbb{E}[D_R(t)|Y(t_c) = 1, N, \mathfrak{H}_{t_c}] = (N(t_c) - N_0(t_c))\,\frac{\lambda_{1,2}[1 - \mathrm{e}^{-(\lambda_{1,2}+\theta)(t-t_c)}]}{\lambda_{1,2} + \theta}\,. \tag{16.22}$$

Altogether, the expected number of events of the patients already randomized is given by

$$\begin{aligned}\mathbb{E}[D_R(t)|N, \mathfrak{H}_{t_c}] = {} & \frac{N_0(t_c)}{\lambda_{0,1} + \lambda_{0,2} - \lambda_{1,2}}\left[\frac{\lambda_{0,1}\lambda_{1,2}}{\lambda_{1,2}+\theta}\left(1 - \mathrm{e}^{-(\lambda_{1,2}+\theta)(t-t_c)}\right)\right.\\ & \left. - \frac{(\lambda_{1,2} - \lambda_{0,2})(\lambda_{0,1} + \lambda_{0,2})}{\lambda_{0,1} + \lambda_{0,2} + \theta}\left(1 - \mathrm{e}^{-(\lambda_{0,1}+\lambda_{0,2}+\theta)(t-t_c)}\right)\right]\\ & + (N(t_c) - N_0(t_c))\,\frac{\lambda_{1,2}[1 - \mathrm{e}^{-(\lambda_{1,2}+\theta)(t-t_c)}]}{\lambda_{1,2} + \theta}\,.\end{aligned} \tag{16.23}$$

So the expected number of events in the subset of patients being already randomized, depends on the distribution parameters of the event times as well as on the number of patients suffering progression until t_c.

16.3.2.2 Calculation of $\mathbb{E}[D_{NR}(t)|N, \mathfrak{H}_{t_c}]$

For determination of the expected number of events of patients not yet randomized, it has to be distinguished between three different scenarios.

Scenario 1 The randomization is finished, t_c is after end of randomization period $a(t_c)$. Thus, no more patients will be recruited after t_c and for $0 \le a(t_c) \le t_c < t$ it is $\mathbb{E}[D_{NR}(t)|N, \mathfrak{H}_{t_c}] = 0$ (Fig. 16.6).

Scenario 2 The randomization is not yet finished, t_c is before end of randomization and the planned time of analysis t is after $a(t_c)$.The expected number of events for $0 \le t_c < a(t_c) \le t$ is

$$\begin{aligned}\mathbb{E}[D_{NR}(t)|N, \mathfrak{H}_{t_c}] &= \int_0^{a(t_c)-t_c} r(t_c, u)\mathbb{P}(T < C, T \in (0, t - t_c - u))\,\mathrm{d}u\\ &= \int_0^{a(t_c)-t_c} r(t_c, u)\left[\int_0^{t-t_c-u} f(s)(1 - G(s))\,\mathrm{d}s\right]\mathrm{d}u\,,\end{aligned} \tag{16.24}$$

with $r(t_c, u) = \frac{N - r \cdot t_c}{a(t_c) - t_c}$, $G(s) = \mathrm{e}^{-\theta s}$ and $f(s)$ the density function of OS which may be derived from (16.12) (Fig. 16.7).

Fig. 16.6 Scenario 1

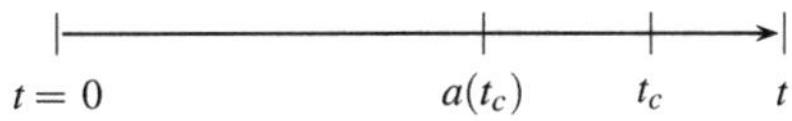

Fig. 16.7 Scenario 2

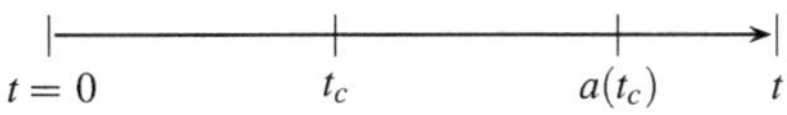

Fig. 16.8 Scenario 3

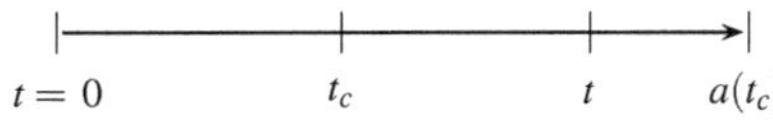

Scenario 3 The randomization is not yet finished. The planned time for interim analysis t is after t_c but before end of randomization (Fig. 16.8).

$$\mathbb{E}[D_{NR}(t)|N,\mathfrak{H}_{t_c}] = \int_0^{t-t_c} r(t_c,t)\left[\int_0^{t-t_c-u} f(s)(1-G(s))\,\mathrm{d}s\right]\mathrm{d}u,$$

with the analogous variables as given in scenario 2.

With regard on the definition of randomization rate (cf. (16.16)), assumption of linear randomization and the above, $\mathbb{E}[D_{NR}(t)|N,\mathfrak{H}_{t_c}]$ may be given in closed form:

$$\mathbb{E}[D_{NR}(t)|N,\mathfrak{H}_{t_c}] = \begin{cases} 0, & \text{if } a(t_c)\le t_c < t\,, \\ \frac{N-N(t_c)}{a(t_c)-t_c}\int_0^{a(t_c)-t_c} \mathbf{P}_i^{f,g}(u,t-t_c)\,\mathrm{d}u, & \text{if } t_c < a(t_c)\le t\,, \\ \frac{N-N(t_c)}{a(t_c)-t_c}\int_0^{t-t_c} \mathbf{P}_i^{f,g}(u,t-t_c)\,\mathrm{d}u, & \text{if } t_c < t \le a(t_c)\,. \end{cases}$$

Insertion of the distribution function of the event and censoring times gives finally

$$\mathbb{E}[D_{NR}(t)|N,\mathfrak{H}_{t_c}] =$$

$$= \begin{cases} 0, & \text{if } 0\le a(t_c)\le t_c < t\,, \\ (N-N(t_c))\left[\frac{\lambda_{0,1}\lambda_{1,2}}{(\lambda_{0,1}+\lambda_{0,2}-\lambda_{1,2})(\lambda_{1,2}+\theta)}\left(1-\frac{e^{-(\lambda_{1,2}+\theta)(t-a(t_c))}-e^{-(\lambda_{1,2}+\theta)(t-t_c)}}{(\lambda_{1,2}+\theta)(a(t_c)-t_c)}\right)-\right. \\ \left.-\frac{(\lambda_{1,2}-\lambda_{0,2})(\lambda_{0,1}+\lambda_{0,2})}{(\lambda_{0,1}+\lambda_{0,2}-\lambda_{1,2})(\lambda_{0,1}+\lambda_{0,2}+\theta)}\left(1-\frac{e^{-(\lambda_{0,1}+\lambda_{0,2}+\theta)(t-a(t_c))}-e^{-(\lambda_{0,1}+\lambda_{0,2}+\theta)(t-t_c)}}{(\lambda_{0,1}+\lambda_{0,2}+\theta)(a(t_c)-t_c)}\right)\right], \\ & \text{if } 0\le t_c < a(t_c)\le t\,, \\ \frac{N-N(t_c)}{a(t_c)-t_c}\left[\frac{\lambda_{0,1}\lambda_{1,2}}{(\lambda_{0,1}+\lambda_{0,2}-\lambda_{1,2})(\lambda_{1,2}+\theta)}\left(t-t_c-\frac{1-e^{-(\lambda_{1,2}+\theta)(t-t_c)}}{\lambda_{1,2}+\theta}\right)-\right. \\ \left.-\frac{(\lambda_{1,2}-\lambda_{0,2})(\lambda_{0,1}+\lambda_{0,2})}{(\lambda_{0,1}+\lambda_{0,2}-\lambda_{1,2})(\lambda_{0,1}+\lambda_{0,2}+\theta)}\left(t-t_c-\frac{1-e^{-(\lambda_{0,1}+\lambda_{0,2}+\theta)(t-t_c)}}{\lambda_{0,1}+\lambda_{0,2}+\theta}\right)\right], \\ & \text{if } 0\le t_c < t \le a(t_c)\,. \end{cases} \tag{16.25}$$

patients is needed. On the other hand, 380 events will occur at observation time of 21.5 months, which saves 1.5 months of study duration.

16.3.5.2 Example 2

Our second example is based on data of a second line non-small cell lung cancer trial with 1,000 patients randomized in total. The last data monitoring committee DMC meeting has to occur after the 800th death event. A non-uniform accrual process is observed for this trial. Eighteen months after start of randomization, the event monitoring for this study is calculated by use of the exponential model as well as the three-state model. Based on the time from randomization, the time from randomization until the 800th death event is estimated for both models. We get stable estimates for both models after about 150 randomized patients and about 40 PFS events and 15 death events observed. The exponential model gives an estimation of 34 months and the three-state model an estimation of 29 months. Since the target number of 800 death events has not been observed so far, we run a simulation using the assumptions of the previous example to investigate until when the both models will estimate the time to the 380th death event and what the expected difference between the estimations is. The target of the 380th death event occurred at 21.5 months (please compare Fig. 16.9).

As seen in Fig. 16.9, 240 patients were randomized, 37 death events, and 80 PFS events were occurred after 6 months from randomization. The exponential model gave an estimation of 25 months, and the three-state model showed an estimation of 21.6 months (please compare Fig. 16.10). This is again a time difference of 4 months. Half of the required death events (190 OS events) were occurred after 14 months from randomization. Most of all patients were randomized and about half of them had a PFS event. Then the exponential model gave a more exact estimation of about 23.2 months, which is still 1.5 months more than the three-state model.

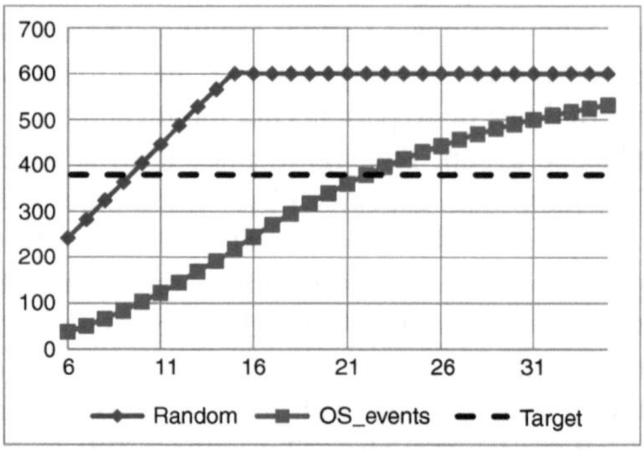

Fig. 16.9 The observed cumulative number of events over the time from randomization. (Results from the simulation example.) The number of patients randomised in (*diamonds - upper line*), the observed number of OS events (*squares - lower line*)

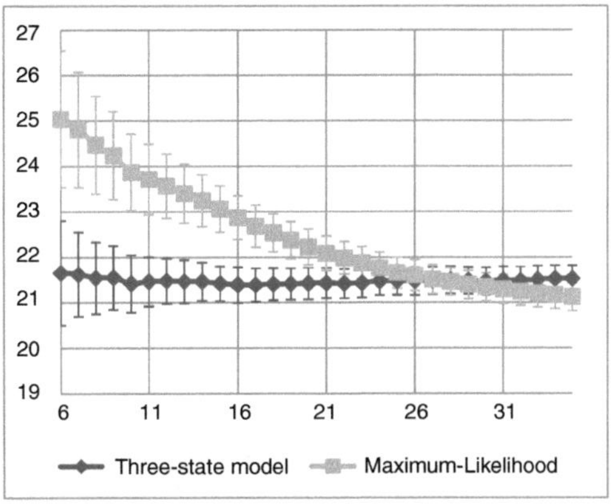

Fig. 16.10 Mean and standard error of the estimations of the time to 380th OS event from all simulations. The *squares* and the *diamonds* represent the 3-state model and the exponential model, respectively

16.3.5.3 Software Available

There is several software available. Multi-state models need specialised software, most of which are written in FORTRAN, R or SAS. The library survival available as part of S-plus and R statistical packages can be used to implement these methods. An R package `msm` was developed in 2002. In addition, an user-friendly R library, `tdc.msm`, was generated for the analysis of multi-state survival data. Technical description of this is provided in the independent article Meira-Machado et al. [25].

Appendix

Derivation of $F_T(.)$ for the disability model by assumption of exponentially distributed state times:

$$\begin{aligned}
F_T(t) &= \mathbb{P}(T \leq t)\\
&= \int_0^t p_{0,0}(0,u)\alpha^*_{0,2}(u,t)\,\mathrm{d}u\\
&= \int_0^t \exp\left\{-\int_0^u (\alpha_{0,1}(v)+\alpha_{0,2}(v))\,\mathrm{d}v\right\}\left[\lambda_{0,2}+\lambda_{0,1}\left(1-\mathrm{e}^{-\lambda_{1,2}(t-u)}\right)\right]\mathrm{d}u\\
&= \int_0^t \exp\left\{-\int_0^u (\lambda_{0,1}+\lambda_{0,2})\,\mathrm{d}v\right\}\left[\lambda_{0,2}+\lambda_{0,1}\left(1-\mathrm{e}^{-\lambda_{1,2}(t-u)}\right)\right]\mathrm{d}u\\
&= \int_0^t \mathrm{e}^{-(\lambda_{0,1}+\lambda_{0,2})u}\left[\lambda_{0,2}+\lambda_{0,1}\left(1-\mathrm{e}^{-\lambda_{1,2}(t-u)}\right)\right]\mathrm{d}u
\end{aligned}$$

$$= \lambda_{0,2} \int_0^t e^{-(\lambda_{0,1}+\lambda_{0,2})u}\, du + \lambda_{0,1} \int_0^t e^{-(\lambda_{0,1}+\lambda_{0,2})u}\, du$$
$$- \lambda_{0,1} e^{-\lambda_{1,2}t} \int_0^t e^{-(\lambda_{0,1}+\lambda_{0,2}-\lambda_{1,2})u}\, du$$
$$= \frac{-\lambda_{0,2}}{\lambda_{0,1}+\lambda_{0,2}} \left(e^{-(\lambda_{0,1}+\lambda_{0,2})t} - 1\right) + \frac{-\lambda_{0,1}}{\lambda_{0,1}+\lambda_{0,2}} \left(e^{-(\lambda_{0,1}+\lambda_{0,2})t} - 1\right)$$
$$+ \frac{\lambda_{0,1} e^{-\lambda_{1,2}t}}{\lambda_{0,1}+\lambda_{0,2}-\lambda_{1,2}} \left(e^{-(\lambda_{0,1}+\lambda_{0,2}-\lambda_{1,2})t} - 1\right)$$
$$= \underbrace{\frac{\lambda_{0,2}}{\lambda_{0,1}+\lambda_{0,2}} + \frac{\lambda_{0,1}}{\lambda_{0,1}+\lambda_{0,2}}}_{=1} - e^{-(\lambda_{0,1}+\lambda_{0,2})t} \underbrace{\left(\frac{\lambda_{0,2}+\lambda_{0,1}}{\lambda_{0,1}+\lambda_{0,2}}\right)}_{=1}$$
$$- \frac{\lambda_{0,1}}{\lambda_{0,1}+\lambda_{0,2}-\lambda_{1,2}} e^{-\lambda_{1,2}t} + \frac{\lambda_{0,1}}{\lambda_{0,1}+\lambda_{0,2}-\lambda_{1,2}} e^{-(\lambda_{0,1}+\lambda_{0,2})t}$$
$$= 1 + e^{-(\lambda_{0,1}+\lambda_{0,2})t} \left(\frac{\lambda_{0,1}-\lambda_{0,1}-\lambda_{0,2}+\lambda_{1,2}}{\lambda_{0,1}+\lambda_{0,2}-\lambda_{1,2}}\right) - \frac{\lambda_{0,1}}{\lambda_{0,1}+\lambda_{0,2}-\lambda_{1,2}} e^{-\lambda_{1,2}t}$$
$$= 1 + \frac{\lambda_{1,2}-\lambda_{0,2}}{\lambda_{0,1}+\lambda_{0,2}-\lambda_{1,2}} e^{-(\lambda_{0,1}+\lambda_{0,2})t} - \frac{\lambda_{0,1}}{\lambda_{0,1}+\lambda_{0,2}-\lambda_{1,2}} e^{-\lambda_{1,2}t} .$$

A slightly different derivation of this formulae is given in Fleischer et al. [16].

References

1. Andersen, P. K., Keiding N.: Multi-state models for event history analysis. *Statistical Methods in Medical Research* **11**, 91–115 (2002)
2. Andersen, P. K.: Multistate Models in Survival Analysis: A Study of Nephropathy and Mortality in Diabetes. *Statistics in Medicine* **7**, 661–670 (1988)
3. Andersen, P. K., Borgan, Ø., Gill, R. D., Keiding, N.: *Statistical models based on counting processes*. Springer, New York (1993)
4. Andersen, P. K., Esbjerg S., Sorensen T. I. A.: Multistate models for bleeding episodes and mortality in liver cirrhosis. *Statistics in Medicine* **19**, 587–599 (2000)
5. Bagiella, E., Heitjan, D. F.: Predicting analysis times in randomized clinical trials. *Statistics in Medicine* **20**, 2055–2063 (2001)
6. Beyersmann, J., Schumacher, M., Allignol, A.: *Competing Risks and Multistate Models with R*. Springer (2012)
7. Birnbaumer, L., Bearer, C. F., Iyengar, R.: A Two-state Model of an Enzyme with an Allosteric Regulatory Site Capable of Metabolizing the Regulatory Ligand. *The Journal of Biological Chemistry* **255**(8), 3552–3557 (1980)
8. Box-Steffensmeier, J. M., De Boef, S.: Repeated events survival models: the conditional frailty model. *Statistics in Medicine* **25**(20), 3518–3533 (2006)
9. Chappell, R.: Competing risk analyses: how are they different and why should you care?. *Clinical Cancer Research* **18**(8), 2127–2129 (2012)

10. Cheson, B. D., Bennett, J. M., Kopecky, K. J., Büchner, T., Willman, C. L., Estey, E. H., Schiffer, C. A., Doehner, H., Tallman, M. S., Lister, A., Lo-Coco, F., Willemze, R., Biondi, A., Hiddemann, W., Larson, R. A., Löwenberg, B., Sanz, M. A., Head, D. R., Ohno, R., Bloomfield, C. D.: Revised Recommendations of the International Working Group for Diagnosis, Standardization of Response Criteria, Treatment Outcomes, and Reporting Standards for Therapeutic Trials in Acute Myeloid Leukemia. *Journal of Clinical Oncology* **21**(24), 4642–4649 (2003)
11. Cheson, B. D., Pfistner, B., Juweid, M. E., Gascoyne, R. D., Specht, L., Horning, S. J., Coiffier, B., Fisher, R. I., Hagenbeek, A., Zucca, E., Rosen, S. T., Stroobants, S., Lister, T. A., Hoppe, R. T., Dreyling, M., Tobinai, K., Vose, J. M., Connors, J. M., Federico, M., Diehl, V.: Revised Response Criteria for Malignant Lymphoma. *Journal of Clinical Oncology* **25**(5), 579–586 (2007)
12. Dantan, E., Poly, P., Dartigues, J.-F., Jacqmin-Gadda, H.: Joint model with latent state for longitudinal and multistate data. *Biostatistics* **12**(4), 723–736 (2011)
13. Deshpande, J. V., Purohit, S. G.: Survival, hazard and competing risks. *Current Science* **80**(9), 1191–1202 (2001)
14. Eisenhauer, E. A., Therasse, P., Bogaerts, J., Schwartz, L. H., Sargent, D., Ford, R., Dancey, J., Arbuck, S., Gwyther, S., Mooney, M., Rubinstein, L., Shankar, L., Dodd, L., Kaplan, R., Lacombe, D., Verweij, J.: New response evaluation criteria in solid tumorus: revised RECIST guidelines (version 1.1). *European Journal of Cancer* **45**, 228–247 (2009)
15. Fine, J. P., Jiang, H., Chappell, R.: On semi-competing risks. *Biometrika* **88**, 907–919 (2001)
16. Fleischer, F., Gaschler-Markefski, B., Bluhmki, E.: A statistical model for the dependence between progression-free survival and overall survival. *Statistics in Medicine* **28**, 2669–2686 (2009)
17. Heng, D. Y. C., Xie, W., Bjarnason, G. A., Vaishampayan, U., Tan, M.-H., Knox, J., Donskov, F., Wood, L., Kollmannsberger, C., Rini, B. I., Choueiri, T. K.: Progression-Free Survival as a Predictor of Overall Survival in Metastatic Renal Cell Carcinoma Treated With Contemporary Targeted Therapy. *Cancer* **117**(12), 2637–2642 (2011)
18. Hougaard, P.: Multi-state Models: A Review. *Lifetime Data Analysis* **5**, 239–264 (1999)
19. Joly, P., Gerds, T. A., Qvist, V., Commenges, D., Keiding, N.: Estimating survival of dental fillings on the basis of interval-censored data and multi-state models. *Statistics in Medicine* **31**, 1139–1149 (2012)
20. Kalbfleisch, J. D., Prentice, R. L.: *The Statistical Analysis of Failure Time Data*. John Wiley & Sons, Inc. (1980)
21. Kvist, K., Andersen, P. K., Angst, J., Kessing, L. V.: Repeated events and total time on test. *Statistics in Medicine* **27**, 3817–3822 (2008)
22. Lan, L., Datta, S.: Non-parametric estimation of state occupation, entry and exit times with multistate current status. *Statistical Methods in Medical Research* **19**, 147–165 (2010)
23. Lawless, J. F.: *Statistical Models and Methods for Lifetime Data*. John Wiley & Sons, Inc. (1982)
24. Nelson, W. B., Prentice, R. L.: *Recurrent Events Data Analysis for Product Repairs, Disease Recurrences, and Other Applications*. ASA-SIAM Series on Statistics and Applied Probability, Philadelphia. (2003)
25. Meira-Machado, L., de Uña-Álvarez, J., Carso-Suárez, C., Andersen, P. K.: Multi-state models for the analysis of time-to-event data. *Statistical Methods in Medical Research* **18**(2), 195–222 (2009)
26. Peng, L., Fin, J. P.: Regression modeling of semicompeting risks data. *Biometrics* **63**, 96–108 (2006)
27. Porta, N., Calle, M. L., Malats, N., Gómez, G.: A dynamic model for the risk of bladder cancer progression. *Statistics in Medicine* **31**, 287–300 (2012)
28. Prentice, R. L., Kalbfleisch, J. D., Peterson, A. V., Flournoy, N., Farewell, V. T., Breslow, N. E.: The Analysis of Failure Times in the Presence of Competing Risks. *Biometrics* **34**, 541–554 (1978)

29. Putter, H., van Houwelingen, H. C.: Frailties in multi-state models: Are they identifiable? Do we need them?. *Statistical Methods in Medical Research*. online version **0**, 1–18 (2011)
30. Rodríguez-Girondo, M., de Uña-Álvarez, J.: Testing Markovianity in the three-state progressive model via future-past association. *Biometrical Journal* **54**(2), 163–180 (2012)
31. Sweeting, M. J., Farewell, V. T., De Angelis, D.: Multi-state Markov models for disease progression in the presence of informative examination times: An application to hepatitis C. *Statistics in Medicine* **29**, 1161–1174 (2010)
32. Schoenfeld, D. A.: Sample-Size Formula for the Proportional-Hazards Regression Model. *Biometrics* **39**(2), 499–503 (1983)
33. Therasse, P., Arbuck, S. G., Eisenhauer, E. A., Wanders, J., Kaplan, R. S., Rubinstein, L., Verweij, J., Van Glabbeke, M., van Oosterom, A. T., Christian, M. C., Gwyther, S. G.: New Guidelines to Evaluate the Response to Treatment in Solid Tumors. *Journal of the National Cancer Institute* **92**(3), 205–216 (2000)
34. Tunes-da-Silva, G., Pedroso-de-Lima, A. C., Sen, P. K.: A semi-Markov multistate model for estimation of the mean quality-adjusted survival for non-progressive processes. *Lifetime Data Analysis* **15**, 216–240 (2009)
35. Uhry, Z., Hédelin, G., Colonna, M., Asselain, B., Arveux, P., Rogel, A., Exbrayat, C., Guldenfels, C., Courtial, I., Soler-Michel, P., Molinié, F., Eilstein, D., Duffy, S. W.: Multi-state Markov models in cancer screening evaluation: a brief review and case study. *Statistical Methods in Medical Research* **19**, 463–486 (2010)

Chapter 17
Review of Designs for Accommodating Patients' or Physicians' Preferences in Randomized Controlled Trials

Afisi S. Ismaila and Stephen D. Walter

Abstract The randomized controlled trial (RCT) is regarded as the principal way to collect scientific data on the efficacy of health interventions. Despite the advantages of RCT design in reducing extraneous variation that may confound interpretation of intervention results, the design may not be suitable for interventions in which patients are likely to have a strong preference for a particular treatment. Some designs incorporating patients or physician preferences by allowing at least a subgroup of them to choose their treatment have been proposed. In this chapter, we review various randomized control trials designs for accommodating participants' and professionals' preferences. Specifically, we discuss the advantages, limitations, applicability, ethical issues and statistical issues of each design. We also discuss the estimation of treatment effect (a measure of the extent to which treatment difference is attributable to treatments); selection effect (a measure of the extent to which treatment response is influenced by self-selection of treatment by patients); and preference effect (a measure of the extent to which treatment difference is caused by an interaction between the patient's choice of treatment and the treatment actually received).

17.1 Introduction

The randomized controlled trial (RCT), defined as an experiment in which the treatments under investigation are allocated by a chance (or random) mechanism, is regarded as the principal way to collect scientific data on the efficacy of health interventions [38]. With proper concealment of allocation and blinding of trial

A.S. Ismaila (✉)
Department of Clinical Epidemiology and Biostatistics, McMaster University, Hamilton, ON, Canada
e-mail: ismailas@mcmaster.ca

S.D. Walter
Department of Medicine, McMaster University, Hamilton, ON, Canada
e-mail: walter@mcmaster.ca

K. van Montfort et al. (eds.), *Developments in Statistical Evaluation of Clinical Trials*, DOI 10.1007/978-3-642-55345-5_17

participants, one of the benefits of random assignment of therapies is that neither the patient nor the outcome assessors knows the assigned treatment before the patient is formally entered in the trial [38]. Furthermore, with sufficient sample sizes, RCTs are capable of providing valid estimates of treatment benefits and controlling extraneous variation [21] RCT designs are widely used with the intention of producing comparable groups of patients who differ only in terms of their exposure to the intervention under study [4, 7, 9].

Despite the advantages of RCT design in reducing extraneous variation that may confound interpretation of intervention results, the design may not be suitable for interventions in which patients are likely to have a strong preference for a particular treatment [7, 8, 12, 34]. Patients' motivation to comply with treatment protocol is likely to be influenced by any preference before treatment began [8]. The potential for a preference effect may be high in RCTs of skilled-based interventions like health education, psychotherapy and surgery. Millat et al. [27] evaluated the impact of RCT on decision-making and therapeutic policies among general and gastrointestinal surgeons in France. They concluded that surgeons are rarely in a state of "equipoise" about surgical interventions. Of the 152 surgeons sampled, 63 % rely on personal beliefs to assess the effectiveness of surgical procedures [27]. Since surgeons are rarely randomized to interventions in conventional RCTs, the likelihood of producing comparable groups with similar surgeon preferences is limited.

Some designs incorporating patients or physician preferences by allowing at least a subgroup of them to choose their treatment have been proposed and reviewed [7, 27]. In general, random allocation of patients to a treatment they do not want may reduce adherence to protocol, increase the dropout rate, restrict generalization of the findings and thus reduce the external validity of the study [5, 6, 17]. Therefore, with random allocation there will always be a risk that the groups will not be matched for motivational factors [8].

The objective of this chapter is to review various randomized control trials designs for accommodating participants' and professionals' preferences. Specifically, we will discuss the advantages, limitations, applicability, ethical issues and statistical issues of each design.

17.2 Background

Suppose an investigator is interested in knowing the relative benefit of two interventions (A or B) in a clinical trial, where A is the experimental treatment and B is the control treatment. The control treatment could be the best standard treatment or a placebo. Let N_E denote the number of eligible patients who are potential recipient of treatment A or B in the clinical trial. Ethically, every study must begin with informed consent.

17.2.1 *Informed Consent*

Informed consent lies at the heart of ethical research involving human subjects [10]. The giving of informed consent is a prerequisite to participation in most RCTs. The informed consent procedure refers to the dialogue, information sharing and general process through which prospective patients choose to participate in research [10]. This procedure requires the study investigators or physicians to inform the patient, in his or her own language, about all risks and benefits associated with the trial, the alternative treatments available and the patient's right to withdraw at any time [10, 38]. The patients should also be informed about the procedure used for treatment assignment [10, 38].

The conventional informed consent procedure is to fully inform the patients prior to treatment allocation and then seek their consent to randomization [22]. Hence, the process of informed consent is likely to generate preferences in patients, even if none existed before [22]. Some researchers have suggested that these patient preferences should be explicitly incorporated into the designs of clinical trial [22]. Two elements are critical to the informed consent process. The first is its timing with respect to treatment allocation; and the second is options it presents to the patients [22]. Schellings et al. [32] distinguished and ranked three types of informed consents: single-consent, incomplete-double-consent and complete-double-consent (or conventional informed consent). In the single-consent, only those in the experimental arm learn about their assigned treatment. In the incomplete-double-consent, all the patients learn about their assigned treatment. In the complete-double-consent, all patients learn about all available interventions in the study. We will discuss some of this later in Sect. 17.3.

17.2.2 *Search Strategy*

We searched Medline, Embase, PsyCINFO, CINAHL, and AMED databases for articles published between 1950 and June 2010 on alternative parallel designs to the conventional RCT for accommodating patients' or physicians' preferences. The search terms include random allocation, clinical trials, preference(s), self selection, choice behavior and patient participation. In addition, we reviewed the references of relevant review articles [7, 26].

17.3 Clinical Trial Designs

In this section, we review the conventional RCT and some of the alternative designs for accommodating patients' or physicians' preferences in randomized controlled trials. We will also discuss the estimation of treatment effect (a measure of the extent to which treatment difference is attributable to treatments); selection effect (a measure of the extent to which treatment response is influenced by self-selection of treatment by patients); and preference effect (a measure of the extent to which treatment difference is caused by an interaction between the patient's choice of treatment and the treatment actually received) [31].

17.3.1 The Conventional Randomized Controlled Trial

The conventional RCT design is illustrated in Fig. 17.1. In this design, all the N_E eligible patients learn about treatment A and B before being asked to consent to random allocation to either treatment. This is described as complete-double-consent [32]. Only those patients who agreed to treatment allocation, say N_R in number (where $N_E \geq N_R$), are enrolled as study participants and randomly assigned to treatment A or B. Eligible patients who do not consent to randomization (say $N_E - N_R$ in number) are not enrolled into the trial. The proportion of N_E patients willing to participate in the conventional RCT is $\pi = N_R/N_E$. For some therapeutic areas π may be as low as 10 % if treatments A and B under evaluation are qualitatively so different (e.g. surgery versus medical intervention) that the proportion of patients having clear intervention preferences are high [29]. Since the N_R patients are usually not a random sample from the eligible N_E patients, the

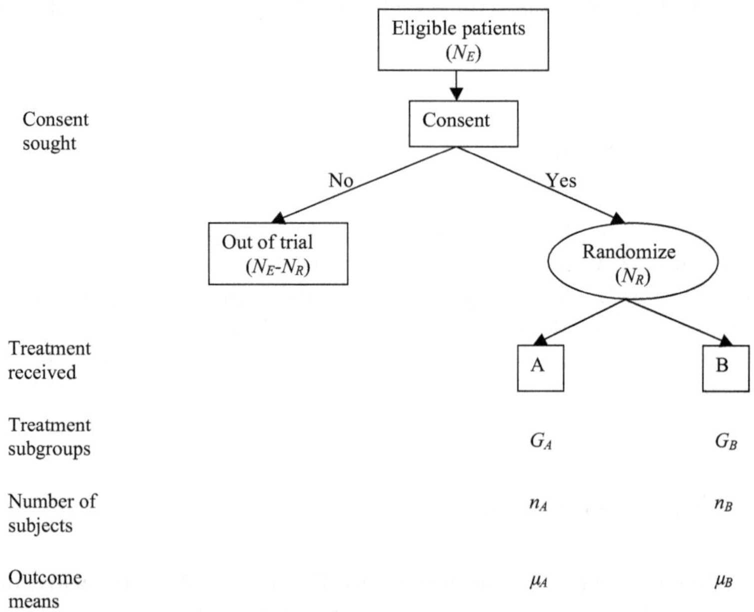

Fig. 17.1 A schematic of the conventional randomization design

generalizability of the results of the trial from the trial participants to the broader patient population may be invalidated.

17.3.1.1 Estimating Treatment, Selection and Preference Effects

Let y_{ik} denote the response for the kth patient ($k = 1, \ldots, n_i$) randomized to the ith treatment ($i = A, B$). Let μ_i denote the mean response of the ith subgroup in the random group. The treatment effect, a measure of the extent to which treatment difference is attributable to treatments themselves is defined by $\Delta_T^{\mathrm{C}} = \mu_A - \mu_B$. The usual estimator of Δ_T^{C} is mean difference based on the sample sizes n_i is given by $\hat{\Delta}_T^{\mathrm{C}} = \overline{y}_A - \overline{y}_B$. The variance of $\hat{\Delta}_T^{\mathrm{C}}$ is given by

$$\mathrm{var}(\hat{\Delta}_T^{\mathrm{C}}) = \sigma^2\left(\frac{1}{n_A} + \frac{1}{n_B}\right).$$

For a balanced design ($n_A = n_B = n$), the variance becomes $\mathrm{var}(\hat{\Delta}_T^{\mathrm{C}}) = 2\sigma^2/n$. This effect can be tested for significance using a z-test or a Student's t-test. Selection effects and preference effects are not directly measurable in a conventional RCT but they are assumed to be equal to zero in expectation.

17.3.2 Comprehensive Cohort Study (CCS)

Olschewski and Scheurlen [28] proposed the comprehensive cohort study as an extension to the conventional randomized controlled trial by allowing patients not consenting to randomization (say $N_E - N_R$ in number) to choose their preferred treatment (Fig. 17.2). This design is essentially a prospective cohort follow-up study with a randomized sub-cohort [29]. The comprehensive cohort design is recommended in clinical trials where full informed consent is mandatory and the proportion of patients with a preference for treatment is high; that is, the proportion of N_E patients willing to participate in the conventional RCT (denoted by π) is low [28].

This design should be prospective and is not applicable to situations in which data from retrospective databases are combined with data from a prospective randomized controlled trial [29]. For example, combining the results of a retrospective cohort study with a randomized controlled trial does not constitute a comprehensive cohort trial or design. The cohort study may not represent a true preference group because some of the patients in the studies might have consented to randomization if offered. Furthermore, heterogeneity may exist in patient selection criteria and study management between the cohort study and clinical trial.

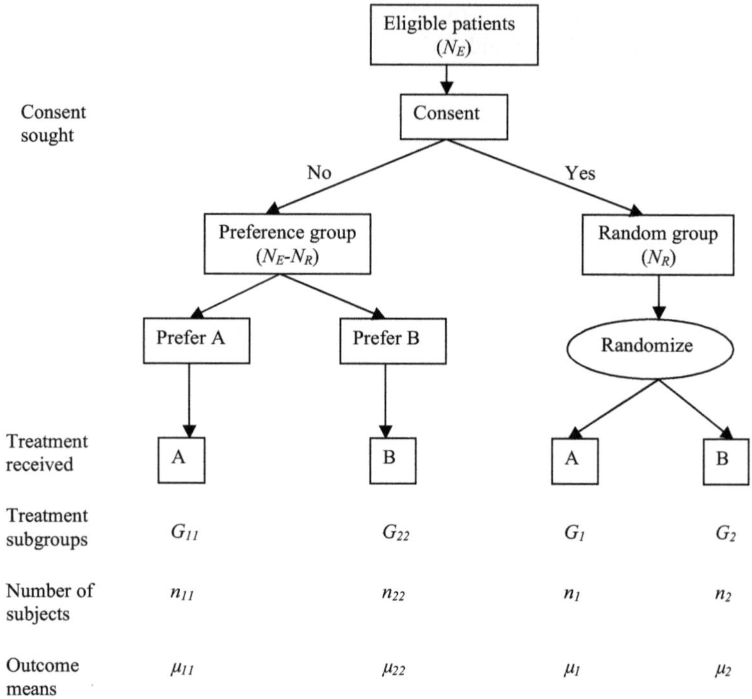

Fig. 17.2 A schematic of the Comprehensive cohort randomized controlled trial [28]

17.3.2.1 Estimating Treatment, Selection and Preference Effects

We will denote the proportion of subjects in the non-random arm (preference group) who expressed preference for intervention A and B by α and β, respectively. So that from Fig. 17.2, the estimates of α and β are given by

$$\hat{\alpha} = \frac{n_{11}}{N_E - N_R}$$

and

$$\hat{\beta} = \frac{n_{22}}{N_E - N_R},$$

where $\hat{\alpha} + \hat{\beta} = 1$. Suppose μ_i ($i = 1$ for A, 2 for B) denotes the mean response of the ith subgroup in the random arm of the design (see Fig. 17.2). The treatment effect for the comprehensive cohort design is denoted by Δ_T^{CC} and defined as $\Delta_T^{CC} = \mu_1 - \mu_2$. The estimate of Δ_T^{CC} based on the observed n_1 and n_2 is $\hat{\Delta}_T^{CC} = \overline{y}_1 - \overline{y}_2$.

This estimate is the same as the one obtained from the conventional randomized controlled trial ($\hat{\Delta}_T^{\mathrm{C}}$). The variance of $\hat{\Delta}_T^{\mathrm{CC}}$ is given by

$$\mathrm{var}(\hat{\Delta}_T^{\mathrm{CC}}) = \frac{S_1^2}{n_1} + \frac{S_2^2}{n_2}.$$

Suppose μ_{ij} denotes the mean response for treatment i ($i = 1$ for A, 2 for B) and preference group j ($j = 1$ (prefer A), 2 (prefer B), and 3 (no preference)). The selection effect and preference effect are not directly measurable from the design. However, the influence of preference on outcome of treatment A and B could be defined as $\Delta_{P*}^{\mathrm{CCa}} = \mu_{11} - \mu_1$ and $\Delta_{P*}^{\mathrm{CCb}} = \mu_{22} - \mu_2$ respectively. The corresponding estimates are $\hat{\Delta}_{P*}^{\mathrm{CCa}} = \overline{y}_{11} - \overline{y}_1$ and $\hat{\Delta}_{P*}^{\mathrm{CCb}} = \overline{y}_{22} - \overline{y}_2$, respectively. When $\hat{\Delta}_{P*}^{\mathrm{CCa}} > 0$ and $\hat{\Delta}_{P*}^{\mathrm{CCb}} > 0$, then the influence of preference on outcome of treatments would be seen as important. Because this estimation involves the preference arm of the design, a regression approach taking all prognostic factors into consideration is always recommended [28, 29]. The goal of this analysis is to adjust for the possible effect of heterogeneity in baseline characteristics of the patients in the preference groups compared to the randomly assigned group.

17.3.2.2 Advantages

The design has all the advantages of a conventional randomized controlled trial. In addition to providing unbiased estimates of the treatment effects, the design allow investigators to measure some influence of preference on outcomes by comparing outcomes in patients not consenting to randomization with those who consented. The external validity of the study is enhanced because almost all eligible patients will enter the study whether or not they consent to randomization [29].

17.3.2.3 Limitations

The design reduces to an observational study if all patients expressed preference for treatments by not consenting to randomization. Hence, an unbiased estimate of treatment effect may be impossible from the preference group because of the potential imbalance in the baseline characteristics of patients. Similarly, the preference-arm of the design will not exist if all patients consented to randomization. The design assumes that every non-consenting patient has a preference for one intervention or the other. As a result it does not explicitly account for the fact that some patients may refuse randomization for reasons other than preference [19]. Finally, the design uses fewer patients to estimate the treatment effect (i.e. only patients in random group).

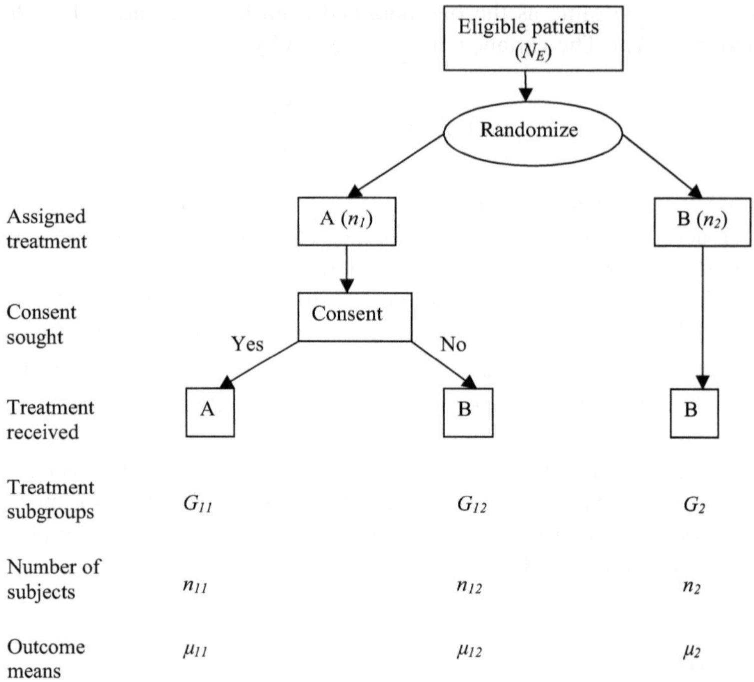

Fig. 17.3 A schematic of the Zelen single consent design [38]

17.3.3 Single Consent Designs

The key feature of these designs is that the N_E eligible patients are randomly assigned to treatments before informed consent is sought. This design is illustrated in Fig. 17.3. Only the patients in the experimental arm (say n_1 in number) are approached for consent. Informed consent is never sought from all the n_2 patients randomized to the standard treatment [38]. As a result these patients are not aware of their inclusion in the trial. The thinking is that patients randomized to the standard treatment are receiving the usual care, for which no consent is needed. Those patients randomized to the experimental treatment that decline informed consent (say n_{12} in number) are offered the standard treatment.

A variant of this design is the "modified single consent design" in which the N_E eligible patients are randomly assigned into two groups, choice and no choice arm, with n_1 and n_2 patients respectively [38, 39]. The patients allocated to the no choice arm receive treatment B (standard treatment). The patients randomized to the choice arm are given the opportunity to choose between treatment A or B after both options are discussed with them.

17.3.3.1 Estimating Treatment, Selection and Preference Effects

Suppose μ_i denotes the mean response for patients randomized to treatment i ($i = 1$ for A, 2 for B). Furthermore, let μ_{ij} denote the mean response for patient randomized to treatment i ($i = 1$ for A, 2 for B) and having preference treatment j ($j = 1$, prefer A; and $j = 2$, prefer B).

Zelen [38, 41] suggests that the treatment effect should be estimated by comparing treatment A to treatment B as randomized irrespective of the treatment they actually received. Therefore, the treatment effect for the single consent design denoted by Δ_T^{SC} is defined as $\Delta_T^{\mathrm{SC}} = \mu_1 - \mu_2$. The estimate of Δ_T^{SC} based on the observed n_1 and n_2 is $\hat{\Delta}_T^{\mathrm{SC}} = \overline{y}_1 - \overline{y}_2$. The variance of $\hat{\Delta}_T^{\mathrm{SC}}$ is given by

$$\mathrm{var}(\hat{\Delta}_T^{\mathrm{SC}}) = \frac{S_1^2}{n_1} + \frac{S_2^2}{n_2}.$$

Selection effect and preference effects are not directly measurable in the single consent designs. However, a comparison of outcomes and characteristics of patients in subgroup G_{12} and G_2 (see Fig. 17.3) may be useful in understanding the possible effect of selection on the experimental treatment B.

Another parameter of interest in the single consent trial is θ_S, the proportion of n_1 patients randomized to the experimental treatment (A) who accepted the treatment when offered [38]. From Fig. 17.3, θ_S can be estimated as $\hat{\theta}_S = n_{11}/n_1$. As the number of patients accepting treatment A increases, θ_S approaches 1 and the single consent design reduces to the conventional randomized design. As θ_S departs from unity, the loss in statistical efficiency of the single consent design becomes apparent [38].

Zelen [38, 41] derived the asymptotic relative efficiency of the single-consent design relative to the conventional RCT as θ_S^2. This derivation assumes that 50 % of the subjects are randomized to each group in both the conventional RCT and the single-consent RCT.

17.3.3.2 Advantages

The single consent design may increase patient enrollment into the study since all the eligible patients (N_E) are allowed to participate in the trial. Furthermore, patients will be aware of the assigned treatment before giving consent or expressing preference [38]. The design is very useful in situations in which experimental intervention is highly attractive to potential subjects and when the control group receives standard intervention [16, 38]. It is very attractive to physicians enrolling their patients or parents enrolling their children because in contrast to the conventional RCT, the single consent trial guarantees that subjects in the experimental arm can receive their preferred treatment [38].

17.3.3.3 Limitations

The design is limited in application because of the need for usual care arm. This does not exist for many therapeutic areas. It is unethical to inform patients about treatment options and randomization only after the results of pre-randomization are known [27, 38]. Furthermore, failure to fully inform the patients in the standard arm of the alternative treatment options in the trial is also an ethical limitation of the design [38]. Because the patients in the standard arm of the trial were not informed of there participation in trial, only routinely collected data during regular clinical visits could be obtained from patients. This limits the applicability of the design to clinical trials in which more data collection and visits may be needed [2]. The design cannot be used when there are important reasons for double-blinding in a randomized controlled trial [38].

The design may result in loss of statistical efficiency compared to conventional RCT if more patients in the experimental arm refuse their assigned treatment [3, 23, 38]. For example, if 70 % of the patients in a single consent design accept the experimental treatment; the efficiency of the design relative to conventional RCT will be 49 %. This means that twice as many subjects are needed to obtain the same sensitivity in single consent design as in a conventional RCT [40]. Hence, the validity of the estimates from the design may be affected by the proportion of patients who accepted the new intervention.

17.3.4 *Double-Consent Design*

The double-consent design was introduced by Zelen as an extension of the single consent design [40, 41]. The design is suitable for comparing two treatments in which there is no best standard treatment. The set of N_E eligible patients are first randomized to treatments A and B with n_1 and n_2 patients respectively (see Fig. 17.4). All the patients are asked if they accept the randomized treatment. If they decline, they are allowed to switch to the alternative treatment or to another treatment not under investigation in the study.

17.3.4.1 Estimating Treatment, Selection and Preference Effects

The double-consent design is an extension of the single consent design. Therefore, the treatment effect for the double consent design denoted by Δ_T^{DC} is defined as $\Delta_T^{\mathrm{DC}} = \mu_1 - \mu_2$. The estimate of Δ_T^{DC} based on the observed n_1 and n_2 is $\hat{\Delta}_T^{\mathrm{DC}} = \overline{y}_1 - \overline{y}_2$. The variance of $\hat{\Delta}_T^{\mathrm{DC}}$ is given by

$$\mathrm{var}(\hat{\Delta}_T^{\mathrm{DC}}) = \frac{S_1^2}{n_1} + \frac{S_2^2}{n_2}.$$

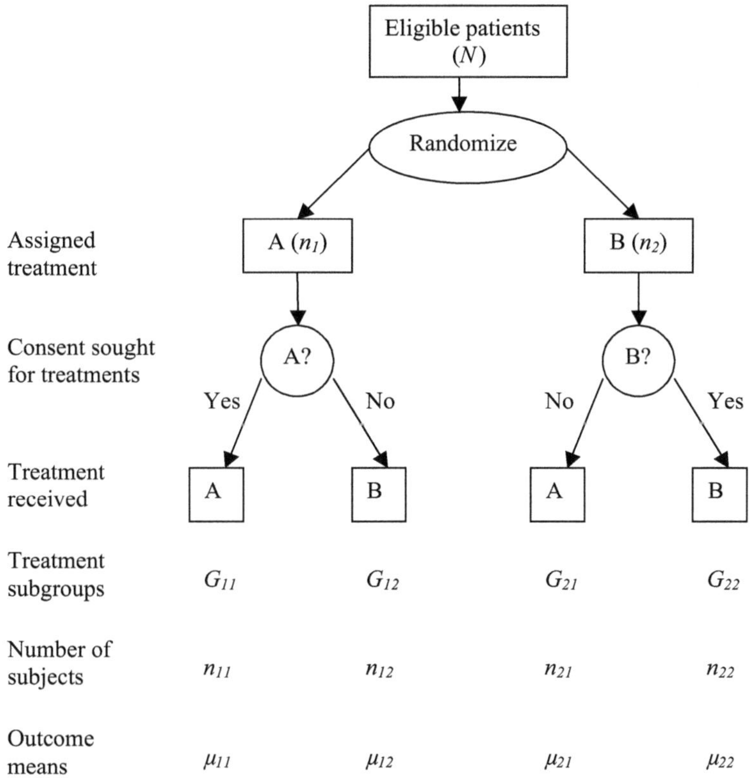

Fig. 17.4 A schematic of the Zelen double-consent design [40, 41]

The selection effect and preference effects are not directly estimable from the design. However, a comparison of patients' characteristics and outcomes between the four treatment subgroups (see Fig. 17.4) can provide some insights into the possible biases associated with selection of treatments [38].

The proportion of patients in the double-consent randomized trial who accepted their assigned treatment is denoted by θ_D. From Fig. 17.4, θ_D can be estimated as $\hat{\theta}_D = (n_{11} + n_{22})/(n_1 + n_2)$. As θ_D approaches unity, the double-consent randomized trial converges to the conventional RCT. Hence, θ_D is a key parameter in understanding of the statistical efficiency of the double-consent RCT relative to conventional RCT. Zelen [41] derived the asymptotic relative efficiency of the single-consent design relative to the conventional RCT as $[2\theta_D - 1]^2$.

17.3.4.2 Advantages

The patient will be aware of the assigned treatment before giving consent or expressing his preference [38]. Furthermore, the design may increase enrolment of patients into the study if more physicians agree to enter patients [18]. The double consent design avoids some of the ethical issues with the single consent design by seeking consents from both arms of the treatment. This design has been widely used in clinical trial across therapeutic areas to minimize cross-over rates in non-placebo controlled trials [1].

17.3.4.3 Limitations

The design cannot be used when there are important reasons for conducting a "double-blind" randomized controlled trial [38]. The design could result in low statistical efficiency relative to conventional RCT if large proportion of subjects rejects the treatment offered [41]. Overall, the sample size for double-consent trial has to be inflated by a factor of $[1/(2\theta_D - 1)]^2$; where θ_D denotes the proportion of patients who accepted their assigned intervention in a double-consent RCT [41].

17.3.5 Two-Stage Clinical Trial Design

Rucker [31] proposed a two-stage randomized clinical trial design for distinguishing treatment effects from those resulting from choosing treatment. The design is illustrated in Fig. 17.5. At the first stage, the N eligible patients are randomly assigned to one of two groups, the preference group and the random group [31]. Unlike the Zelen designs, informed consent is sought at the first stage of randomization. At the second stage, patients in the random group are further randomized to treatment A or B after informed consent. Patients in the preference group are allowed to choose their preferred treatments. The undecided patients in the preference group are further randomized to treatment A or B.

17.3.5.1 Estimating Treatment, Selection and Preference Effects

Rucker [31] used a linear model approach to estimate the effects of selection, preference and treatment. Rucker denotes the expected preference rates for treatments A and B by α and β, respectively. Therefore, $1 - \alpha - \beta$ is the proportion of patients expressing no preference. These rates are directly estimable in the preference arm of the design (see Fig. 17.5). Their estimates are $\hat{\alpha} = n_{11}/(N - m)$ and $\hat{\beta} = n_{22}/(N - m)$. Suppose μ_i ($i = 1$ for A, 2 for B) denotes the mean response of the ith subgroup in the random arm of the design. The treatment effect for the two-stage design is denoted by Δ_T^{TS} and defined as $\Delta_T^{\text{TS}} = \mu_1 - \mu_2$. The estimate of Δ_T^{TS} based on the observed n_1 and n_2 is $\hat{\Delta}_T^{\text{TS}} = \overline{y}_1 - \overline{y}_2$.

Let μ_{ij} denote the mean response for treatment i ($i = 1$ for A, 2 for B) and preference group j ($j = 1$ (prefer A), 2 (prefer B), and 3 (no preference)). Only

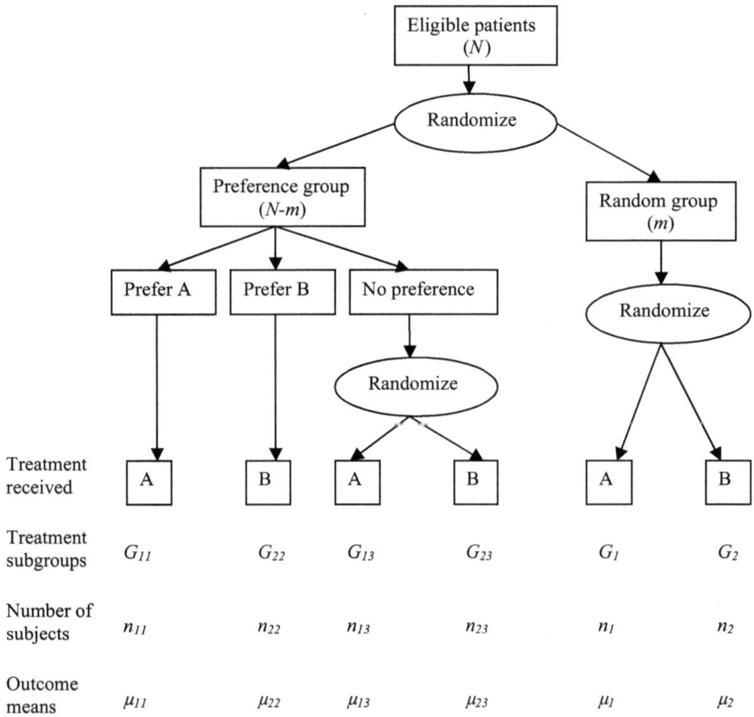

Fig. 17.5 A schematic of the two-stage trial design [31]

four (μ_{11}, μ_{13}, μ_{22}, μ_{23}) of the six response means are directly estimable from the preference arm as $\hat{\mu}_{11} = \overline{y}_{11}$, $\hat{\mu}_{13} = \overline{y}_{13}$, $\hat{\mu}_{22} = \overline{y}_{22}$ and $\hat{\mu}_{23} = \overline{y}_{23}$. The mean response of patients who received treatment A but prefer treatment B (μ_{12}) and the mean response of patients who received treatment B but prefer treatment A (μ_{21}) are not directly observable but defined by Rucker [31] as

$$\mu_{12} = \frac{\mu_1 - \alpha\mu_{11} - (1-\alpha-\beta)\mu_{13}}{\beta}, \quad \mu_{21} = \frac{\mu_2 - \beta\mu_{22} - (1-\alpha-\beta)\mu_{23}}{\alpha}.$$

Selection effect for the two-stage design (Δ_S^{TS}) was defined as [31]

$$\Delta_S^{\mathrm{TS}} = \frac{(\mu_{11} + \mu_{21}) - (\mu_{12} + \mu_{22})}{2}.$$

If higher values of μ_{ij} indicate better response, then $\Delta_S^{\mathrm{TS}} > 0$ implies that there is selection effect in favor of patients preferring treatment A. That is, patients who self-select treatment A tend to have better outcomes than those who selected treatment B [31]. The estimator of Δ_S^{TS} is given by

$$\hat{\Delta}_S^{\mathrm{TS}} = \frac{(\overline{y}_{11} + \overline{y}_{21}) - (\overline{y}_{12} + \overline{y}_{22})}{2}.$$

The preference effect for the two-stage design (Δ_P^{TS}) is defined as [31]

$$\Delta_P^{\mathrm{TS}} = \frac{(\mu_{11} + \mu_{22}) - (\mu_{12} + \mu_{21})}{2}.$$

This is the difference between the outcomes of patients receiving their preferred intervention and patients not receiving their preferred intervention. Hence, Δ_P^{TS} implies that patients who received their preferred treatment benefit greatly than others [31]. The estimator of Δ_P^{TS} is given by

$$\hat{\Delta}_P^{\mathrm{TS}} = \frac{(\overline{y}_{11} + \overline{y}_{22}) - (\overline{y}_{12} + \overline{y}_{21})}{2}.$$

17.3.5.2 Advantages

The design preserved randomization and allows an unbiased estimation of treatment benefits. The design may increase trial enrolment because all eligible patients are enrolled. Thus the external validity is enhanced. The design is capable of distinguishing treatment effects from selection effects and preference effects.

17.3.5.3 Limitations

The design may be expensive to implement [25, 31]. A simulation result has shown that at least 100 patients are required to ensure the reliability of the statistical tests [31]. External validity may reduce because only patients accepting randomization will enter the study [19]. Patients who have strong preference for one treatment may refuse participation or elect to be in the preference arm of the trial [19, 31]. Hence, the design may not be suitable for a clinical trial comparing surgical and medical interventions [31]. Internal validity is enhanced because all patients are randomized [19]. However, comparison of preference vs. random arm is subject to confounding because patients characteristics may determine choice of treatment [19]. Because of the random arm, all the problems associated with conventional RCT designs like treatment refusal and non-compliance are still present in the design. Furthermore, the design needs twice as many patients to estimate the treatment effect.

17.3.6 Preference-Conventional RCT Design

Wennberg [36] and Wennberg et al. [37] proposed a clinical trial design in which the N eligible patients who consent to randomization are assigned randomly to one

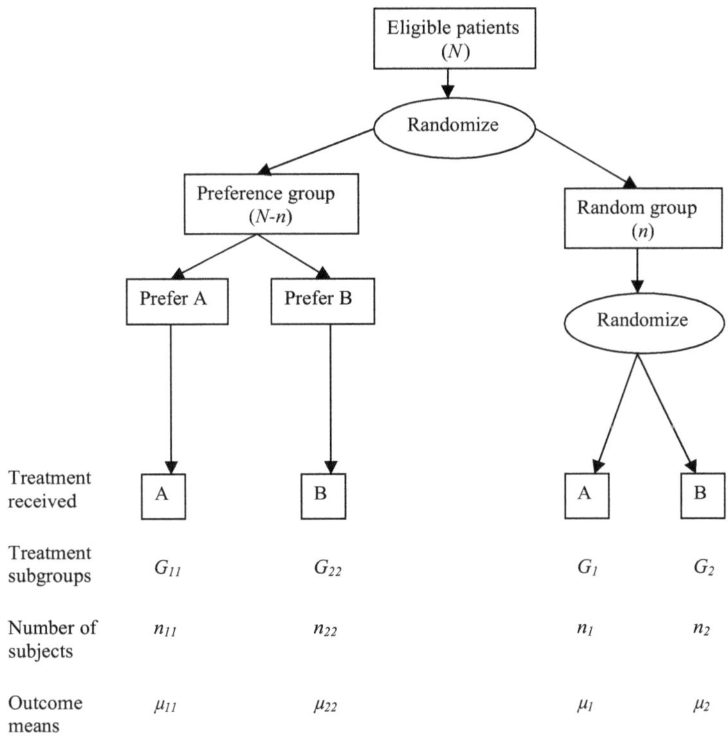

Fig. 17.6 A schematic of the Preference-conventional RCT design [36,37]

of two groups; the preference group and the random group. The design is illustrated in Fig. 17.6. Patients in the random group are further randomized to treatment A or B while patients in the preference group are allowed to choose their preferred treatments. Those requiring additional information or advice before deciding on a treatment are counseled by physicians operating under standard protocol who do not administer treatments.

17.3.6.1 Estimating Treatment, Selection and Preference Effects

The preference-conventional RCT could be seen as a special case of the two stage trial design in which all the patients randomized to the preference arm express preference for one of the treatment and hence no undecided patient ($\mu_{13} = \mu_{23} = 0$ and $\alpha + \beta = 1$). Therefore, the treatment effect (Δ_T^{PC}) is defined as $\Delta_T^{\text{PC}} = \mu_1 - \mu_2$. The estimate of Δ_T^{PC} based on the observed n_1 and n_2 is $\hat{\Delta}_T^{\text{PC}} = \overline{y}_1 - \overline{y}_2$.

Selection effect (Δ_S^{PC}) is defined as $\Delta_S^{\text{PC}} = \big((\mu_{11} + \mu_{21}) - (\mu_{12} + \mu_{22})\big)/2$. Substituting μ_{12} and μ_{21} implies that

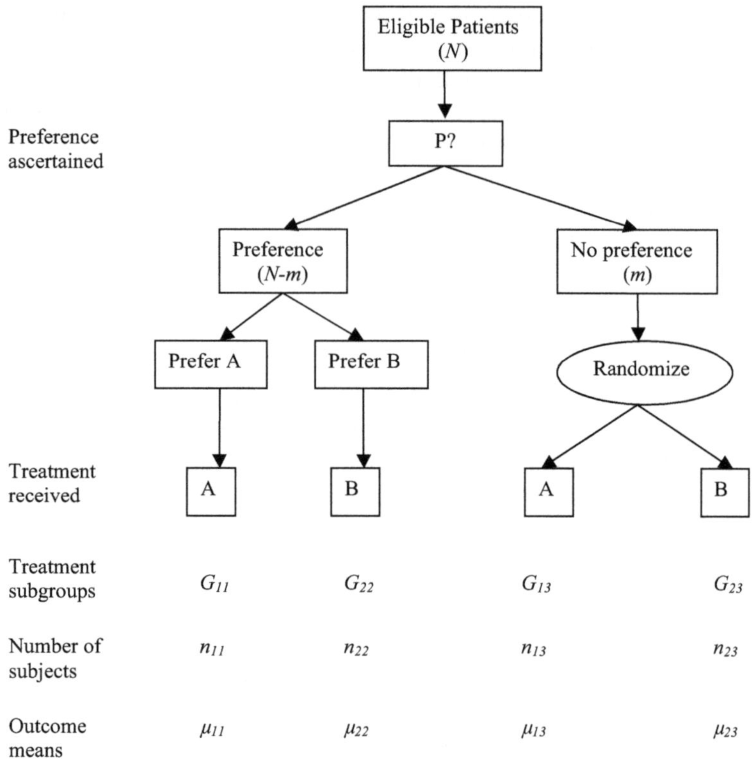

Fig. 17.7 A schematic of the partially randomized patient-preference design [6, 8]

design. However, a comparison of the outcomes and patient characteristics across four treatment subgroups (see Fig. 17.7) could provide some information about preferences in relation to each treatment.

17.3.7.2 Advantages

The design may increase enrollment into the study and thus achieve a high degree of representation [8, 17, 18]. This is because physicians will have less explanation to do as the treatment choices are more open [22]. Furthermore, motivational factors will be optimized by letting patients select their preferred treatment [8]. The design may yield results that are more relevant to decision making in a clinical setting [18].

17.3.7.3 Limitations

If patients are allowed to choose, by preference, the treatment arm they join, rather than a random assignment, the differences in outcome may be explained by the differences in the baseline characteristics of patients in the randomized and non-randomized groups [17, 26]. Hence, uncontrolled confounding factors may bias the main results because patients are only partially randomized [15]. This weakens the internal validity of the study and makes it impossible to measure the true treatment effect [7, 18].

Partially randomized preference design is not feasible if all patients choose a treatment [7, 18]. Furthermore, if only a small proportion of patients eligible for the study accepts randomization, then the evaluation effectively becomes an observational study [22]. The design is not applicable to many trials in which participant blinding is important because two of the four arms are open label [15].

Patient preferences may change over time, both during the trial and subsequently [5, 17]. Hence, preference arms in trials may not reflect true, informed, rigorously assessed preferences [5]. The results of the analyses of the effects of preferences on outcome may be ambiguous [5].

17.3.8 Design with Clinician-Preferred Treatment

One of the ethical bases for conducting RCTs is that the clinicians delivering the treatments are in a state of equipoise—unsure about which treatment is better [20, 30, 33]. This may be difficult to ascertain in a placebo controlled trial or an RCT of new treatment against a standard treatment. Korn and Baumrind [20, 21] identified two potential problems of not explicitly incorporating clinician preferences into clinical trials. First, it may be difficult to obtain agreement on the eligibility criteria if clinicians have personal preference for one treatment [20]. Second, how unsure about the desirability of the one treatment over the other does a clinician has to be for appropriate randomization of his patients [20].

On the basis of these two limitations of conventional RCTs, Korn and Baumrind [20, 21] proposed a design that takes clinician preferences into account (see Fig. 17.8). First, patients are assessed for study eligibility. Second, informed consent is obtained from all eligible patients (say N in size) to participate in the trial. Third, medical history of each eligible patient is reviewed independently by each clinician or panel of clinician to determine optimal treatment options for the patient. Where there is a consensus, the patient receives the clinician-preferred treatment (A or B). However, when the clinicians disagree on the choice of treatment, the patient is randomly assigned to either treatment A or B and treated by a clinician who preferred the treatment.

17.3.8.1 Estimating Treatment, Selection and Preference Effects

The RCT with clinician-preferred treatment could be seen as a variant of the preference clinical trial in which physicians are making the choices rather than patients. Hence four response means (μ_{11}, μ_{13}, μ_{22}, μ_{23}) are directly estimable

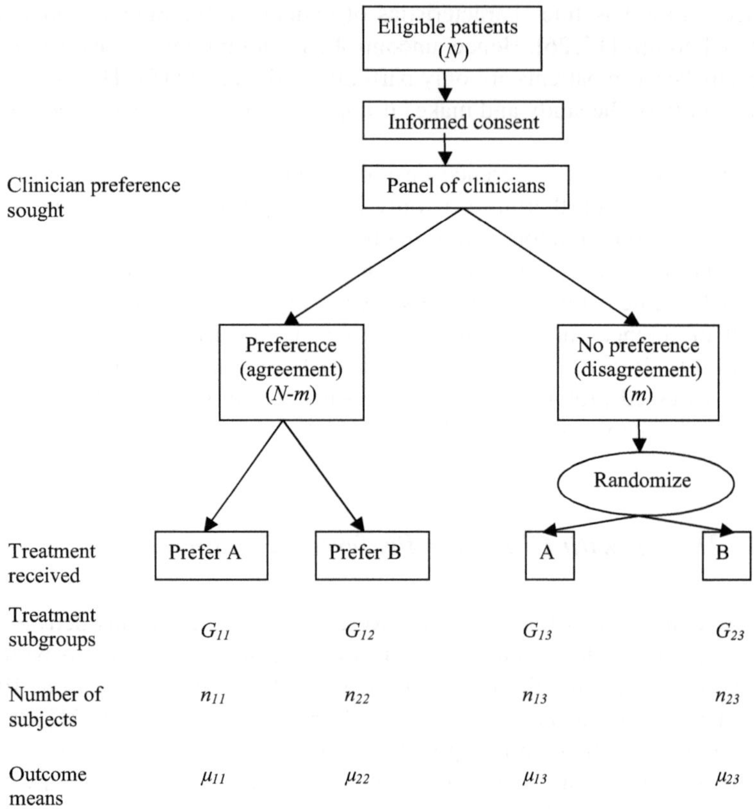

Fig. 17.8 A simplified schematic of a clinician-preferred treatment design [20]

from the design. We define the treatment effects as $\Delta_T^{\mathrm{CPT}} = \mu_{13} - \mu_{23}$. Therefore, Δ_T^{CPT} is the treatment effect among patients whom clinicians disagree on the choice of treatment. The estimate of Δ_T^{CPT} based on the observed n_{13} and n_{23} is defined as $\hat{\Delta}_T^{\mathrm{CPT}} = \overline{y}_{13} - \overline{y}_{23}$.

Selection effect and preference effect are not directly measurable in the design. However, a comparison of the outcomes and patient characteristics across four treatment subgroups (see Fig. 17.8) could provide some information about preferences in relation to each treatment.

17.3.8.2 Advantages

The design ensures that patients receive the best available care possible in the trial because of the need for consensus among study physicians. Furthermore, clinicians are allowed to participate only in the treatment arm that they preferred. This may

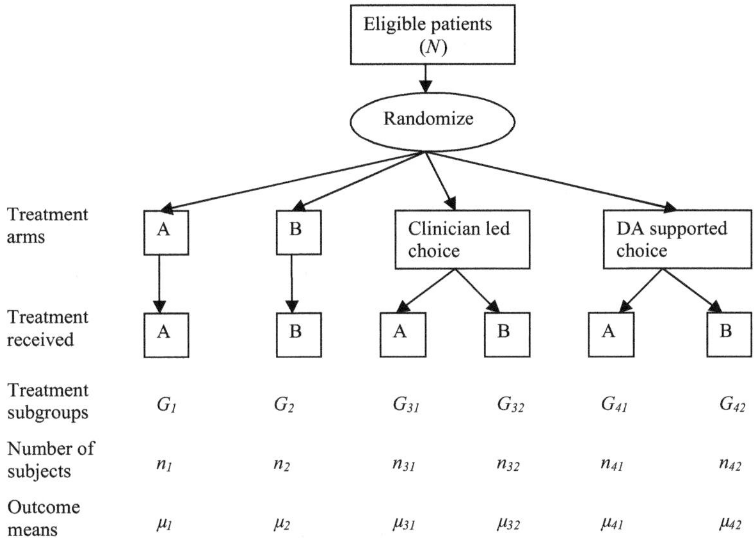

Fig. 17.9 A schematic of the Four-arm, decision-support, randomized controlled trial design [24]

be beneficial in surgical trials where surgeons may have a strong preference for participating in one arm of the trial.

17.3.8.3 Limitations

External validity of the study may be jeopardized. The applicability of the results of the trials is limited because of the requirement for disagreement on treatment before randomization [20, 21]. The design reduces to an observational study if there is agreement on the optimal choice of treatment for patients by clinicians. Hence, an unbiased estimate of the treatment effect may be impossible because it may be confounded by clinicians' preferences. The design is not applicable in studies requiring double-blinding.

17.3.9 Four-Arm, Decision-Support, Randomized Controlled Trial Design

McCaffery, Irwig and Bossuyt [24] proposed a design for evaluating the long term health impact of decision aids in which patients are randomly assigned to four arms design (see Fig. 17.9). This design could be seen as a variant of the two-stage design [31]. In the first two arms patients received their assigned treatments (A or B) in a similar manner to conventional RCT. The last two arms represent the

17.4 Discussion and Conclusion

In this paper, we reviewed various randomized controlled trials for accommodating patients and physician preferences. Ten designs were identified. We provided a description of each design and methods of estimating various measures of treatment, selection and preference effects. Furthermore, we looked at the merits and demerits of each design in comparison with the conventional randomized controlled trial. The results are summarized in Table 17.1.

Nine of the designs incorporated a choice arm, in which patients are allowed to choose their preferred treatment on their own or in consultation with a clinician. In one of the designs (design with clinician preference), the clinicians are the ones selecting treatments for patients.

For all of the designs reviewed an unbiased estimate of the treatment effect is possible. Four of the nine designs have a random group similar to a conventional RCT. These designs are the comprehensive cohort design, two stage clinical trial design, preference-conventional RCT and the four-arm decision-support, randomized controlled trial design. As a result, estimates of treatment effects are equivalent to the estimates from a conventional RCT with two arms. Two other designs (partially randomized patient preference trial and design with clinician or patient-preferred treatment) randomized patients who have no preference for either intervention under study. While the estimate of treatment benefit from the two designs may be unbiased, the generalization to the overall eligible patients may be questionable. The remaining two designs (single-consent RCT and double-consent RCT) allow patients randomized to a particular intervention to accept or switch to an alternative intervention (sometimes the standard treatment). While an unbiased estimate of the treatment effect is possible under the intention-to-treat principle, the designs raise some important issues on whether such analysis is the best for evaluating the efficacy of new intervention. Of particular concern is the potential for loss of statistical power of the trial to detect a real treatment difference due to dilution effect if a large proportion of patients reject their allocated treatments [2,35]. Some studies have shown that this problem could be compounded if the extent of crossover differs considerably between the intervention arms [3,23].

Of the nine designs that allow the incorporation of the preference arms, only two designs (two-stage clinical trial design, preference-conventional RCT) provide a good framework for estimating the selection and preference effects. In the other six designs, some forms of preference measures are possible but mostly confounded within treatments.

There are two main ethical issues: (1) type of informed consent and (2) timing of informed consent. A complete-double-consent design in which all patients learn about all available interventions in the study is the ethical gold standard of informed consent. From Table 17.1, seven of the designs reviewed obtained complete informed consent from patients prior to enrolling them into trials. In two of the designs, informed consents were obtained post-randomization.

Table 17.1 Summary of RCT designs for accommodating patients' or physicians' preferences

Designs	Initial allocation to groups	Type of informed consent	Timing of informed consent	Treatment arms	Preference expressed by		Effects directly estimable
					Patients	Clinicians	
Conventional randomized controlled trial	Random	Complete-double	Before randomization	Random arms only	No	No	Treatment
Comprehensive cohort study	Non-random	Complete-double	Before randomization	Random and non-random	Yes	No	Treatment, preference
Single-consent randomized trial	Random	Single	After randomization	Random arms only	Yes	No	Treatment
Double-consent randomized trial	Random	Incomplete-double	After randomization	Random arms only	Yes	No	Treatment
Two stage clinical trial	Random	Complete double	Before randomization	Random and non-random	Yes	No	Treatment, selection, preference
Preference conventional randomized controlled trial	Random	Complete-double	Before randomization	Random and non-random	Yes	No	Treatment, selection, preference

(continued)

Table 17.1 (continued)

Designs	Initial allocation to groups	Type of informed consent	Timing of informed consent	Treatment arms	Preference expressed by		Effects directly estimable
					Patients	Clinicians	
Partially randomized patient preference trial	Non-random	Complete-double	Before randomization	Random and non-random	Yes	No	Treatment
Design with clinician-preferred treatment	Non-random	Complete-double	Before randomization	Random and non-random	No	Yes	Treatment
Four-arm, decision-support, randomized controlled trail design	Random	Complete-double	Before randomization	Random and non-random	Yes	No	Treatment
Design with clinician or patient-preferred treatment	Random	Incomplete-double	After randomization	Random and non-random	Yes	Yes	Treatment

Some of the designs are related to the two stage design. For example, the preference-conventional RCT could be seen as a special case of the two stage trial design in which all the patients randomized to the preference arm expressed preference for one of the treatments and hence, no undecided patients. Similarly, the partially randomized patient-preference design could be seen as the choice arm of the two stage design or as a two stage design with no random arm. Furthermore, the design with clinician-preferred treatment could be seen as a special case of the partially randomized patient-preference design in which preferences are clinician-based rather than patient-based. The four-arm, decision-support, randomized controlled trial design is also a special case of the two stage trial design. Of all the designs reviewed, the two-stage design provides a comprehensive framework for estimating treatment effect, selection effect and preference effect while allowing a segment of eligible patients to choose the intervention of their choice.

There are some debates on whether preference arms in the clinical trials truly represent patients' or physicians' preferences [5]. This is because individual preferences for treatment are affected by several factors including how the questions for eliciting preferences were framed; participants' understanding of concept of risk; numeric description of risks and benefits; and varying expectations [5, 13]. Furthermore, education and socio-economic status of patients may affect his or her preferences for treatment [19]. King et al. [19] find that well educated and employed patients are more likely to refuse randomization because of preference. Hence the methods for eliciting preference from patients have to be methodologically rigorous. Compared to the traditional RCTs, the availability of the random arm could reduce enrollment into the random arm of the design. This may lead to loss of power and reduce external validity of the study. There are some conflicting evidences from the literature regarding whether patients' or physicians' preferences affects outcomes [17, 19]. This means that designs incorporating preferences may not be ideal for evaluating all interventions. It should at least be considered in all RCTs in which treatments and mode of administration are very different and where blinding of patients may not be possible.

References

1. Adamson, J., Cockayne, S., Puffer, S., Torgerson, D.J.: Review of randomized trials using the post-randomized consent (Zelen's) design. *Contemporary Clinical Trials* **27**(4), 305–319 (2006)
2. Altman, D.G., Whitehead, J., Parmar, M.K., Stenning, S.P., Fayers, P.M., Machin, D.: Randomized consent designs in cancer clinical trials. *European Journal of Cancer* **31A**(12), 1934–1944 (1995)
3. Anbar, D.: The relative efficiency of Zelen's prerandomization design for clinical trials. *Biometrics* **39**(3), 711–718 (1983)
4. Armitage, P.: The role of randomization in clinical trials. *Statistics in Medicine* **1**(4), 345–352 (1982)
5. Bowling, A., Rowe, G.: "You decide doctor". What do patient preference arms in clinical trials really mean? *Journal of Epidemiology and Community Health* **59**(11), 914–915 (2005)

6. Bradley, C.: Clinical trials–time for a paradigm shift? *Diabetic Medicine* **5**(2), 107–109 (1988)
7. Bradley, C.: Designing medical and educational intervention studies. A review of some alternatives to conventional randomized controlled trials. *Diabetes Care* **16**(2), 509–518 (1993)
8. Brewin, C.R., Bradley, C.: Patients' preferences and randomized clinical trials. *British Medical Journal* **289**, 313–315 (1989)
9. Byar, D.P., Simon, R.M., T., F.W., Schlesselman, J.J., DeMets, D.L., Ellenberg, J.H., Gail, M.H., Ware, J.H.: Randomized clinical trials. Perspectives on some recent ideas. *New England Journal of Medicine* **295**(2), 74–80 (1976)
10. Canadian Institutes of Health Research (CIHR), Natural Sciences and Engineering Research Council of Canada (NSERC), Social Sciences and Humanities Research Council of Canada (SSHRC): *Tri-Council Policy Statement: Ethical Conduct for Research Involving Humans, December 2010* URL http://www.pre.ethics.gc.ca/pdf/eng/tcps2/TCPS_2_FINAL_Web.pdf. (Last accessed January 3, 2011)
11. Cooper, K.G., Grant, A.M., Garratt, A.M.: The impact of using a partially randomized patient preference design when evaluating alternative managements for heavy menstrual bleeding. *British Journal of Obstetrics and Gynaecology* **104**(12), 1367–1373 (1997)
12. Coward, D.D.: Partial randomization design in a support group intervention study. *Western Journal of Nursing Research* **24**(4), 406–421 (2002)
13. Edwards, A., Elwyn, G.: Understanding risk and lessons for clinical risk communication about treatment preferences. *Quality in Health Care* **10**(Suppl I), i9–i13 (2001)
14. Feine, J.S., Awad, M.A., Lund, J.P.: The impact of patient preference on the design and interpretation of clinical trials. *Community Dentistry and Oral Epidemiology* **26**(1), 70–74 (1998)
15. Halpern, S.D.: Evaluating preference effects in partially unblinded, randomized clinical trials. *Journal of Clinical Epidemiology* **56**(2), 109–115 (2003)
16. Homer, C.S.: Using the Zelen design in randomized controlled trials: debates and controversies. *Journal of Advanced Nursing* **38**(2), 200–207 (2002)
17. Howard, L., Thornicroft, G.: Patient preference randomized controlled trials in mental health research. *British Journal of Psychiatry* **188**, 303–304 (2006)
18. Janevic, M.R., Janz, N.K., Dodge, J.A., Lin, X., Pan, W., Sinco, B.R., Clark, N.M.: The role of choice in health education intervention trials: a review and case study. *Social Science and Medicine* **56**(7), 1581–1594 (2003)
19. King, M., Nazareth, I., Lampe, F., Bower, P., Chandler, M., Morou, M., Sibbald, B., Lai, R.: Impact of participant and physician intervention preferences on randomized trials: a systematic review. *Journal of the American Medical Association* **293**(9), 1089–1099 (2005)
20. Korn, E.L., Baumrind, S.: Randomized clinical trials with clinician-preferred treatment. *Lancet* **337**(8734), 149–152 (1991)
21. Korn, E.L., Baumrind, S.: Clinician preferences and the estimation of causal treatment differences. *Statistical Science* **13**(3), 209–227 (1998)
22. Lambert, M.F., Wood, J.: Incorporating patient preferences into randomized trials. *Journal of Clinical Epidemiology* **53**(2), 163–166 (2000)
23. Matts, J., McHugh, R.: Randomization and efficiency in Zelen's single-consent design. *Biometrics* **43**(4), 885–894 (1987)
24. McCaffery, K., Irwig, L., Bossuyt, P.: Patient decision aids to support clinical decision making: evaluating the decision or the outcomes of the decision. *Medical Decision Making* **27**(5), 619–625 (2007)
25. McPherson, K., Britton, A.R., Wennberg, J.E.: Are randomized controlled trials controlled? Patient preferences and unblind trials. *Journal of the Royal Society of Medicine* **90**(12), 652–656 (1997)
26. Millat, B., Borie, F., Fingerhut, A.: Patient's preference and randomization: new paradigm of evidence-based clinical research. *World Journal of Surgery* **29**, 596–600 (2005)
27. Millat, B., Fingerhut, A., Flamant, Y., Hay, J.M., L., F.P., Farah, A., Duron, J.J., Courchevel, J.M.: Survey of the impact of randomized clinical trials on surgical practice in France. *European Journal of Surgery* **165**(2), 87–94 (1999)

28. Olschewski, M., Scheurlen, H.: Comprehensive cohort study: an alternative to randomized consent design in a breast preservation trial. *Methods of Information in Medicine* **24**(3), 131–134 (1985)
29. Olschewski, M., Schumacher, M., Davis, K.B.: Analysis of randomized and non-randomized patients in clinical trials using the comprehensive cohort follow-up study design. *Controlled Clinical Trials* **13**, 226–239 (1992)
30. Piantadosi, S.: *Clinical trials: a methodologic perspective*. John Wiley and Sons, New York (1997)
31. Rucker, G.: A two-stage trial design for testing treatment, self-selection and treatment preference effects. *Statistics in Medicine* **8**(4), 477–485 (1989)
32. Schellings, R., Kessels, A.G., Ter Riet, G., Knottnerus, J.A., Sturmans, F.: Randomized consent designs in randomized controlled trials: systematic literature search. *Contemporary Clinical Trials* **27**(4), 320–332 (2006)
33. Shaw, L.W., Chalmers, T.C.: Ethics in cooperative clinical trials. *Annals of the New York Academy of Sciences* **169**(2), 487–495 (1970)
34. Silverman, W.A., Altman, D.G.: Patients' preferences and randomized trials. *Lancet* **347**(8995), 171–17 (1996)
35. Snowdon, C., Elbourne, D., Garcia, J.: Zelen randomization: attitudes of parents participating in a neonatal clinical trial. *Controlled Clinical Trials* **20**(2), 149–171 (1999)
36. Wennberg, J.E.: What is outcomes research? In: A.C. Gelijns (ed.) *Medical Innovation at the Crossroads, Vol. I: Modern Methods of Clinical Investigation*, pp. 33–46. National Academy Press, Washington, DC (1990)
37. Wennberg, J.E., Barry, M.J., Fowler, F.J., Mulley, A.: Outcomes research, PORTs, and health care reform. *Annals of the New York Academy of Sciences* **703**, 52–62 (1993)
38. Zelen, M.: A new design for randomized clinical trials. *New England Journal of Medicine* **300**(22), 1242–1245 (1979)
39. Zelen, M.: Alternatives to classic randomized trials. *Surgical Clinics of North America* **61**(6), 1425–1432 (1981)
40. Zelen, M.: Strategy and alternate randomized designs in cancer clinical trials. *Cancer Treatment Reports* **66**(5), 1095–1100 (1982)
41. Zelen, M.: Randomized consent designs for clinical trials: an update. *Statistics in Medicine* **9**(6), 645–656 (1990)

Chapter 18
Dose Finding Methods in Oncology: From the Maximum Tolerated Dose to the Recommended Phase II Dose

Xavier Paoletti and Adélaide Doussau

Abstract Phase I oncology clinical trials are designed to identify the optimal dose that will be recommended for phase II trials. This dose is typically defined as the dose associated with a certain probability of dose limiting toxicity (DLT) during the first cycle of treatment, although toxicity is repeatedly measured over cycles on an ordinal scale. We present the main dose finding methods developed in the era of cytotoxic agents. We illustrate their properties and limitations in different scenarios. We also explore different implementations of these methods that have been proposed in a Bayesian or likelihood framework or that can rely on several dose-toxicity models. We highlight the fact that the binary nature of the primary outcome (DLT or no DLT) drastically limits the performances of any methods. We then present adaptive dose-finding designs that use toxicity measurements at all cycles of treatment and not only the first one; some authors have proposed to consider the DLT as a time to event variable while others have analyzed the longitudinal measurements of toxic side events. This however raises the delicate issue of the definition of the optimal dose. These approaches are illustrated on two dose finding phase I trials; data are reanalysed and results are compared and discussed. Integration of richer information appears appealing in phase I dose-finding trials, as it gives more accurate estimates of the risk of toxicity and increases the ability of selecting the correct dose. Use of longitudinal data in addition allows for detecting cumulative or delayed effects of strong magnitude. Model-based methods give a flexible framework for using more complete data.

X. Paoletti (✉)
Department of Biostatistics/INSERM U900, Institut Curie, Paris, France
e-mail: xavier.paoletti@curie.fr

A. Doussau
USMR, Bordeaux University-Hospital, ISPED Centre INSERM
U897-Epidemiologie-Biostatistique, Bordeaux, France
e-mail: adelaide.doussau@isped.u-bordeaux2.fr

K. van Montfort et al. (eds.), *Developments in Statistical Evaluation of Clinical Trials*, DOI 10.1007/978-3-642-55345-5_18

18.1 Introduction

Phase I oncology clinical trials are designed to evaluate the toxicity profile of several doses of a new treatment and to identify a dose that can be safely recommended for phase II trials (RPIID). For decades, the fundamental underlying assumption in oncology was "more is better". According to this assumption, the "optimal" dose, recommended for phase II, is defined as the maximum tolerated dose (MTD). The main endpoint is toxicity. Severity of toxicity in cancer clinical trials is graded according to the Common Terminology Criteria for Adverse Events from the National Cancer Institute, which ranges from 1 (mild adverse event) to 5 (death) [33]. Classically, the MTD is a dose associated with a predefined probability of severe grade 3 or 4 toxicity, called dose-limiting toxicity (DLT) evaluated on the first cycle of treatment. This target probability of toxicity ranges from 20 to 30 % [53]. Two families of methods have been developed to find this dose, sometimes called algorithmic and model-based dose escalation designs [26]. O'Quigley et al. [37] proposed a continual reassessment method, an adaptive design based on continuous reestimation of the dose-DLT probability using Bayesian (CRM) or likelihood inference (CRML) [39]. Numerous extensions have been published [19]; all allow the trial to be started at the lowest dose with sequentially increasing dose levels. As these designs (dose allocation, sample size) are modified on the basis of previous observations, they are often described as adaptive methods.

Although model-based methods have been repeatedly shown to have better operating characteristics than algorithm-based methods, the overall probability to identify the correct dose remains low [42]. The performances of these trials including small sample sizes are limited by the elementary binomial variability of the main outcome. The design of these studies needs to be improved, as errors in identification of the MTD are a major cause of failure of subsequent development of new agents in oncology [4, 44]. Since the early 2000s, new classes of molecules have been developed that raise specific issues that cannot be efficiently tacked with basic methods. There is then a convergent need from both the statisticians and the physicians to improve dose finding studies.

Molecularly targeted agents (MTA) have various mechanisms of action as they target different signaling pathways specific to cancer cells. Among their various specificities, it should be noted that: (i) an increasing relationship between dose and activity has not been clearly established for most of agents [16]; an activity plateau above a certain dose is likely; (ii) the toxicity profile is different, with more milder non hematological toxic side events [24]; (iii) they are administered over long periods (even until disease progression for treatment of advanced stages).

The limits of the usual definition of the MTD based on the occurrence of severe toxicity during the first cycle of treatment have been frequently highlighted [57]: moderate toxic side events, repeated measurements of toxicity throughout the trial, activity endpoints are not considered. Not all forms of severe toxicity have the same impact on the possibility to continue the treatment. Phase I trials provide much more information than a simple binomial outcome. Methodological research has been very active; important axes of statistical developments include optimization of the CRM, joint modeling of toxicity and activity endpoints, integration of time into assessment of the endpoint.

It would be impossible to provide a comprehensive review of all methods and we rather focus on the introduction of the time dimension n the dose finding process; we will introduce two motivating examples in Sect. 18.2 and certain notations and we will describe standard methods for dose finding in a single dimension (that is when all doses can be ordered according to increasing toxicity) in Sect. 18.3; in Sect. 18.3.4 we will try to provide an overview of the competing performances of the various methods with particular emphasis on the CRM and the maximum performances we can obtain; we describe methods accounting for temporal aspects in Sect. 18.4 and the approaches used to incorporate longitudinal data and ordinal outcomes in Sect. 18.5. Finally, in Sect. 18.6, we present the application of methods using data obtained after the first cycle on motivating examples. We try to show that although the MTD is still an important notion for clinical development, the dose recommended for phase II should make better use of data collected during phase I.

18.2 Motivating Examples

Two motivating examples of phase I clinical trials of targeted agents are described in this section and will be reanalyzed in Sect. 18.6.

18.2.1 The Erlotinib-Radiotherapy (RT) Trial

The European innovative therapies for children with cancer (ITCC) consortium carried out a phase I trial of erlotinib, a tyrosine kinase inhibitor, in combination with radiotherapy in children with glioblastoma [15]. An adaptation of the CRML was used to identify the dose associated with a 20 % probability of DLT during the first 6 weeks of treatment. A cycle was defined as a 21-day period. Twenty children were evaluated at three increasing doses of erlotinib ranging from 75 to 125 mg/m^2. Two DLTs were observed over the two first cycles: fatal grade 5 seizures, grade 3 skin rash and pruritus. The probability of DLT at 125 mg/m^2 after all patients had been included was 16 % (95 % CI: 4–45 %), and this dose was recommended for phase II studies. A total of 96 cycles were delivered to 20 patients; 12 children completed the 6th cycle of treatment. Six children (26 cycles) received 75 mg/m^2, six children (34 cycles) received 100 mg/m^2 and eight children (36 cycles) received 125 mg/m^2. Nineteen cases of grade 2 toxicity and seven cases of grade 3-5 toxicity were recorded during the first six cycles of treatment, including six cases of grade 3-5 after the first cycle.

18.2.2 The R-Viscum Trial

The European organization for research and treatment of cancer (EORTC) carried out a phase I trial of intravenous aviscumine, an Escherichia coli-derived recombinant type II ribosome-inactivating protein, in adult patients with solid tumors [49]. The CRML was used to identify the dose associated with a 20 % probability of DLT during the first 3 weeks cycle of treatment [40]. Forty-one patients were evaluated at 14 increasing doses ranging from 10 to 6,400 ng/kg. Four DLTs were observed: one case of fatigue and three of hepatitis. The dose recommended for phase II studies was 5,600 ng/kg. One patient was included at each of the following doses levels: 10, 20, 40, 100, 200, 400, 800, 1,600, 2,400 ng/kg; 4 to 10 patients were included at 3,200, 4,000, 4,800, 5,600, 6,400 ng/kg. Ninety-seven cycles were administered; three patients completed six cycles of treatment. We considered clinical toxicities deemed related to treatment according to the investigators as well as laboratory toxicities that worsened from baseline. The worst grades experienced at each cycle were grade 2 in 38 cycles and grade 3 in 22 cycles, including 12 cases of grade 3 toxicity that occurred after the first cycle.

In both trials, data collected after cycle 1 were not formally included in the process to recommend a dose for phase II. The impact of this complementary information will be investigated.

18.3 Notations and Standard Methods

18.3.1 General Notations

Let us assume that n patients are to be sequentially enrolled in a dose finding trial with K dose levels, $d_1, \ldots, d_K$. A patient j is treated at the dose $X_j = d_k$ and he experiences the binary outcome Y_j taking value 1 in case of DLT and 0 otherwise. Each dose level d_k is associated with a probability of toxicity $R_k = \Pr(Y_j = 1|X_j = d_k)$ that increases with the dose. Dose levels can be combination of several agents but one assumes that they can be ordered according to increasing toxicity. The MTD is defined as the dose with a probability of DLT closest to some predefined percentile, denoted τ. Note that in the following discussion, we use the "American" definition of MTD, i.e. the dose with the highest *acceptable* probability of toxicity. This should not be confused with the definition used in some European countries where the MTD is the lowest dose with an *unacceptable* probability of toxicity, typically greater than 33 %.

18.3.2 Algorithm-Based Methods

18.3.2.1 3 + 3 Designs

The most common method used in phase I clinical trials is the so-called 3+3 design. Derived from up and down procedures developed by Robbins and Monro [46], it has been adapted as follows: patients are sequentially entered in groups (often called cohorts) of three patients. The outcomes of a group guide the escalation for the next three patients; suppose dose d_k has been administered to three patients:

- If 0 DLT, escalate to d_{k+1}
- If 1 DLT, expand the dose level and recommend d_k
- If more than 2 DLT out of 3 or 6 patients, recommend d_{k-1}

The trial comes to a halt and d_k is identified as the MTD when two DLT have been observed at d_{k+1} and at most one out of 6 patients at d_k. This method and some variations have been investigated by several authors ([21, 22, 45, 52] among others).

The main conclusions are that the operating characteristics investigated both in simulations and in probabilistic computations are disappointing with low statistical efficiency, poor accuracy of the final recommendation, too many patients treated at excessively low doses deemed to be ineffective, high risk of being conservative in case of a DLT below the MTD [45].

18.3.2.2 Accelerated Titration Designs

This method has been further developed by Simon et al. [51]. An accelerated escalation is obtained by including only one patient per dose level as long as only mild (grade 0 or grade 1) adverse events are observed. As soon as moderate or severe toxicity is experienced, one switches back to the 3 + 3 design. Simon also proposed including intra-patient dose escalation after the first cycle to maximize the chance that a patient would receive an adequate dose. However information after the first cycle is not taken into account in estimation of the MTD and the accelerated titration design eventually cools down to one of the designs investigated by Storer [52]. In these two papers, operating characteristics investigated in simulations were similar to the 3 + 3; the probability of defining the correct dose was not modified except when large number of doses were escalated before reaching the MTD, in which case the accelerated titration design was more efficient. The risk of over toxicity was slightly higher with the accelerated design, but the distribution of allocated patients was more favorable with fewer patients allocated to very low dose levels. The overall duration of the trial was not modified.

18.3.3 Continual Reassessment Method (CRM) and Extensions

The basic principle behind CRM is to reassess the estimate of the dose toxicity relationship after each observation or group of observations to allocate the dose for which the current estimated probability of DLT is closest to τ. Continual reassessment of the dose-DLT relationship is obtained by fitting a model of this relationship to all available data at the current timepoint. The usual CRM uses an underparameterized working model; more specifically, in the case of a homogeneous sample, a one-parameter working model: $P(Y_j = 1|X_j = d_k) = \psi(d_k, a)$ where a is the unknown parameter over the parameter space A. This model, which is an increasing function of d and a monotonic function of a, must be rich enough to ensure that for all $\tau \in (0, 1)$ there exists a value a such that $\psi(d, a) = \tau$. The model is under-parameterized and may not provide an accurate global fit to the true dose-toxicity relations. The only requirement is that it has sufficient flexibility to provide a local fit, limited only by some simple technical conditions which are described in Shen and O'Quigley [8, 50]. Several proposals include the logistic model with one of the parameter fixed, the probit model or the power model. They will be discussed in Sect. 18.3.4; reasons for not working with a richer two-parameter model are outlined in Shen and O'Quigley [50], Cheung [8].

Originally, a Bayesian estimation procedure was proposed. Let us denote $g(a)$ the prior distribution and $L_j(a)$ the likelihood after the inclusion of j patients for whom the paired data $\Omega_j = \{(x_1, y_1), \ldots, (x_j, y_j)\}$ have been observed.

$$L_j(a) = \frac{1}{j} \prod_{\ell=1}^{j} \left[\psi(x_\ell, a)^{y_\ell} \times (1 - \psi(x_\ell, a))^{1-y_\ell}\right] .$$

Posterior distribution of a, $f_j(a)$ is then

$$f_j(a|\Omega_j) = \frac{L_j(a)g(a)}{\int_A L_j(u)g(u)du}$$

O'Quigley and colleagues [37] proposed to compute either the mean posterior distribution $\int_A \psi(d_k, a) f(a)da$ for all k or the less calculation intensive $\psi(d_k, \tilde{a})$ with $\tilde{a} = \int_A af(a)da$. The dose allocated to the next patient or group of patients is the dose closest to the target, i.e. the current estimate of the MTD, denoted, γ: $\hat{\gamma} = |\,\psi(d_k, \tilde{a}) - \tau\,|$. The prior $\exp(-a)$ was first introduced together with gamma or normal prior depending on the parametrization of the model.

Alternatively, parameter estimate, $\hat{a}$, can be obtained from the maximum likelihood estimator [39]; the next recommended dose level then minimizes $|\,\psi(d_k, \hat{a}) - \tau\,|$.

The derivative of the log-likelihood can be written:

$$I_j(a) = \frac{1}{j}\sum_{\ell=1}^{j}\left[y_\ell \frac{\psi'}{\psi}(x_\ell, a) + (1-y_\ell)\frac{-\psi'}{1-\psi}(x_\ell, a)\right] \tag{18.1}$$

$$= \frac{1}{j}\sum_{k=1}^{K}\left[t_k \frac{\psi'}{\psi}(d_k, a) + (n_k - t_k)\frac{-\psi'}{1-\psi}(d_k, a)\right]. \tag{18.2}$$

where t_k denotes the number of DLT observed among the n_k patients allocated at d_k. It is obvious that the maximum of the likelihood arises on the boundary of the parameter space provided at least one DLT and one non DLT have been observed. This is generally called the heterogeneity requirement. The design then consists in two steps with a pre-DLT stage (or run-in) that can be driven either using a Bayesian inference or a simple escalation rule; Paoletti et al. [40] used intermediate grade to calibrate this pre-DLT stage so that doses are rapidly escalated after each new patient when only mild toxicity are observed but more dose escalation is slowed down when moderate (grade 2) events is observed. After the first DLT, dose allocations are derived from the model estimates.

Shen and O'Quigley [50], followed by Cheung et al. [8] showed the good asymptotic properties of the method under model misspecification. In particular, the recommended dose X_{n+1} converges to the true MTD, and the estimate of the risk of DLT $\hat{\gamma}$ converges to its true value γ when n goes to infinity.

Although the decision to stop a trial includes considerations not integrated into the statistical methods (toxicity profile, pharmacokinetic parameters etc.), several stopping rules have been proposed that use either the converging property [38], accuracy of confidence intervals [60] or fixed sample sizes at a dose. We refer the reader to these publications for details.

18.3.3.1 Escalation with Overdose Control (EWOC)

Other authors [2] have suggested controlling for the expected risk that a patient might receive a dose higher than the MTD, by selecting doses so that the posterior probability of overdosing does not exceed some predefined value α. In practice, a 2-parameter logistic model is used, $\psi(ad_k + b)$ re-parameterized in terms of the MTD (γ) and the probability of DLT at d_1, $p_1 = \Pr(Y = 1|X = d_1)$:

$$\gamma = d_1 + \frac{\psi^{-1}(\tau) - \psi^{-1}(p_1)}{a}.$$

The probability of DLT at d_k is then expressed as:

$$\psi(\tau, \gamma, d_k) = \psi\left(\frac{-\gamma \log\left(\frac{p_1}{1-p_1}\right) + d_1 \log\left(\frac{\tau}{1-\tau}\right)}{d_1 - \gamma} + \frac{\log\left(\frac{p_1}{1-p_1}\right) - \log\left(\frac{\tau}{1-\tau}\right)}{d_1 - \gamma} d_k\right) \tag{18.3}$$

The posterior density function (PDF) of (p_1, γ) is denoted $P(p_1, \gamma|\Omega)$. The marginal PDF of the MTD given the accumulated data until patient j is:

$$\pi(\gamma|\Omega_j) = \int P(p_1, \gamma|\Omega_j)dp_1$$

and the marginal cumulative distribution of the MTD at level $d_k \in (d_1, \ldots, d_K)$ is

$$\pi_j(d_k) = \int_{d_1}^{d_k} \pi(\gamma|\Omega_j)d\gamma$$

In practice, the dose selected for the next patient has a posterior probability of exceeding the MTD equal to α: $x_{j+1} = \pi_j^{-1}(\alpha)$.

By extending this idea, Neuenschwander [34] computed the credibility intervals for the probability of toxicity at each dose and identified the probability that a dose may be too low, close to the target, higher or unacceptably high. The dose recommended for the next patient should then have a high probability of being close to the target and a low probability of overdosing and underdosing. However, these two approaches require more flexible models and in both publications, a two-parameter logistic model was used with slightly informative priors to counterbalance the cost for estimating an extra parameter. Simulations show the good operating characteristics with a tendency to be more conservative than the CRM and to more often pick up dose lower than the MTD. In addition as pointed out by Zhang et al. [59], prior distribution is difficult to calibrate in the context of first in man trials for which no information on the toxicity profile is available. CRM can become sensitive to the variance selected for the prior distribution, especially when a small number of patients are enrolled or a large range of doses is explored.

18.3.4 Operating Characteristics of the Competing Methods

18.3.4.1 Softwares

Several software packages that implemented the CRM and 3 + 3 are freely available from websites. A non exhaustive list is provided below.

- NP1: Kramar et al. [23] at *A-Kramar@o-lambret.fr*
- EPCT: Machin et al. at *ukccsg@le.ac.uk* or *epct@cteru.comsg*
- Piantadosi S at http://www.cancerbiostats.onc.jhmi.edu/software.cfm
- MD Anderson at http://biostatistics.mdanderson.org/SoftwareDownload/
- `dfCRM` (R package): K Cheung at http://cran.r-project.org/.

Nevertheless, statisticians are encouraged to develop their own programs as software packages generally implement a single model for the dose-toxicity relationship, a limited set of prior distributions, allocation or stopping rules etc.; none of the softwares cover all of the various developments.

18.3.4.2 Simulation Evaluations

Although asymptotic convergence properties could be drawn for the CRM [50] in contrast with the 3 + 3 method, evaluations with small sample sizes typical of phase I trials are mandatory. Due to the adaptive nature of these designs, analysis and dose allocation are intertwined; reanalysis of collected data using a competing scheme is consequently impossible without further assumptions. Simulations are therefore commonly used. Performances are described by the distribution of the final recommended doses as well as the mean distribution of the allocated doses, the risk of DLT, the distribution of duration of trials etc. As recalled in [9], these parameters depend on the location of the recommended dose (the higher the MTD the lower the probability to pick up the right dose), the shape of the dose toxicity curve (flatter relations lead to poorer results), the sample size, the targeted probability. Comprehensive review of all results is not possible as many authors have compared different approaches [1, 17, 19, 22, 41, 48]. We will focus on the (i) the respective merits of the methods described above, (ii) the operating characteristics of CRM based on 2-parameter models versus 1-parameter models, the maximum value of any method.

18.3.4.3 CRM vs 3 + 3

Iasonos et al [19] conducted simulations that showed that CRM-based methods outperform the standard method by accurately finding the true MTD and by treating more patients at optimal dose levels, which is consistent with the literature [1]. This finding was reported for all scenarios, except when the first levels corresponded to the MTD. Otherwise, CRM-based methods may reach the MTD in fewer patients than the standard method by treating fewer patients at sub-optimal low doses. However, even in situations in which the standard method comes to an halt with a small number of patients, the risk of selecting the wrong dose is very high [38]. The standard method results in very similar number of DLT compared to the CRM. This was confirmed by a review of phase I trials of targeted agents [25]. Across the eight scenarios investigated by Iasonos, the absolute accuracy of CRM (that is the probability to identify the correct dose) was higher by 7–20 % than the standard method. More patients were systematically treated at the optimal dose. Although, CRM-based methods reached the MTD faster, this did not imply that these methods result in earlier termination of the trial. Patients accrual could be optimized by using prospective decision tree, but the benefit in terms of trial duration would be only limited.

18.3.4.4 CRM Implementations

Inference methods (Bayesian or likelihood) for the CRM does not have strong impact on performances provided the prior was correctly specified [39]. Cheung [8] and Paoletti [41] found that prior distribution such as the exponential distribution, may become overly conservative when more than 4 or 5 dose levels must be escalated. It is crucial to carefully elicit the prior. Morita et al. [32] developed a method to construct non informative priors in various scenarios.

The choice of model is a delicate issue that goes beyond the scope of this chapter. The power model received lots of attention as a model with good operating characteristics in this context;

$$\psi(d_k, a) = \alpha_k^a$$

in which $\alpha_k \in (0, 1)$ is a recoding of the dose d_k with $\alpha_1 < \alpha_k < \alpha_K$. The adequate choice of recoding to obtain consistent designs is described in Cheung [9] and the impact on operating characteristics is studied in Paoletti and Kramar [41]. We refer the reader to these documents for comprehensive review of the model properties. The four points that we would like to emphasize are:

- A 1-parameter model (typically a power model or a logistic model with a fixed slope) outperforms a 2-parameter model (such as a logistic model) when using likelihood inference [41]; moreover in a substantial fraction of simulations, a 2-parameter model was not identifiable due to lack of data (DLTs at 2 dose levels are necessary);
- A 2-parameter model must be accompanied by Bayesian inference in order to introduce some information and to facilitate the estimate of the risk of DLT. This makes any comparison difficult as the amount of introduced information may strongly bias the evaluation depending on the (mis)specification of the prior;
- If a logistic one-parameter model is chosen, it is preferable to fix the slope and estimate the intercept as shown in [41], especially when the doses are increased according to relative increments (a higher dose is a percentage of the previous one);
- Although asymptotic convergence is obtained with a wide class of models provided that the model is locally not excessively misspecified [50], operating characteristics are strongly influenced by the choice of model [41]. Roughly, models with a flat slope are associated with more rapid dose escalation, but with more oscillations and slower convergence rate; conversely models with a steeper slope are more stable but lead to more conservative escalation steps. No model is uniformly superior; the choice of the model must be decided after evaluating several scenarios in line with the agent under development. Cheung [9] proposed some tools to help in building the model.

18.3.4.5 Maximum Performances

The question "how well does any approach do?" begs another: "how well is it possible to do?". Constructing optimal designs has been extensively investigated using either information criterion (based on the variance-covariance matrix [6, 30, 58]). Resulting designs are usually not put into practice in oncology due to the risk of overdosing in absence of sequential escalation. They are more often applied in other medical fields [6] for dose ranging phase II studies.

Alternatively, a so-called *optimal* method has been proposed by Paoletti and O'Quigley [36, 42] to serve as a benchmark in simulations studies to quantify the best performances that can be obtained given a dose toxicity relation and a sample size. This approach relies on a notion of 'complete' information, as if a patient could be independently treated at each dose level. Complete information can be summarized with the lowest dose at which the toxicity outcome takes the value 1. This threshold can be derived from a continuous random variable, which is similar to a latent variable. Complete information can be generated if the true probability of toxicity is known. Hence the method cannot be put into practice, but it can serve in a simulation setting. Complete information for the full sample provides observed frequency of DLT at each level d_k. From this efficient non parametric estimate of the discrete dose-toxicity relation, one can select the dose closest to the target as the best estimate of the MTD given a sample size and given a dose-toxicity relation.

This *optimal* method shows the limits of the dose finding process. In a large set of scenarios, the highest probability to pick up the right dose without any informative prior after 25 patients was below 60 % [11]. Paoletti and Kramar [41] underlined that the CRM with likelihood inference produce performances very close to the ones of the *optimal* method. Any method of Phase I clinical trials is limited by the simple binary variability induced by the outcome of interest. It is somehow surprising that so much data is collected to be eventually so dramatically reduced.

Improvement of dose finding will result from the incorporation of more information rather than refining existing methods.

18.4 Accounting for Temporal Aspects

18.4.1 Definition of the Phase II Recommended Dose

Current approaches use only a small fraction of the information collected. Traditional DLT definition, based on grade 3–4 toxicity data from cycle 1 only, has been designed for cytotoxic chemotherapy, and may not be appropriate for new molecularly targeted agents and chronic administration, for which late or moderate toxicities also deserve attention. When the outcome may occur at different treatment cycles, several definitions of the RPIID can be drawn. The time to occurrence of DLT is a possible endpoint; the target is then to identify a dose associated with a predefined risk of *cumulative* toxicity over a given period T. A risk of DLT *per cycle* is an alternative measure of the toxicity of a compound; the target is then to identify a dose associated with a predefined risk of DLT per cycle. Furthermore, the time-course of the risk of toxicity as measured by an ordinal variable over time may be a quantity of interest in order to detect potential late or cumulative effects.

The definition of the RPIID may depend on the expected toxicity profile of the compound under study. Definitions based on the probability per cycle may be relevant for reversible adverse events while irreversible events are probably better described by cumulative measures. We explore some methods adapted to this context.

18.4.2 Methods for Time to Event Endpoints

18.4.2.1 Extending the Evaluation Period with the CRM

The risk of late or repeated toxic side events with non cytotoxic compounds is increasingly feared [57]. The first attempt was to increase the duration of the DLT assessment period; instead of 1 cycle (typically 3 weeks), 2 cycles of treatment were used to define DLT. However, this approach raises numerous issues. The waiting period before accruing new patients is twice as long as that of a 1-cycle assessment; the second issue relates to the risk of missing (or censored) data; due to progressive disease, 50 % of the patients go off-study after completing two cycles of treatment [43]. Data are not missing completely at random as disease progression is expected to depend on the dose (and probably also on baseline characteristics such as the disease type). Simply replacing the patient with missing DLT evaluation results in biased estimates; specific analysis methods are therefore required.

18.4.2.2 Time to Event Continual Reassessment Method (Tite-CRM)

Consider that the binary outcome Y_j denoting DLT is now measured over a period T irrespective of the cycle duration. At a given timepoint $t < T$ of the trial, Y_j may be viewed as censored if no DLT has been observed. Cheung and Chappell [7] proposed to extend the CRM by considering a model of the dose-toxicity relation weighted by the individual follow-up of each patient without DLT; $\phi(d_k, w, a)$ is now a monotone increasing function in w with the constraint that $\phi(d_k, 0, a) = 0$ and $\phi(d_k, 1, a) = \psi(d_k, a)$. The authors investigated a simple linear weight function

$$\phi(d_k, w, a) = w\psi(d_k, a) \text{ with } w(t, T) = \frac{t}{T} \tag{18.4}$$

that assumes that the hazard of DLT is uniform over T. Weighted likelihood can be easily obtained from (18.1). Maximization provides the parameter estimate once some heterogeneity in the response has been observed. Bayesian inference using a prior distribution for parameter a can also be implemented. Design and analysis are then very similar to those of CRM, except that incomplete data can be used and new patients can be included even when some patients are still under study. New allocations will rely on all available information collected up to the current timepoint. The main assumption with this weight function is that the hazard is constant over time. Adaptive weighting schemes independent of the dose effect are presented as an alternative:

$$w(t, T) = \frac{\kappa}{z+1} + \frac{1}{z+1}\left(\frac{t - t_{(\kappa)}}{t_{(\kappa+1)} - t_{(\kappa)}}\right)$$

where z is the total number of toxic observations, $t_{(0)} < t_{(1)} \cdots < t_{(z+l)} \leq T$ are the ordered failure times, and $\kappa = \max_{0 \leq j \leq z}\{j : t \geq t_{(j)}\}$. If most toxic responses occur near the end of the follow-up period, less weight will be given to patients free of DLT but with short follow-up. DLT is always given a weight 1. This approach follows the underlying idea behind CRM that a simple, possibly under-parameterized, working model is efficient on very scarce data with adequate sampling. Cheung and Chappell demonstrate the asymptotic convergence for weight functions independent of the parameter a.

The authors initially presented the operating characteristics for finite sample sizes of 25 patients allocated to a maximum of 6 dose levels and followed for up to $T =$ 6 months. These operating characteristics were compared to those of CRM with a 6-month evaluation period. The MTD was the dose having a cumulative risk of DLT of 20 %. No censoring before T was implemented. The authors reported probabilities of picking up the correct dose comparable to the CRM, or slightly worse depending on the scenario. Nevertheless the duration of the trial was dramatically reduced compared to the CRM. Results were fairly robust to the choice of failure time distributions in the absence of censoring at time earlier than T. If the censoring rate is high, then the probability of toxicity would be more accurately estimated by dynamic weighting [10]. As for the CRM evaluation, the likelihood inference compared favorably to the Bayesian inference. In particular, with a sample size of 25, the inflexibility of the prior distribution limited the possibility of exploring the highest dose levels.

The R-package `dfcrm` introduced earlier can be used to run simulations and to conduct a trial.

18.4.2.3 Time to Event Escalation With Overdose Control (Tite-EWOC)

As previously described, the EWOC method attempts to control the proportion of patients receiving a dose higher than the MTD. Mauguen et al. [29] explored Cheung's proposal [7] to combine the EWOC method with the weighting approach used for time to event data, in order to decrease the duration of dose-finding trial, without impairing the overdose control ability. The likelihood function after j patients is then weighted as in (18.4)

$$L_j(p_1,\gamma) = \prod_{\ell} w\psi(p_1,\gamma,d_1)^{y_\ell}[1 - w\psi(p_1,\gamma,d_1)]^{1-y_\ell}$$

where $\psi(p_1,\gamma,d_1)$ is given in (18.3).

In a simulation study, the authors found similar performances when using EWOC and Tite-EWOC, for various mean inter-patient arrival times and scenarios. In the situations in which Tite-EWOC was slightly inferior to EWOC method, Tite-EWOC more frequently recommended the dose immediately lower than the MTD. The magnitude of reduction of trial duration was directly related to the mean inter-patient arrival time.

18.5 Methods for Longitudinal Ordinal Data

18.5.1 Mixed Effect Proportional Odds Models

An alternative approach using repeated measurements of adverse events has also been developed [12]. As follow-up is very regular in these first-in-man trials, the treatment cycle was used as time scale. The patient was evaluated for toxic side events at each cycle. Let Y_{ij} denote an ordinal variable with three levels g, representing the severity of the worst toxic side events occurring for patient i at cycle j of treatment. Y_{ij} takes value 1 if no toxicity or grade 1 toxicity is observed, 2 for moderate grade 2 toxicity and 3 for severe grade 3–5 toxicity. $Y_{i1} = 3$, the severe toxicity at cycle 1, then corresponds to the usual definition of DLT. Let $p_{2+}(d_k,t_{ij})$ and $p_3(d_k,t_{ij})$ be respectively the probability of outcome 2 or 3 and the probability of outcome 3 at time t_{ij} and at dose d_k. The dose is considered to be constant for a patient throughout the trial. $p_{2+}(d_k,t_{ij})$ and $p_3(d_k,t_{ij})$ are monotonically increasing functions of the dose and are assumed to be related by a proportional odds model (POM). The logistic proportional odds mixed effect regression model (POMM) therefore constitutes a natural candidate [13]. A random intercept u_i is introduced to account for the expected correlation between repeated measurements for a given patient treated at the same dose for several cycles, leading to the following model:

$$\text{logit}\, p_g(d_k,t_{ij}) = \text{logit}(\Pr(Y_{ij} \le g | X_i = d_k)) = \alpha_g - \beta_1 \times d_k - u_i \qquad (18.5)$$

where $u_i \sim \mathcal{N}(0,\sigma_0^2)$ and $g = 1,2$. We denote θ the vector of the four parameters: $\theta = (\boldsymbol{\alpha_1},\boldsymbol{\alpha_2},\boldsymbol{\beta_1},\boldsymbol{\sigma_0})$. According to the odds proportional assumption, the association between the dose and the risk of severe toxicity is the same as the association between the dose and the risk of moderate or severe toxicity. Of note, we first assume that there is no time effect on the risk of toxicity. The marginal probability of event g over all administered cycles can therefore be denoted by $p_{2+}(d_k)$ and $p_3(d_k)$.

Given the observations of the dose and event outcomes (x_i, y_{ij}), at a given timepoint of the trial, the likelihood for the parameter vector θ, is

$$L(y_{ij}|\theta) = \prod_{i,j} (p_1(x_i|u_i))^{I_{y_{ij}=1}} \times (p_2(x_i|u_i))^{I_{y_{ij}=2}} \times (p_3(x_i|u_i))^{I_{y_{ij}=3}} \qquad (18.6)$$

where $I_{[y_{ij}=g]}$ takes value 1 if $Y_{ij} = g$ and 0 otherwise. As the random effect is unknown, evaluation of L must integrate the random effect distribution. No closed form is available and maximization is obtained using Laplace approximation and adaptive Gauss-Hermite quadrature.

Maximization of the full likelihood provides unbiased estimates in the case of data missing at random according to Rubin's classification [31]. In fact, patients usually go off-study after severe toxicity or when their cancer progresses resulting in missing data; if we assume that disease progression is largely independent of the risk of toxicity given the dose level, the missing data are expected to be missing at random.

Following the principle of adaptive design for dose finding trials, estimates $\hat{p}_g(d_k)$ are used to conduct dose allocation; we call this method POMM-CRML. A possible decision criterion consists of minimizing $|\hat{p}_3(d_k) - \tau|$. Patients are sequentially enrolled in the trial starting at the lowest dose. A new patient can only be included when the previous patients have completed at least one cycle of treatment. Extension to grouped inclusions is straightforward. Before each new inclusion,

1. Fit a POMM to all collected data, i.e. to the outcomes at all cycles for all patients previously included available at the time of the new inclusion. Simpler models can be fitted when estimates of the model (18.5) cannot be obtained.
2. Evaluate the decision criteria and identify the dose whose estimate of the risk of severe toxicity per cycle is closest to τ.
3. The new patient is treated at this current recommended dose.
4. The trial is terminated when the maximum allowable number of patients has been treated or after certain stopping rules have been verified [38, 60].

Time trend can be further investigated using a model with a time covariate.

$$\text{logit}(P(Y_{ij} \leq g|X_i = d_k, t_{ij})) = \alpha_g - \beta_1 d_k - \beta_2 t_{ij} - u_i \qquad (18.7)$$

where $u_i \sim \mathcal{N}(0, \sigma_0^2)$ and $g = 1, 2$. Probabilities of outcomes are then assumed to be related by a proportional odds model for both time and dose. Model (18.7) can be used to test time effects as reflected by a significant β_2 parameter. As this situation turns out to be very rare [43,51], it was proposed to test for this effect only at the end of the trial to avoid decreasing the test power. Alternatively, a sequential probability ratio test may be implemented. In case of increasing time trend, definition of the RPIID can be challenging. Other characteristics of the compound would then be used to define the dose that should be further investigated, if any; they include the pharmacokinetics, the possibility of changing the schedule, the toxicity profile etc.

18.5.2 *Operating Characteristics*

18.5.2.1 Identifying the RPIID

Operating characteristics were investigated in a simulation study assuming fixed sample sizes of 30 patients who could receive up to 6 cycles and various dose-toxicity relations; scenarios assumed eight dose levels (60, 120, 200, 300, 400, 600, 800, 1,000 mg) transformed on the log scale with increasing risk of toxicity following either a proportional odds model or not; the MTD was either level 2, 4 or 6. Missing data were assumed to be due to severe toxicity or progression; time to progression was independent of the risk of toxicity and of the dose [16]. Results from one scenario are provided in Table 18.1. In all scenarios studied, fitting a mixed effect POM appeared to be feasible for the sample sizes typically used in phase I trials (models could be fitted in more than 97 % of simulations). The CRML correctly identified the MTD based on the first cycle only in less than 50 % of the simulations, regardless the explored dose-toxicity relation. In absence of increasing risk of toxicity with time, the adaptive POMM-CRML allowed the probability of a correct recommendation to be increased to more than 62 %. The mean number of patients treated at the correct dose was systematically increased. These good performances were also observed when using a retrospective longitudinal analysis after all data had been collected using the CRML, resulting in the same level of correct recommendations. Of note, in some simulations, overly toxic doses were recommended after longitudinal reanalysis that would not have been recommended with CRML (<1 % of simulations). The two missing data processes displayed similar results. When the dose-toxicity relationship violated the proportional odds assumption, the results remained fairly robust. Even when the slope of the dose-toxicity relationship was higher for G2-5 compared to G3 toxicities, it did not result in an increased risk of recommendation of higher doses than the true RPIID or in an increased risk of overtreatment.

18.5.2.2 Detection of Time Trend

The model including both time and dose could be estimated in more than 92 % of simulations. The power to identify a time trend after treating 30 patients increased with the strength of the time effect from 46 % (for $OR = 1.33$ at cycle i compared to cycle $i - 1$) to 93 % ($OR = 1.79$, i.e. receiving an additional cycle of treatment was roughly equivalent to receiving the next higher dose level). The false-positive rate in the absence of a time effect was 5 % of simulations, indicating an adequate type I error rate control despite the limited sample size.

Table 18.1 Results of competing strategies using the CRM on the first cycle (CRM_{30}), the CRM on the first cycle together with a reanalysis of longitudinal data (CRM_{30} with POMM), and the proposed method using longitudinal data for dose allocation (POMM-CRML); data may be complete, or missing after G3+ toxicity or after progression

Dose level d_k	1	2	3	**4**	5	6	7	8	N cycles
$\log(d_k)$	4.1	4.8	5.3	**5.7**	6	6.4	6.9,	7.2	mean (sd)
Probability G3	0.02	0.06	0.14	**0.24**	0.37	0.55	0.71	0.83	
Probability G2+	0.07	0.21	0.4	**0.58**	0.71	0.84	0.91	0.95	
Cycle 1									
$CRML_{30}$									
Mean number of pts / dose	1.8	3.9	7.8	**9.3**	5.4	1.4	0.3	0.05	30 (0)
Distribution DR (%)	0.3	5.1	29.2	**46.7**	17.9	0.8	0	0	
Longitudinal data									
Complete data									
Retrospective POMM analysis									
Mean number of pts / dose	1.8	3.9	7.8	**9.3**	5.4	1.4	0.3	0.05	180 (0)
Distribution DR (%)		0.7	13.3	**73.4**	11.1	0.9	0.3	0	
POMM-CRML adaptive design									
Mean number of pts / dose	1.7	3.0	6.2	**12.9**	4.6	1.2	0.3	0.5	180 (0)
Distribution DR (%)		0.5	12.8	**76.5**	10.1	0.1	0	0	
Missing data after the first severe toxicity									
Retrospective POMM analysis									
Mean number of pts / dose	1.8	3.9	7.8	**9.3**	5.4	1.4	0.3	0.05	111.3 (19.3)
Distribution DR (%)	0.2	1.9	18.5	**62.3**	15.4	1.0	0.0	0.6	
POMM-CRML adaptive design									
Mean number of pts / dose	1.8	3.4	7.0	**11.1**	4.9	1.2	0.3	0.2	110.1 (14.8)
Distribution DR (%)	0.2	1.3	20.9	**62.7**	14.5	0.1	0.2	0.1	
Missing data after progression									
Retrospective POMM analysis									
Mean number of pts / dose	1.8	3.9	7.8	**9.3**	5.4	1.4	0.3	0.05	111.5 (19.3)
Distribution DR (%)	0.2	0.8	18.6	**63.0**	14.7	1.4	0.4	0.9	
POMM-CRML adaptive design									
Mean number of pts / dose	1.8	3.0	6.7	**12.2**	4.7	1.1	0.3	0.2	119.3 (9.1)
Distribution DR (%)	0.1	1.0	17.2	**69.6**	12.2	0	0	0	

G3: Severe toxicity, G2+: Moderate or severe toxicity. *DR* dose recommended at the end of the simulated trial, *Bold entries* correspond to the target dose

18.6 Applications

Data from the trials described in Sect. 18.2 were reanalyzed using both time to event models (for the ITCC trial 1) and a proportional odds (PO) mixed effect model for both examples. As reanalysis of adaptive designs is not directly feasible and requires further assumptions, only estimates of the probability of toxicity at different time points of the trials are provided, bearing in mind that if other methods had been

Table 18.2 Final analysis of the erlotinib + radiotherapy with CRML

Dose level d_k	d_1	d_2	d_3	d_4
Dose in mg/m^2	75	100	125	150
α_k	0.07	0.2	0.35	0.50
♯ pts at d_k	6	6	8	0
♯ G3 at cycle 1	1	0	1	0
$\hat{p}_3(d_k)$	0.02	0.06	0.16	0.35

G3: severe toxicity. α_k is the code for d_k

used, the dose allocation would not have been the same; analyzing data deriving from a different design results in a certain degree of loss of efficiency.

18.6.1 *The ITCC/Erlotinib + RT Trial*

18.6.1.1 CRML

Results of the trial conducted with the CRML are summarized in the Table 18.2. The final recommended dose was dose level 3 (125 mg/m^2) with an estimated risk of DLT of 0.16, 95 % confidence interval $(0.04, 0.45)$. After 20 patients had been evaluated, the confidence interval was too large to provide any meaningful information and could not be used to guide the decision to stop the trial. Nevertheless, the probability to maintain the same dose level (d_2) for the next 5 patients was higher than 85 % and the trial was stopped.

18.6.1.2 Tite-CRM

The cumulative risk of DLT over $T = 6$ cycles was estimated using the time-to event CRM. A power 1-parameter model was chosen; the three dose levels of erlotinib and radiotherapy were coded as previously: $\alpha_1 = 0.07, \alpha_2 = 0.2$ and $\alpha_3 = 0.35$. A Bayesian inference was used with a non informative exponential prior distribution.

Table 18.3 gives three snapshots of the estimated risk of severe toxicity: on 28th June 2006, just before patient 11 was to be included, the estimates ranged from $\hat{p}_3(d_1) = 16\,\%$ to $\hat{p}_3(d_3) = 49\,\%$. On 16th July, the same patients had longer follow-up without DLT resulting in lower estimated risk of DLT. The last reanalysis after all data have been collected indicated a much larger cumulative risk of DLT: probability at d_1 was 20 % whereas probability at d_3 was as high as 54 %. This should be compared to the results of the estimates based on the first cycle only where d_3 appeared to be tolerable with $\hat{p}_3(d_3) = 0.16$. At completion, 13 patients received 6 cycles or experienced DLT and had weight 1, whereas 7 had weights ranging from 0.33 to 0.83. Dynamic weighting gave fairly similar results.

Table 18.3 Erlotinib + radiotherapy reanalysis with the Tite-CRM: after six patients have completed evaluation and three are under evaluation at two different timepoints; and after trial completion

Patient j	1	2	3	4	5	6	7	8	9
Dose level d_k	1	1	1	1	1	1	2	2	2
Y_j	0	0	1	0	0	0	0	0	0
The 28th of June									
♯cycles	6	6	2	6	4	2	3	2	1
w_j	1	1	1	1	0.67	0.33	0.5	0.33	0.17
$\hat{p}_3(d_k)$	$\hat{p}_3(d_1) = 0.16$						$\hat{p}_3(d_2) = 0.33$		
The 16th of July									
♯cycles	6	6	2	6	4	2	4	3	2
w_j	1	1	1	1	0.67	0.33	0.67	0.5	0.33
$\hat{p}_3(d_k)$	$\hat{p}_3(d_1) = 0.12$						$\hat{p}_3(d_2) = 0.31$		

Final reanalysis after trial completion				
Dose level d_k	d_1	d_2	d_3	d_4
♯pat at d_k	6	6	8	0
♯$G3$ at any cycle	2	1	3	0
$\hat{p}_3(d_k)$	0.21	0.38	0.54	0.70

G3: severe toxicity; #pat: number of patients; #cycles: number of cycles

18.6.1.3 POMM-CRM

Data were then reanalyzed to identify the RPIID, defined as the dose associated with 20 % of DLT *per cycle* and to detect a trend time. Only retrospective analysis of the data using a PO mixed effect model is considered. Models (18.5) and (18.7) were estimated by adjusting for the log of the dose. Estimates of fixed intercepts, time and dose were then $\hat{\theta} = (\alpha_1 = 4.64, \alpha_2 = 6.19, \beta_1 = 0.80, \beta_2 = -0.03)$; variance of the random effect, σ_0^2 could not be estimated as it appeared to be excessively large; time trend was not significant ($p = 0.82$); estimates of the model with dose only gave $\hat{\theta} = (\alpha_1 = 4.69, \alpha_2 = 6.24, \beta_1 = 0.80)$. The predicted probabilities of toxicity per cycle at each dose are shown in Table 18.4. The risk of severe toxicity at each cycle using a PO mixed effect model appears lower compared to the estimates on the first cycle only using a logistic model. The risk of toxicity at the highest visited dose d_3 might have encouraged the investigators to explore d_4.

In conclusion, incorporating evaluation of toxic side effects obtained from further cycles of treatment can lead to radically different recommendations depending on how we define the DLT and the RPIID. Were cumulative risk of DLT be the main endpoint then the treatment appears quite toxic and lower dose levels should be recommended. Conversely, the risk of severe toxicity per cycle is in line with what had been observed on the first cycle. Some toxic side events are reversible and manageable, while others are not, which may guide the modeling choice. Typically, risk of skin rash observed with erlotinib + radiotherapy is probably well described

Table 18.4 Erlotinib + radiotherapy reanalysis with POMM: observed and predicted per cycle probability of graded toxicity, according to the log of the dose

Dose level d_k	d_1	d_2	d_3
Number of pat	6	6	8
Number of cy.	26	34	36
Number of G2	4	8	7
Number of G3	2	1	4
Obs. G3 (per cy.), in %	7.7	2.9	11.1
Obs. G2+ (per cy.), in %	23.1	26.5	30.6
Pred. G3 (per cy.),in %	5.8	7.2	8.5
Pred. G2+ (per cy.),in %	22.6	26.9	30.5

G3: Severe toxicity, G2+: Moderate or severe toxicity, cy.: cycle, Pred: Predicted, Obs: Observed. The four smallest dose levels were collapsed in one column

as a risk per cycle. On the contrary, hemorrhages, another adverse event of this combination, would be more appropriately assessed by cumulative risk. Combining both modelings is a promising field of research.

18.6.2 The EORTC/R-Viscum Trial

18.6.2.1 CRML

Results of the trial conducted with the CRML are summarized in Table 18.5. 11 dose levels were escalated after each new patient tolerated the treatment before the first DLT, fatigue grade 3, was observed at $d_{11} = 4{,}000$ ng/kg. The levels were coded as follows: $0.0035, 0.005, 0.009, 0.015, 0.024, 0.035, 0.05, 0.07, 0.11, 0.2, 0.33, 0.48, 0.62, 0.74$. As no limit in the dose increase had been defined in the protocol, the coding was set up when the first DLT was observed: d_{10}, the dose at which the first DLT was observed, was coded 0.2; other codes were constructed so that if d_k is the current estimate of the MTD (that is the dose closest to 20 %) then $\psi(d_{k+1}, \hat{a})$ was close to 0.33; in other words, the estimated slope around the MTD was stable wherever the MTD was located.

After 27 patients have been included, the sponsor amended the protocol and switched to the 3 + 3 design. Final estimates using the CRM model are displayed in Table 18.5; columns of the first six doses levels have been collapsed; the final recommended dose was d_{13} (5,600 mg/kg) with an estimated risk of DLT of 0.16, 95 % confidence interval $(0.07, 0.37)$. The probability to maintain the same dose level for the next five patients was 90 %, had we continued with the CRM.

Table 18.5 Final analysis of the R-Viscum with the CRML

Dose level d_k	$d_1 - d_8$	d_9	d_{10}	d_{11}	d_{12}	d_{13}	d_{14}
Dose in mg/kg	10–1,600	2,400	3,200	4,000	4,800	5,600	6,400
α_k	0.0035–0.07	0.11	0.20	0.33	0.48	0.62	0.74
♯pat at d_k	8	1	4	4	10	7	5
♯$G3$ at cycle 1	0	0	0	1	1	0	2
$\hat{p}_3(d_k)$	0.0–0.0	0.0	0.0	0.02	0.06	0.16	0.31

G3: severe toxicity, #pat: number of patients

Table 18.6 r-Viscum trial reanalysis: observed and predicted probability of graded toxicity per cycle, according to the dose

	Dose (ng/kg)										
	10–100	200	400	800	1,600	2,400	3,200	4,000	4,800	5,600	6,400
Number of pat	4	1	1	1	1	1	4	6	10	7	5
Number of cy.	15	2	1	2	2	2	13	11	25	13	8
Number of G2	3	0	0	0	1	2	4	5	12	7	4
Number of G3	2	0	1	0	1	0	0	3	9	2	4
Obs. G3 (per cy.) %	20	0	100	0	50	0	0	27	36	15	50
Obs. G2+ (per cy.) %	13	0	100	0	100	100	30	72	84	69	100
Pred. G3 (per cy.) %	4	4	4	5	5	6	8	14	24	37	50
Pred. G2+ (per cy.) %	43	45	46	48	50	55	64	76	86	92	95

G3: Severe toxicity, G2+: Moderate or severe toxicity, #pat: number of patients, #cy.: number of cycles. The four smallest dose levels were collapsed in one column

18.6.2.2 POMM-CRM

The data introduced in Sect. 18.2 were reanalysed to identify the RPIID and detect a time trend. The targeted probability of severe toxicity per cycle was set at 20 %, to match the target used in the trial. As before, retrospective analysis of the data using a POMM is presented. Models (18.5) and (18.7) were estimated. Estimates of fixed intercepts, time and dose and variance of the random intercept were $\mathring{\theta} = (\alpha_1 = 0.35, \alpha_2 = 3.33, \beta_1 = 4.48, \beta_2 = 0.04, \sigma_0^2 = 3.44)$; time trend was not significant (p=0.88); estimates of the model with dose as the only covariate gave $\theta = (\alpha_1 = 0.31, \alpha_2 = 3.28, \beta_1 = 4.42, \sigma_0^2 = 3.35)$; the dose effect as well as the random intercept were both significant in this model ($p = 0.01$ and $p < 0.001$ respectively). The predicted probabilities of toxicity per cycle at each dose are shown in Table 18.6. The first four columns were collapsed to form one column labeled "10–100". According to the POMM, the recommended dose would have been d_{12} (4,800 ng/kg).

Analysis of all collected data improved the accuracy of the estimate of the risk of DLT. As all toxic side effects were reversible, a longitudinal model is appealing. The additional information from all cycles of treatment can then be easily interpreted and are consistent with the usual estimates from the first treatment cycle.

18.7 Conclusions

Following the introduction of the CRM in 1990, dose finding methods have been the subject of numerous statistical developments. The main endpoint of the trial was a binary variable (presence or absence of dose limiting toxicity) measured on the first cycle of treatment; the objective of the trial was formalized as identifying a pre-defined percentile of an unknown dose-toxicity relation. The introduction of adaptive designs using simple "working" models to estimate the risk of DLT improved the performance of the dose finding process and the flexibility of the design. Building on this model, it was possible to account for heterogeneity using covariates [42], to target different levels of risk etc. however, the operating characteristics of these methods are very close to the maximum performance that can be achieved. Performances are limited by the amount of information contained in the binary variable used as primary endpoint.

In the last decade, new classes of agents have emerged. The very severe toxicity at the first cycle may no longer be the most appropriate endpoint to identify the dose recommended for phase II.

The approaches presented in Sect. 18.4 explore repeated measurements of toxic side effects expressed either as cumulative risk or risk per cycle. Time to event does not improve the ability to identify the correct dose, but reduces the overall duration of a trial and more importantly provides a valid tool to account for late toxic side events. Conversely, the longitudinal approach was associated with a net improvement in terms of probability to identify the correct dose for a set of different scenarios, compared to the CRML. A similar improvement would be obtained by simply fitting the proportional odds mixed effect model to the final sample collected by using the CRML. The conclusions were robust across scenarios, even when the dose-toxicity relationship violated the proportional odds assumption. In addition, this methods provided a simple tool to assess the time trend of the risk of toxicity. Estimating the model was not straightforward, especially early in the trial. A Bayesian approach, with prior information on the variance parameter for the random effect, may increase the accuracy of the estimation.

Proportional odds models appear to be an interesting framework for modeling graded toxicity; it is likely that the same mechanism is responsible for severe toxicities and moderate toxicities, which makes the proportional odds assumption reasonable. A hypothesis of proportionality over repeated cycles of treatment was proposed by Simon, who used a proportional odds Kmax model to detect late toxicity in NCI phase I trials of old agents [51] but he did not develop adaptive dose finding methods in this framework. This was done by Legezda and Ibrahim [27] who simplified the model of Simon that was not identifiable with limited sample sizes. They proposed a mixed effect model for binary endpoint where the risk of DLT at a given cycle results from the administered dose plus the cumulative dose received since treatment initiation weighted by the clearance of the agent under study. They further hypothesize that the clearance is known at the start of the trial, and may vary from patients to patients (random effect). This model is directly drawn from PK-PD models developed for cytotoxic agents; this therapeutic class, administered intravenously, commonly induces hematological toxicity as a consequence of the concentration of the agent in the blood. The authors reported good operating characteristics after 10 cycles of treatment. As they allowed intra-patient dose adaptation (a patient may receive different doses throughout the trial) and did not implement end of treatment after severe toxicity, their results cannot be directly compared to those from Doussau et al. described above. More recently, the PO model applied on the first cycle of treatment was evaluated but it failed to improve the chance to pick up the correct level [56], probably due to the additional parameters that must be estimated and due to the fact that the MTD is still defined as a probability of DLT. It is therefore much more promising to use the ordinal scale to increase the power to investigate time trends. Under the realistic assumption that any time trend would be observed on the risk of both moderate and severe toxicity, graded toxicity becomes a more informative variable.

Another very important axis of methodological research to exploit richer information involves incorporating multiple endpoints into the dose finding process. In particular, the joint distribution of the risk of severe toxicity as well as the chance of clinical response, also measured as a binary [54] or a ternary endpoint (tumor shrinkage, stabilization progression) [18] can serve to identify the dose with the best trade-off between acceptable toxicity rate and response rate. This approach transforms dose-finding trials from simple identification of the MTD to the more stimulating objective of recommending the phase II dose. This topic, opened by Thall and Russel [55], has been frequently explored: authors relied on conditional probabilities [35] or copula [5] to obtain this joint estimation. Other authors have taken advantage of the natural ordering between a toxic dose without and with activity and a dose active without toxicity leading to the use of ordinal models [28]. The two major limitations to implement these methods in clinical practice are (i) the statistical complexity requiring the use of priors and Bayesian inference as well as extensive computations and (ii) the lack of sensitivity of the clinical outcome used to measure agent activity; in fact, less than 5 % of patients in first-in-man trials of new agents have a tumor response measured according to the usual RECIST [20,47] making it challenging to model dose-activity relationship. More recent contributions have explored the possibility of using continuous markers of activity [3, 14].

Although the transfer of these innovative designs to clinical practice has given disappointing results up to now, model-based designs are tailored to the complexity of dose finding trials of targeted agents and should meet the expectations of clinicians and sponsors.

Acknowledgements The authors are indebted to Pr Rodolphe Thiebaut (Bordeaux University Hospital/ISPED, France) who actively contributed to the developments on longitudinal data presented in Sect. 18.5. The text benefited from careful review of an external reviewer and from Dr Antony Saul for English edition. We would like also to express our gratitude to Pr Patrick Schöefski (University hospital Leuven, Belgium), Pr Pierre Fumoleau (Dijon Cancer Center and University, France), Elisa Rizzo and Laurence Collette (EORTC-HQ Brussels, Belgium) who provided the individual patient data for the EORTC R-Viscum study. Likewise, we are grateful to Dr Birgit Geoerger and Dr Marie-Cécile le Deley (Institut Gustave Roussy, Villejuif, France) for providing data for the ITCC erlotinib trial. This work was partly supported by a grant from the French NCI (INCa) *# 2010-1-PL SHS-06-IC-1 / optidose*.

References

1. Ahn, C.: An evaluation of phase I cancer clinical trial designs. *Statistics in medicine* **17**(14), 1537–1549 (1998). PMID: 9699228
2. Babb, J., Rogatko, A., Zacks, S.: Cancer phase I clinical trials: efficient dose escalation with overdose control. *Statistics in medicine* **17**(10), 1103–1120 (1998). PMID: 9618772
3. Bekele, B.N., Shen, Y.: A Bayesian approach to jointly modeling toxicity and biomarker expression in a phase I/II dose-finding trial. *Biometrics* **61**(2), 343–354 (2005). DOI 10.1111/j.1541-0420.2005.00314.x. PMID: 16011680
4. Booth, C., Calvert, A., Giaccone, G., Lobbezoo, M., Seymour, L., Eisenhauer, E.: Endpoints and other considerations in phase I studies of targeted anticancer therapy: recommendations from the task force on methodology for the development of innovative cancer therapies (mdict). *European Journal of Cancer* **44**(1), 19–24 (2008)
5. Braun, T.: The bivariate continual reassessment method. extending the CRM to phase I trials of two competing outcomes. *Control Clinical Trials* **23**(3), 240–56 (2002)
6. Bretz, F., Dette, H., Pinheiro, J.C.: Practical considerations for optimal designs in clinical dose finding studies. *Statistics in medicine* **29**(7–8), 731–742 (2010). DOI 10.1002/sim.3802. PMID: 20213708
7. Cheung, Y., Chappell, R.: Sequential designs for phase I clinical trials with late-onset toxicities. *Biometrics* **56**(4), 1177–1182 (2000). PMID: 11129476
8. Cheung, Y., Chappell, R.: A simple technique to evaluate model sensitivity in the continual reassessment method. *Biometrics* **58**(3), 671–674 (2002). PMID: 12230003
9. Cheung, Y.K.: *Dose finding by the continual reassessment method*. CRC Press, Boca Raton (2011)
10. Chevret, S.: *Statistical Methods for Dose-Finding Experiments*. Wiley (2006)
11. Doussau, A., Asselain, B., Le Deley, M.C., Geoerger, B., Doz, F., Vassal, G., Paoletti, X.: Dose-finding designs in pediatric phase I clinical trials: comparison by simulations in a realistic timeline framework. *Contemporary clinical trials* **33**(4), 657–665 (2012). DOI 10.1016/j.cct.2011.11.015. PMID: 22521954
12. Doussau, A., Thiébaut, R., Paoletti, X.: Dose-finding design using mixed effect proportional odds model for longitudinal graded toxicity data in phase i oncology clinical trials. *Statistics in Medicine* **32**(30), 5430–5447 (2013). DOI 10.1002/sim.5960
13. Ezzet, F., Whitehead, J.: A random effects model for ordinal responses from a crossover trial. *Statistics in medicine* **10**(6), 901–906; discussion 906–907 (1991). PMID: 1876780
14. Fedorov, V., Wu, Y., Zhang, R.: Optimal dose-finding designs with correlated continuous and discrete responses. *Statistics in medicine* **31**(3), 217–234 (2012). DOI 10.1002/sim.4388. PMID: 22162014
15. Geoerger, B., Hargrave, D., Thomas, F., Ndiaye, A., Frappaz, D., Andreiuolo, F., Varlet, P., Aerts, I., Riccardi, R., Jaspan, T., Chatelut, E., Le Deley, M.C., Paoletti, X., Saint-Rose, C., Leblond, P., Morland, B., Gentet, J.C., Méresse, V., Vassal, G.: Innovative therapies for children with cancer pediatric phase I study of erlotinib in brainstem glioma and relapsing/refractory brain tumors. *Neuro-oncology* **13**(1), 109–118 (2011). DOI 10.1093/neuonc/noq141. PMID: 20974795
16. Gupta, S., Hunsberger, S., Boerner, S.A., Rubinstein, L., Royds, R., Ivy, P., LoRusso, P.: Meta-analysis of the relationship between dose and benefit in phase I targeted agent trials. *Journal of the National Cancer Institute* **104**(24), 1860–1866 (2012). DOI 10.1093/jnci/djs439. PMID: 23169991

17. Heyd, J.M., Carlin, B.P.: Adaptive design improvements in the continual reassessment method for phase I studies. *Statistics in medicine* **18**(11), 1307–1321 (1999). PMID: 10399198
18. Houédé, N., Thall, P., Nguyen, H., Paoletti, X., Kramar, A.: Utility-based optimization of combination therapy using ordinal toxicity and efficacy in phase I/II trials. *Biometrics* **66**(2), 532–540 (2010). DOI 10.1111/j.1541-0420.2009.01302.x. PMID: 19673865
19. Iasonos, A., Wilton, A.S., Riedel, E.R., Seshan, V.E., Spriggs, D.R.: A comprehensive comparison of the continual reassessment method to the standard 3 + 3 dose escalation scheme in phase I dose-finding studies. *Clinical trials (London, England)* **5**(5), 465–477 (2008). DOI 10.1177/1740774508096474. PMID: 18827039
20. Italiano, A., Massard, C., Bahleda, R., Vataire, A.L., Deutsch, E., N, M., Pignon, J.P., Vassal, G., Armand, J.P., Soria, J.C.: Treatment outcome and survival in participants of phase I oncology trials carried out from 2003 to 2006 at institut gustave roussy. *Annals of oncology* **19**(4), 787–792 (2008). DOI 10.1093/annonc/mdm548. PMID: 18042834
21. Ivanova, A., Montazer-Haghighi, A., Mohanty, S., Durham, S.D.: Improved up-and-down designs for phase I trials. *Statistics in medicine* **22**(1), 69–82 (2003). DOI 10.1002/sim.1336. PMID: 12486752
22. Korn, E.L., Midthune, D., Chen, T.T., Rubinstein, L.V., Christian, M.C., Simon, R.M.: A comparison of two phase I trial designs. *Statistics in medicine* **13**(18), 1799–1806 (1994). PMID: 7997713
23. Kramar, A., Houédé, N., Paoletti, X.: np1: a computer program for dose escalation strategies in phase I clinical trials. *Computer methods and programs in biomedicine* **88**(1), 8–17 (2007). DOI 10.1016/j.cmpb.2007.06.006. PMID: 17719124
24. Le Tourneau, C., Diéras, V., Tresca, P., Cacheux, W., Paoletti, W.: Current challenges for the early clinical development of anticancer drugs in the era of molecularly targeted agents. *Targeted oncology* **5**(1), 65–72 (2010). DOI 10.1007/s11523-010-0137-6. PMID: 20361265
25. Le Tourneau, C., Gan, H.K., Razak, A.R.A., Paoletti, X.: Efficiency of new dose escalation designs in dose-finding phase I trials of molecularly targeted agents. *PloS one* **7**(12), e51,039 (2012). DOI 10.1371/journal.pone.0051039. PMID: 23251419
26. Le Tourneau, C., Lee, J., Siu, L.: Dose escalation methods in phase I cancer clinical trials. *Journal of the National Cancer Institute* **101**(10), 708–720 (2009)
27. Legezda, A.T., Ibrahim, J.G.: Longitudinal design for phase I clinical trials using the continual reassessment method. *Controlled Clinical Trials* **21**, 574–88 (2000)
28. Mandrekar, S., Qin, R., Sargent, D.: Model-based phase I designs incorporating toxicity and efficacy for single and dual agent drug combinations: methods and challenges. *Statistics in Medicine* **29**(10), 1077–83 (2010)
29. Mauguen, A., Le Deley, M.C., Zohar, S.: Dose-finding approach for dose escalation with overdose control considering incomplete observations. *Statistics in medicine* **30**(13), 1584–1594 (2011). DOI 10.1002/sim.4128. PMID: 21351289
30. McLeish, D., Tosh, D.: Sequential designs in bioassay. *Biometrics* **46**, 103–16 (1990)
31. Molenberghs, G., Verbeke, G.: *Models for Discrete Longitudinal Data*. Springer (2005)
32. Morita, S., Thall, P., Müller, P.: Determining the effective sample size of a parametric prior. *Biometrics* **64**(2), 595–602 (2008). DOI 10.1111/j.1541-0420.2007.00888.x. PMID: 17764481
33. National Cancer Institute, U.N.I.o.H.C.T.E.P.: Common terminology criteria for adverse events (CTCAE) v4.03. (2010). URL http://ctep.cancer.gov/protocolDevelopment/electronic_applications/ctc.htm
34. Neuenschwander, B., Branson, M., Gsponer, T.: Critical aspects of the Bayesian approach to phase I cancer trials. *Statistics in medicine* **27**(13), 2420–2439 (2008). DOI 10.1002/sim.3230. PMID: 18344187
35. O'Quigley, J., Hughes, M., Fenton, T.: Dose-finding designs for HIV studies. *Biometrics* **57**(4), 1018–24 (2001)
36. O'Quigley, J., Paoletti, X., Maccario, J.: Non-parametric optimal design in dose finding studies. *Biostatistics* **3**(1), 51–56 (2002). DOI 10.1093/biostatistics/3.1.51. PMID: 12933623
37. O'Quigley, J., Pepe, M., Fisher, L.: Continual reassessment method: a practical design for phase I clinical trials in cancer. *Biometrics* **46**(1), 33–48 (1990). PMID: 2350571

38. O'Quigley, J., Reiner, E.: Miscellanea. a stopping rule for the continual reassessment method. *Biometrika* **85**(3), 741–748 (1998). DOI 10.1093/biomet/85.3.741. URL http://biomet.oxfordjournals.org/content/85/3/741.full.pdf+html
39. O'Quigley, J., Shen, L.Z.: Continual reassessment method: a likelihood approach. *Biometrics* **52**(2), 673–684 (1996). PMID: 8672707
40. Paoletti, X., Baron, B., Schöffski, P., Fumoleau, P., Lacombe, D., Marreaud, S., Sylvester, R.: Using the continual reassessment method: lessons learned from an EORTC phase I dose finding study. *European journal of cancer* **42**(10), 1362–1368 (2006). DOI 10.1016/j.ejca.2006.01.051. PMID: 16740385
41. Paoletti, X., Kramar, A.: A comparison of model choices for the continual reassessment method in phase I cancer trials. *Statistics in medicine* **28**(24), 3012–3028 (2009). DOI 10.1002/sim.3682. PMID: 19672839
42. Paoletti, X., O'Quigley, J., Maccario, J.: Design efficiency in dose finding studies. *Computational Statistics & Data Analysis* **45**(2), 197–214 (2004). DOI 10.1016/S0167-9473(02)00323-7. URL http://www.sciencedirect.com/science/article/pii/S0167947302003237
43. Postel-Vinay, S., Gomez-Roca, C., Molife, L.R., Anghan, B., Levy, A., Judson, I., De Bono, J., Soria, J.C., Kaye, S., Paoletti, X.: Phase I trials of molecularly targeted agents: should we pay more attention to late toxicities? *Journal of Clinical Oncology* **29**(13), 1728–1735 (2011). DOI 10.1200/JCO.2010.31.9236. PMID: 21444876
44. Ratain, M., Humphrey, R., Gordon, G., Fyfe, G., Adamson, P., Fleming, T., Stadler, W., Berry, D., Peck, C.: Recommended changes to oncology clinical trial design: revolution or evolution. *European Journal of Cancer* **41**(1), 8–11 (2008)
45. Reiner, E., Paoletti, X., O'Quigley, J.: Operating characteristics of the standard phase I clinical trial design. *Computational Statistics and Data Analysis* **30**(3), 303–315 (1999). DOI 10.1016/S0167-9473(98)00095-4
46. Robbins, H., Monro, S.: A stochastic approximation method. *Annals of Mathematical Statistics* **22**, 400–407 (1951)
47. Roberts, T.G.J., Goulart, B.H., Squitieri, L., Stallings, S.C., Halpern, E.F., Chabner, B.A., Gazelle, G.S., Finkelstein, S.N., Clark, J.W.: Trends in the risks and benefits to patients with cancer participating in phase I clinical trials. *Journal of the American Medical Association* **292**(17), 2130–2140 (2004). DOI 10.1001/jama.292.17.2130. PMID: 15523074
48. Rosenberger, W., Haines, L.: Competing designs for phase I clinical trials: a review. *Statistics in medicine* **21**(18), 2757–2770 (2002). DOI 10.1002/sim.1229. PMID: 12228889
49. Schöffski, P., Riggert, S., Fumoleau, P., Campone, M., Bolte, O., Marreaud, S., Lacombe, D., Baron, B., Herold, M., Zwierzina, H., Wilhelm-Ogunbiyi, K., Lentzen, H., Twelves, C., European Organization for Research and Treatment of Cancer New Drug Development Group: Phase I trial of intravenous aviscumine (rViscumin) in patients with solid tumors: a study of the european organization for research and treatment of cancer new drug development group. *Annals of oncology* **15**(12), 1816–1824 (2004). DOI 10.1093/annonc/mdh469. PMID: 15550588
50. Shen, L., O'Quigley, J.: Consistency of continual reassessment method under model misspecification. *Biometrika* **83**(2), 395–405 (1996). DOI 10.1093/biomet/83.2.395. URL http://biomet.oxfordjournals.org/content/83/2/395.full.pdf+html
51. Simon, R., Freidlin, B., Rubinstein, L., Arbuck, S.G., Collins, J., Christian, M.C.: Accelerated titration designs for phase I clinical trials in oncology. *Journal of the National Cancer Institute* **89**(15), 1138–1147 (1997). PMID: 9262252
52. Storer, B.: Design and analysis of phase I clinical trials. *Biometrics* **45**(3), 925–937 (1989). PMID: 2790129
53. Storer, B.: Small-sample confidence sets for the MTD in a phase I clinical trial. *Biometrics* **49**(4), 1117–1125 (1993). PMID: 8117905
54. Thall, P., Cook, J.: Dose-finding based on efficacy-toxicity trade-offs. *Biometrics* **60**(3), 684–93 (2004)

55. Thall, P., Russell, K.: A strategy for dose-finding and safety monitoring based on efficacy and adverse outcomes in phase I/II clinical trials. *Biometrics* **54**(1), 251–64 (1998)
56. Van Meter, E., Garrett-Mayer, E., Bandyopadhyay, D.: Proportional odds model for dose-finding clinical trial designs with ordinal toxicity grading. *Statistics in medicine* **30**(17), 270–280 (2011)
57. Verweij, J., Disis, M.L., Cannistra, S.A.: Phase I studies of drug combinations. *Journal of clinical oncology: official journal of the American Society of Clinical Oncology* **28**(30), 4545–4546 (2010). DOI 10.1200/JCO.2010.30.6282. PMID: 20855831
58. Whitehead, J., Williamson, D.: Bayesian decision procedures based on logistic regression models for dose-finding studies. *Journal of biopharmaceutical statistics* **8**(3), 445–467 (1998). DOI 10.1080/10543409808835252. PMID: 9741859
59. Zhang, J., Braun, T., Taylor, J.: Adaptive prior variance calibration in the bayesian continual reassessment method. *Statistics in Medicine* **32**(13), 2221–34 (2013). DOI 10.1002/sim.5621
60. Zohar, S., Chevret, S.: The continual reassessment method: comparison of bayesian stopping rules for dose-ranging studies. *Statistics in Medicine* **20**(19), 2827–43 (2001). DOI 10.2515/therapie/2011042

MIX
Papier aus verantwortungsvollen Quellen
Paper from responsible sources
FSC® C105338

If you have any concerns about our products,
you can contact us on
ProductSafety@springernature.com

In case Publisher is established outside the EU,
the EU authorized representative is:
Springer Nature Customer Service Center GmbH
Europaplatz 3, 69115 Heidelberg, Germany

Printed by Libri Plureos GmbH
in Hamburg, Germany